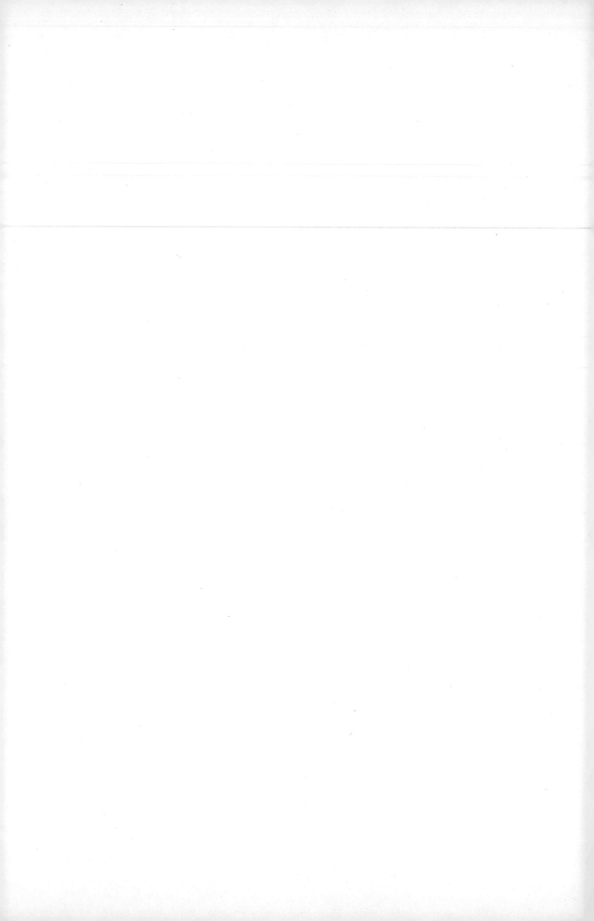

FLEXIBLE MULTIBODY DYNAMICS

EFFICIENT FORMULATIONS AND APPLICATIONS

Arun K. Banerjee

Formerly Principal Research Scientist
Lockheed Martin Advanced Technology Center
Palo Alto, CA
USA

WILEY

Registered office
John Wiley & Sons Ltd, The Atrium, Southern Gate, Chichester, West Sussex, PO19 8SQ, United Kingdom

For details of our global editorial offices, for customer services and for information about how to apply for permission to reuse the copyright material in this book please see our website at www.wiley.com.

Library of Congress Cataloging-in-Publication Data

Banerjee, Arun K., author.
Flexible multibody dynamics : efficient formulations and applications / Arun K. Banerjee.
pages cm
Includes bibliographical references and index.
ISBN 978-1-119-01564-2 (hardback)
1. Machinery, Dynamics of. 2. Multibody systems–Mathematical models. I. Title.
TJ173.B235 2015
621.8′11–dc23

2015033623

A catalogue record for this book is available from the British Library.

Cover images: Red rescue helicopter moving in blue sky: © aragami123345/iStock; Robotic probe in deep space, computer illustration: Victor Habbick Visions/Science Photo Library/Getty

Set in 10/12pt Times by Aptara Inc., New Delhi, India
Printed and bound in Malaysia by Vivar Printing Sdn Bhd

1 2016

This book is dedicated to:

Alpona, my wife who suffered over thirty-five years as I worked at my job and on weekends and nights for publications, and urged me to write this book after I retired, but did not live to see it published;

and

Professor Thomas Kane, my mentor, from whose magnificent books I learned to *do* dynamics, and who started me on the path to published research that this book embodies;

and

Our daughter, Onureena, and son, Abheek, who have been my blessings;

and

Lockheed Martin Space Systems Company and its managers, particularly Dr. Ron Dotson, who gave me a free hand to work on the algorithms reported here, which led to a flexible multibody dynamics code developed with major help from my colleague, Mark Lemak, and several independent formulations.

Contents

Preface

This book is based on my published research on deriving computationally efficient equations of motion of multibody systems with rotating, flexible components. It reflects my work of over thirty-five years done mostly at Lockheed Missiles & Space Company, and also at Martin Marietta and Northrop Corporations. The cover of the book depicts two examples of flexible multibody systems: the Galileo spacecraft, which was sent to Jupiter, with its rotating antenna dish on an inertially-fixed base with a deployed truss, and a helicopter in flight. Other examples of multibody systems, apart from the human body itself, are robotic manipulators, a space shuttle deploying a tethered subsatellite, and a ship reeling out a cable to a vehicle doing sea floor mine searches. Formulation of equations of motion is the first step in their simulation-based design.

In this book, I choose to use Kane's method of deriving equations of motion, for two reasons: efficiency in reducing labor of deriving the equations, and simplicity of the final equations due to a choice of variables that the method allows. However, the contribution of the book goes beyond a direct formulation of Kane's equations to more computationally efficient algorithms like block-diagonal and order-*n* formulations. Another major contribution of this book is in compensating for errors of premature linearization, inherent with the use of vibration modes in large overall motion problems, by using geometric stiffness due to inertia loads.

A highlight of this book is the application of the theory to complex problems. In Chapter 1, I explain Kane's method, first with a simple example and then by applying it to a realistic problem of the dynamics of a three-axis controlled spacecraft with fuel slosh. Presented separately are Kane's method of direct linearization of equation of motion and a method of a posteriori compensation for premature linearization by adding geometric stiffness due to inertia loads; in the Appendix, a guideline for choosing variables that simplify equations of motion is provided. In Chapter 2, Kane's method is used to derive nonlinear dynamical equations for tethered satellite deployment, station-keeping and retrieval, and a problem of impact dynamics of a nose cap during ejection of a parachute for recovery of a booster launching a satellite. The next two chapters cover large overall motion of beams and plates that illustrate the application of Kane's method of direct linearization. Chapter 5 gives a derivation of equations of large overall motion of an arbitrary flexible body, with a method of redeeming prematurely linearized equations by adding motion-induced geometric stiffness. Chapter 6 incorporates the motion-induced geometric stiffness into the dynamics of a system of flexible bodies in large overall motions. Chapter 7 is a review material from structural dynamics, based mainly on the book by Craig, with some additional work on mode selection done at Lockheed. Chapter 8 produces an algorithm for dynamical equations, with block-diagonal mass matrices, used for the Hubble and Next Generation Space telescopes, and to systems with and without structural loops, comparing results with test data for an antenna deployment. Chapter 9 illustrates the power of efficient motion variables in a block-diagonal algorithm, treating multiple loops. Chapter 10 simplifies the block-diagonal formulation to an order-*n* method for a system of spring-connected rigid rods, to simulate large bending of beams in large overall motion, comparing results with the finite element method; a Fortran code for the formulation is in Appendix B of this book. Chapter 11 uses a variable-*n* order-*n* algorithm for deploying a boom from a spacecraft, and a cable from a ship to an underwater vehicle. Chapter 12 covers flexible rocket dynamics.

This book is for readers with backgrounds in rigid body dynamics and structural dynamics. In writing it I was helped by Prof. Paul Mitiguy at Stanford (on efficient variables), Prof. Arun Misra at McGill (on formation flying of tethered satellites), and Dr. John Dickens of Lockheed (on modal truncation vectors and geometric stiffness issues). I thank my Lockheed colleagues: Mark Lemak, who developed a multibody dynamics code from the algorithms given here and produced the results in Chapters 6–9; and David Levinson, whose high praise was a booster for me to write this book. John Dickens provided the structural dynamics codes. Dr. Ron Dotson, a manager at Lockheed, gave me a free hand to develop the algorithms. Dr. Tushar Ghosh of L-3 Communications advised me on current practice, and meticulously edited the book.

1

Derivation of Equations of Motion

1.1 Available Analytical Methods and the Reason for Choosing Kane's Method

In this book we derive equations of motion for a system of rigid and flexible bodies undergoing large overall motion. Various choices of analytical methods are available for this task, such as Newton-Euler methods, and methods based on D'Alembert's principle together with the principle of virtual work, Lagrange's equations, Hamilton's equations, Boltzmann-Hamel equations, Gibbs equations, and Kane's equations. The most recent among these is Kane's method, based on a paper published in 1965 by Kane and Wang [1], and the method was given detailed exposition, with extensive applications, by Kane [2], Kane and Levinson [3], and Kane, Likins, Levinson [4]. Likins [5] also did a comparison of these various analytical methods for deriving equations of motion in a comprehensive report that also considered applications to flexible spacecraft.

In a survey paper, Kane and Levinson [6] took up a fairly complex example, of an 8 degree-of-freedom (dof) system consisting of a spacecraft containing a four-bar linkage to show the difference between seven analytical methods. To summarize, their conclusion was that (a) D'Alembert's method is less laborious than a method using conservation of momentum, with both requiring introduction and elimination of constraint forces; (b) Lagrange's equations require no introduction of workless constraint forces, but the labor to derive the equations is prohibitive; (c) Lagrange's equations in quasi-coordinates use variables that simplify the equations of motion but require order-n^3 computations for certain terms for an n-dof system, and the process of getting the final equations is formidable; (d) Gibbs equations is somewhat better, using quasi-coordinates but requiring one to form terms with n^2 computations for an n-dof system. With an exposition of Kane's method, they showed that Kane's method is superior to the rest of the methods, on the basis of two crucial considerations: (1) operational simplicity, meaning reduced labor in the derivation of the equations of motion either by hand or in terms of computer operations via symbol manipulation; and (2) simplicity of the final form of the equations, simplicity giving rise to reduction in computational time; simplicity is achievable depending on whether a method allows a choice of motion variables such as quasi-coordinates, or what Kane calls generalized speeds. An exposition of Kane's method is given later.

Flexible Multibody Dynamics: Efficient Formulations and Applications, First Edition. Arun K. Banerjee.

1.2 Kane's Method of Deriving Equations of Motion

Consider a system of particles and rigid bodies whose configuration in a Newtonian reference frame N is characterized by generalized coordinates, $q_1, q_2, \cdots, q_n$. Let $u_1, u_2, \ldots, u_n$ be motion variables, called *generalized speeds* by Kane, introduced as linear combinations of $\dot{q}_1, \dot{q}_2, \ldots, \dot{q}_n$, where an overdot indicates time derivative, that are kinematical differential equations of the form,

$$u_i = \sum_{j=1}^{n} W_{ij}\dot{q}_j + X_i \quad (i = 1, \ldots, n) \tag{1.1}$$

Here W_{ij} and X_i are functions of the generalized coordinates and time t, for an n-dof system. W_{ij} and X_i are chosen so that Eq. (1.1) can be uniquely solved for $\dot{q}_1, \dot{q}_2, \ldots, \dot{q}_n$. Typically prescribed motion terms appear in X_i. The angular velocity of any rigid body and the velocity of any material point of the system can always be expressed uniquely as a linear function of the generalized speeds, $u_1, u_2, \ldots, u_n$. Thus, for a particle P_k in a system, Kane [3] has shown that its velocity in a Newtonian, meaning inertial, reference frame N, defined as the inertial time-derivative of the position vector of P_k from a point O fixed in N, can always be split in two groups of terms:

$$^N\mathbf{v}^{P_k} = \frac{^N d\mathbf{p}^{OP_k}}{dt} = \sum_{i=1}^{n} {^N\mathbf{v}_{\mathrm{i}}^{P_k}} u_{\mathrm{i}} + {^N\mathbf{v}_{\mathrm{t}}^{P_k}} \tag{1.2}$$

Here ${^N\mathbf{v}_{\mathrm{i}}^{P_k}}$, ${^N\mathbf{v}_{\mathrm{t}}^{P_k}}$ are vector functions of the generalized coordinates, $q_1, q_2, \ldots, q_n$. Kane [3] calls the vector, ${^N\mathbf{v}_{\mathrm{i}}^{P_k}}$, that is the coefficient of the ith generalized speed u_i in Eq. (1.2), the ith partial velocity of the point P_k. Similarly for a rigid body B_k, the velocity of its mass center B_k^* and the angular velocity of B_k in N for a system can always be expressed as

$$\begin{aligned} {^N\mathbf{v}^{B_k^*}} &= \sum_{i=1}^{n} {^N\mathbf{v}_{\mathrm{i}}^{B_k^*}} u_i + {^N\mathbf{v}_{\mathrm{t}}^{B_k^*}} \\ {^N\boldsymbol{\omega}^{B_k}} &= \sum_{i=1}^{n} {^N\boldsymbol{\omega}_{\mathrm{i}}^{B_k}} u_i + {^N\boldsymbol{\omega}_{\mathrm{t}}^{B_k}} \end{aligned} \tag{1.3}$$

Again, ${^N\mathbf{v}_i^{B_k^*}}$ is the ith partial velocity of B_k^*, and ${^N\boldsymbol{\omega}_i^{B_k}}, {^N\boldsymbol{\omega}_t^{B_k}}$ are vector functions of the generalized coordinates, and Kane calls the vector ${^N\boldsymbol{\omega}_i^{B_k}}$, the coefficient of u_i in ${^N\boldsymbol{\omega}^{B_k}}$ Eq. (1.3), the ith partial angular velocity of the body B_k in N. Typically, ${^N\mathbf{v}_t^{P_k}}, {^N\mathbf{v}_t^{B_k^*}}, {^N\boldsymbol{\omega}_t^{B_k}}$ in Eqs. (1.2) and (1.3) are remainder terms associated with prescribed velocity and angular velocity. Partial velocities and partial angular velocities are crucial items in Kane's method, and throughout this book we will see their central roles in the formulation of equations of motion. Once the velocities of points with mass and of mass centers and angular velocities of rigid bodies are expressed in some vector basis fixed in B_k, inertial acceleration of those points and mass centers, as well as angular acceleration of those bodies, can be obtained by differentiating these vector expressions in a Newtonian reference frame N. This is done by appealing to the rule for differentiation of a vector in two reference frames, expressed as:

$$
\begin{aligned}
{}^{N}\mathbf{a}^{P_k} &= \frac{{}^{N}d^{N}\mathbf{v}^{P_k}}{dt} = \frac{{}^{B_k}d^{N}\mathbf{v}^{P_k}}{dt} + {}^{N}\boldsymbol{\omega}^{B_k} \times {}^{N}\mathbf{v}^{P_k} \\
{}^{N}\mathbf{a}^{B_k^*} &= \frac{{}^{N}d^{N}\mathbf{v}^{B_k^*}}{dt} = \frac{{}^{B_k}d^{N}\mathbf{v}^{B_k^*}}{dt} + {}^{N}\boldsymbol{\omega}^{B_k} \times {}^{N}\mathbf{v}^{B_k^*} \\
{}^{N}\boldsymbol{\alpha}^{B_k} &= \frac{{}^{N}d^{N}\boldsymbol{\omega}^{B_k}}{dt} = \frac{{}^{B_k}d^{N}\boldsymbol{\omega}^{B_k}}{dt}
\end{aligned}
\tag{1.4}
$$

The first equality indicates a definition, and the second equality sign provides a basic kinematic relationship between differentiation of a vector in two reference frames, and it assumes that the frame B_k, or equivalently, rigid body B_k is different from frame N.

Kane's equations of motion are stated in terms of what Kane calls generalized inertia forces and generalized active forces. For an n-dof system consisting of NR number of rigid bodies and NP number of particles, the ith generalized inertia force is defined by the following dot-products with the ith partial velocities and partial angular velocities:

$$
\begin{aligned}
F_i^* = &-\sum_{j=1}^{NR}\left[m_j{}^{N}a^{B_j^*}\cdot{}^{N}\mathbf{v}_i^{B_j^*} + (\mathbf{I}^{B_j/B_j^*}\cdot{}^{N}\boldsymbol{\alpha}^{B_j} + {}^{N}\boldsymbol{\omega}^{B_j}\times\mathbf{I}^{B_j/B_j^*}\cdot{}^{N}\boldsymbol{\omega}^{B_j})\cdot{}^{N}\boldsymbol{\omega}_i^{B_j}\right] \\
&-\sum_{j=1}^{NP} m_j{}^{N}\mathbf{a}^{P_j}\cdot{}^{N}\mathbf{v}_i^{P_j} \qquad (i = 1,\ldots\ldots, n)
\end{aligned}
\tag{1.5}
$$

Here ${}^{N}\mathbf{a}^{\mathbf{B}_j^*}, {}^{N}\mathbf{a}^{\mathbf{P}_j}$ are the Newtonian frame accelerations of the mass centers B_j^* of the body B_j and particle P_j, respectively; $\mathbf{I}^{B_j/B_j^*}$ is the inertia dyadic of B_j about B_j^*; and ${}^{N}\boldsymbol{\alpha}^{B_j}$ is the angular acceleration of B_j in N. The ith generalized active force for this n-dof system of NR number of rigid bodies and NP number of particles is given by the following dot-products with partial velocities and partial angular velocities:

$$
F_i = \sum_{j=1}^{NR}\left[\mathbf{F}^{\mathbf{B}_j^*}\cdot{}^{N}\mathbf{v}_i^{\mathbf{B}_j^*} + \mathbf{T}^{B_j}\cdot{}^{N}\boldsymbol{\omega}_i^{B_j}\Big)\right] + \sum_{j=1}^{NP}\mathbf{F}^{\mathbf{P}_j}\cdot{}^{N}\mathbf{v}_i^{\mathbf{P}_j} \qquad (i = 1,\ldots,n) \tag{1.6}
$$

Here the resultant of all contact and body forces on body B_j are $\mathbf{F}^{B_j^*}$ at B_j^* together with a couple of torque $\mathbf{T}^{B_j}$, and the resultant of external and contact forces on particle P_j is $\mathbf{F}^{P_j}$. Note that all non-working interaction forces are automatically eliminated by taking the sum in Eq. (1.6) over bodies and particles, with actions and reactions canceling, as generalized active forces are formed. Some special cases of generalized active force that are covered by Eq. (1.6) are those due to elastic-dissipative mechanical systems, by "conservative" forces derivable from a potential function $V(q_1, \ldots, q_n, t)$ and dissipative forces from a dissipation function $D(u_1, \ldots, u_n)$.

$$F_i^c = -\frac{\partial V}{\partial q_i} - \frac{\partial D}{\partial u_i} \qquad (i = 1, \ldots, n) \tag{1.7}$$

1.2.1 Kane's Equations

Kane's Equations for an n-dof system can now be written by adding up the generalized active and inertia forces, and setting them equal to zero, as

$$F_i + F_i^* = 0, \quad i = 1, \ldots, n \quad \text{or,} \quad -F_i^* = F_i, \quad i = 1, \ldots, n \tag{1.8}$$

These dynamical equations of motion, together with the kinematical equation of Eq. (1.1) can be written as two sets of n coupled, nonlinear, differential equations in matrix form:

$$\begin{aligned} [M(q)]\{\dot{U}\} &= \{C(q, U, t)\} + \{F(q, U, t)\} \\ [W(q)]\{\dot{q}\} &= \{U - X(q, t)\} \end{aligned} \tag{1.9}$$

Here $M(q)$ is called the nxn "mass matrix," $C(q, U, t)$ the nx1 "Coriolis and centrifugal inertia force matrix," and $F(q, U, t)$ the nx1 "generalized force" matrix. Equation (1.9) completely describes the dynamics of the system. Note that the algebra involved in forming Eqs. (1.5) and (1.6) can be quite massive for a complex mechanical system, as may be checked by an analyst deriving equations of motion by hand. That is why a computerized symbol manipulation code, Autolev, was developed by Levinson and Kane [7] to derive the equations of motion. Finally, it should be mentioned that Kane had originally [2] called Eq. (1.8) the Lagrange's form of D'Alembert's Principle, because just as Lagrange's equations can be derived by dot-multiplying the D'Alembert force equilibrium equations by the components of virtual displacement in a virtual work principle, Kane obtains his equations by dot-multiplying the D'Alembert equilibrium equations by the partial velocities and partial angular velocities, to represent what may be thought of as a virtual power principle.

1.2.2 Simple Example: Equations for a Double Pendulum

Figure 1.1 shows a planar double pendulum. Consider the links OP, PQ as massless rigid rods, each of length l, with lumped mass m at the end of each rod acted on only by gravity. Configuration of the pendulum is defined by two generalized coordinates, q_1, q_2, as shown.

To use Kane's method we may choose as generalized speeds, following Eq. (1.1):

$$u_1 = \dot{q}_1; \quad u_2 = \dot{q}_1 + \dot{q}_2 \quad \text{or} \quad \begin{Bmatrix} \dot{q}_1 \\ \dot{q}_2 \end{Bmatrix} = \begin{Bmatrix} u_1 \\ u_2 - u_1 \end{Bmatrix} \tag{1.10}$$

The velocity of P in the Newtonian reference frame N can be written in terms of the angular velocity of the link OP in N, $u_1\mathbf{n}_3$ ($\mathbf{n}_3$ being perpendicular to the plane in Figure 1.1) as

$$^N\mathbf{v}^P = u_1\mathbf{n}_3 \times l(\cos q_1\mathbf{n}_1 + \sin q_1\mathbf{n}_2) = lu_1(-\sin q_1\mathbf{n}_1 + \cos q_1\mathbf{n}_2) \tag{1.11}$$

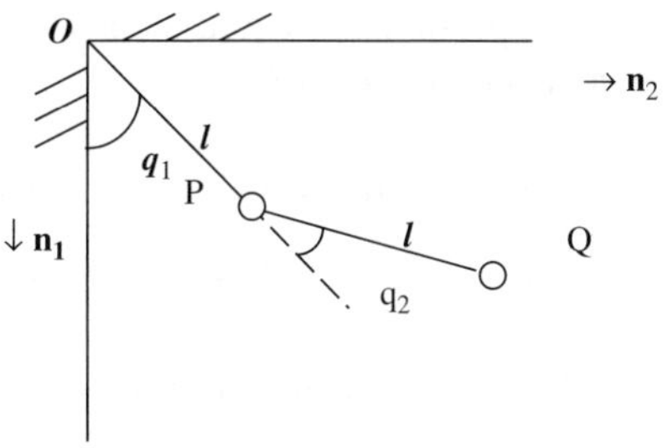

Figure 1.1 A Planar Double Pendulum.

Here $\mathbf{n}_1, \mathbf{n}_2$ are unit vectors in N directed downward and to the right, respectively, as shown in Figure 1.1. The velocity of Q in N is given in terms of the velocity of P in N by

$$\begin{aligned} {}^N\mathbf{v}^Q &= {}^N\mathbf{v}^P + u_2\mathbf{n}_3 \times l[(\cos(q_1+q_2)\,\mathbf{n}_1 + \sin(q_1+q_2)\,\mathbf{n}_2] \\ &= \mathbf{n}_1 l[-u_1\sin q_1 - u_2\sin(q_1+q_2)] + \mathbf{n}_2 l[u_1\cos q_1 + u_2\cos(q_1+q_2)] \end{aligned} \tag{1.12}$$

Here $u_2\mathbf{n}_3$ is the angular velocity in N of the link PQ. Now we form partial velocities of P and Q, coefficients of generalized speeds, u_1, u_2 in Eqs. (1.11), (1.12) shown in Table 1.1.

Accelerations of P and Q in N are obtained by differentiating the velocity vectors in N:

$${}^N\mathrm{a}^P = \frac{{}^N d\,{}^N\mathbf{v}^P}{dt} = l\left[\dot{u}_1(-\sin q_1\mathrm{n}_1 + \cos q_1\mathrm{n}_2) + u_1^2(-\cos q_1\mathrm{n}_1 - \sin q_1\mathrm{n}_2)\right] \tag{1.13}$$

$$\begin{aligned} {}^N\mathbf{a}^Q = \frac{{}^N d\,{}^N\mathbf{v}^Q}{dt} &= l\left[\dot{u}_1(-\sin q_1\mathbf{n}_1 + \cos q_1\mathbf{n}_2) + u_1^2(-\cos q_1\mathbf{n}_1 - \sin q_1\mathbf{n}_2)\right] \\ &+ l\left\{\dot{u}_2\left[-\sin(q_1+q_2)\,\mathbf{n}_1 + \cos(q_1+q_2)\,\mathbf{n}_2\right] + u_2^2[-\cos(q_1+q_2)\,\mathbf{n}_1 - \sin(q_1+q_2)\,\mathbf{n}_2]\right\} \end{aligned} \tag{1.14}$$

Negatives of the generalized inertia forces in two generalized speeds are formed by Eq. (1.5):

$$-F_1^* = ml^2\left[2\dot{u}_1 + \dot{u}_2\cos q_2 - u_2^2\sin q_2\right] \tag{1.15}$$

$$-F_2^* = ml^2\left[\dot{u}_1\cos q_2 + u_1^2\sin q_2 + \dot{u}_2\right] \tag{1.16}$$

These equations would be more complex had we chosen $u_i = \dot{q}_i, i = 1, 2$. Relative simplicity of the form of Eqs. (1.15), (1.16) is due to an efficient choice of generalized speeds, which

Table 1.1 Partial Velocities for the Double Pendulum Example.

r	${}^N\mathbf{v}_r^P$	${}^N\mathbf{v}_r^Q$
1	$l(-\sin q_1\mathbf{n}_1 + \cos q_1\mathbf{n}_2)$	$l(-\sin q_1\mathbf{n}_1 + \cos q_1\mathbf{n}_2)$
2	0	$l[-\sin(q_1+q_2)\,\mathbf{n}_1 + \cos(q_1+q_2)\,\mathbf{n}_2$

is described later in this chapter's appendix. The expressions for the generalized active forces due to gravity on P and Q are obtained via Eq. (1.6).

$$F_1 = \text{mg}\, \mathbf{n}_1 \bullet {}^N\mathbf{v}_1^P + mg\mathbf{n}_1 \cdot \bullet {}^N\mathbf{v}_1^Q = -2\,\text{mgl}\sin q_1 \tag{1.17}$$

$$F_2 = \text{mg}\mathbf{n}_1 \bullet {}^N\mathbf{v}_2^P + mg\mathbf{n}_1 \bullet {}^N\mathbf{v}_2^Q = -\text{mgl}\ \sin(q_1 + q_2) \tag{1.18}$$

Substitution of Eqs. (1.15)–(1.18) in Kane's equation, Eq. (1.8), yields the dynamical equations:

$$ml^2 \left[2\dot{u}_1 + \dot{u}_2 \cos q_2 - u_2^2 \sin q_2\right] = -2\,mgl \sin q_1 \tag{1.19}$$

$$ml^2 \left[\dot{u}_1 \cos q_2 + u_1^2 \sin q_2 + \dot{u}_2\right] = -mgl \sin(q_1 + q_2) \tag{1.20}$$

These dynamical equations are written in matrix form as:

$$\begin{bmatrix} 2ml^2 & ml^2 \cos q_2 \\ ml^2 \cos q_2 & ml^2 \end{bmatrix} \begin{Bmatrix} \dot{u}_1 \\ \dot{u}_2 \end{Bmatrix} = (ml^2 \sin q_2) \begin{Bmatrix} u_2^2 \\ -u_1^2 \end{Bmatrix} - \begin{Bmatrix} 2mgl \sin q_1 \\ mgl \sin(q_1 + q_1) \end{Bmatrix} \tag{1.21}$$

Note that the multiplier matrix of the column matrix for the derivatives of the generalized speeds, the so-called mass matrix, is symmetric. This is true for all rigid body systems. Dynamical equations, Eq. (1.21), together with the kinematical equations, Eq. (1.10), complete the equations of motion for the double pendulum.

1.2.3 Equations for a Spinning Spacecraft with Three Rotors, Fuel Slosh, and Nutation Damper

Consider a more complex spacecraft example, for which equations of motion were derived by the author and reported in Ref. [8]. Figure 1.2 shows a spinning spacecraft with three-axis control, a nutation damper, and a thruster with thrust fuel sloshing in a tank. The spacecraft is a gyrostat G, meaning a rigid body with three fixed-axis reaction control rotors W_1, W_2, W_3; sloshing fuel is represented as a spherical pendulum with a massless rod with an end point mass m_P, and the nutation damper has a point mass m_Q, at point Q, with $\hat{Q}$ being the location in the spacecraft body G when the nutation damper spring is unstretched. The slosh pendulum attachment point is located from the mass center G^* of G by the position vector $z_0\mathbf{g}_3$. We show in Figure 1.3 two angles Ψ_1, Ψ_2 to orient the slosh pendulum, and let σ denote the nominal length plus the elastic stretch of the spring at the nutation damper of mass m_Q. This describes a 12 dof system, for which we define the generalized speeds, u_i, $(i = 1, \dots, 12)$, as follows:

$$u_i = {}^N\mathbf{v}^{G*} \cdot \mathbf{g}_i, \quad i = 1, 2, 3 \tag{1.22}$$

$$u_{3+i} = {}^N\boldsymbol{\omega}^G \cdot \mathbf{g}_i, \quad i = 1, 2, 3 \tag{1.23}$$

$$u_{6+i} = {}^G\boldsymbol{\omega}^{W_i} \cdot \mathbf{g}_i, \quad i = 1, 2, 3 \tag{1.24}$$

$$u_{9+i} = \dot{\psi}_i, \quad i = 1, 2 \tag{1.25}$$

$$u_{12} = \dot{\sigma} \tag{1.26}$$

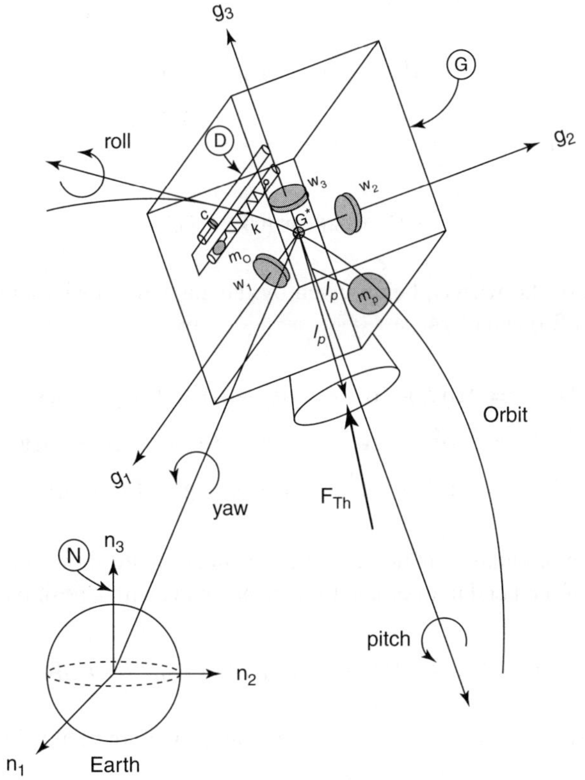

Figure 1.2 Spacecraft with Three Reaction Wheels for Attitude Control, a Thruster with Thrust Fuel Sloshing in a Tank Represented by a Spherical Pendulum, and a Nutation Damper. Roll, pitch, and yaw angles shown are about the orbit frame, not the body frame. Ayoubi *et al.* [8]. Reproduced with permission of Springer.

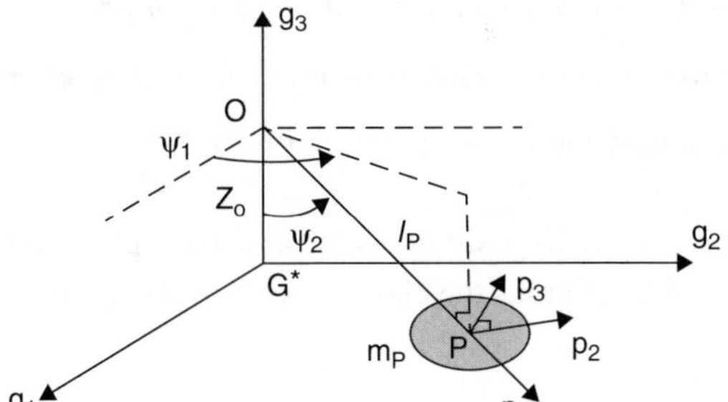

Figure 1.3 Attachment Point and Orientation of the Pendulum Representing Fuel Slosh. Ayoubi *et al.* [8]. Reproduced with permission of Springer.

The velocity of G* follows from Eq. (1.22):

$$ {}^{N}\mathbf{v}^{G*} = u_1\mathbf{g}_1 + u_2\mathbf{g}_2 + u_3\mathbf{g}_3 \tag{1.27} $$

The angular velocity of G, from Eq. (1.23) is:

$$ {}^{N}\boldsymbol{\omega}^{G} = u_4\mathbf{g}_1 + u_5\mathbf{g}_2 + u_6\mathbf{g}_3 \tag{1.28} $$

Angular velocities of the wheels, being given by the angular velocity addition theorem [3], follow from Eqs. (1.23) and (1.24), as:

$$ {}^{N}\boldsymbol{\omega}^{W_1} = {}^{N}\boldsymbol{\omega}^{G} + {}^{G}\boldsymbol{\omega}^{W_1} = (u_4 + u_7)\,\mathbf{g}_1 + u_5\mathbf{g}_2 + u_6\mathbf{g}_3 \tag{1.29} $$

$$ {}^{N}\boldsymbol{\omega}^{W_2} = {}^{N}\boldsymbol{\omega}^{G} + {}^{G}\boldsymbol{\omega}^{W_2} = u_4\mathbf{g}_1 + (u_5 + u_8)\mathbf{g}_2 + +u_6\mathbf{g}_3 \tag{1.30} $$

$$ {}^{N}\boldsymbol{\omega}^{W_3} = {}^{N}\boldsymbol{\omega}^{G} + {}^{G}\boldsymbol{\omega}^{W_3} = u_4\mathbf{g}_1 + u_5\mathbf{g}_2 + (u_6 + u_9)\mathbf{g}_3 \tag{1.31} $$

The slosh pendulum orientation angles and attachment point are shown in Figure 1.3. Let the particle at the end of the pendulum of length l_P be named P, with a position vector given by:

$$ \mathbf{p}^{G*P} = [l_P s\psi_2(c\psi_1\mathbf{g}_1 + s\psi_1\mathbf{g}_2)] + (z_0 - l_P c\psi_2)\mathbf{g}_3 \tag{1.32} $$

Here we have denoted $c\psi_1 = \cos\psi_1, s\psi_2 = \sin\psi_2$, and so on. Inertial velocity of the particle at P is:

$$ {}^{N}\mathbf{v}^{P} = {}^{N}\mathbf{v}^{G*} + \frac{{}^{G}d\mathbf{p}^{G*P}}{dt} + {}^{N}\boldsymbol{\omega}^{G} \times \mathbf{p}^{G*P} \tag{1.33} $$

Equation (1.33) is expanded as follows:

$$ \begin{aligned} {}^{N}\mathbf{v}^{P} = {} & \mathbf{g}_1[u_1 + u_5(z_0 - l_P c\psi_2) - u_6 l_P s\psi_1 s\psi_2 + l_P(u_{11}c\psi_1 c\psi_2 - u_{10}s\psi_1 s\psi_2)] \\ & + \mathbf{g}_2[u_2 + u_6 l_P c\psi_1 s\psi_2 - u_4(z_0 - l_P c\psi_2) + l_P(u_{10}c\psi_1 s\psi_2 + u_{11}s\psi_1 c\psi_2)] \\ & + \mathbf{g}_3[u_3 + l_P(u_4 s\psi_1 s\psi_2 - u_5 c\psi_1 s\psi_2 + u_{11}s\psi_2)] \end{aligned} \tag{1.34} $$

The nutation damper mass particle position vector in Figure 1.2 is, $\mathbf{p}^{\mathbf{Q}} = \sigma\mathbf{g}_1 + \underline{\mathbf{z}}_{\mathrm{Q}}\mathbf{g}_3$ (recall that σ denotes the nominal length plus the elastic stretch of the spring). Thus, inertial velocity of Q is:

$$ \begin{aligned} {}^{N}\mathbf{v}^{Q} & = {}^{N}\mathbf{v}^{G*} + {}^{N}\boldsymbol{\omega}^{G} \times \mathbf{p}^{G*Q} + \frac{{}^{G}d\mathbf{p}^{G*Q}}{dt} \\ & = \mathbf{g}_1(u_1 + u_5 z_Q + u_{12}) + \mathbf{g}_2(u_2 + u_6\sigma - u_4 z_Q) + \mathbf{g}_3(u_3 - u_5\sigma) \end{aligned} \tag{1.35} $$

With all the velocities and angular velocities expressed, a table of partial velocities and partial angular velocities can be formed from Eqs. (1.27), (1.28)–(1.31), (1.34), and (1.35) as the coefficients of the generalized speeds in these equations. In the Table 1.2 of partial velocities, the first column refers to the index of the generalized speed and the other columns describe corresponding particular partial velocity or partial angular velocity.

Inertial angular accelerations of G, W_1, W_2, W_3 are obtained by differentiating the angular velocity expressions in Eqs. (1.28)–(1.31) in the N-frame.

$$ {}^{N}\boldsymbol{\alpha}^{G} = \dot{u}_4\mathbf{g}_1 + \dot{u}_5\mathbf{g}_2 + \dot{u}_6\mathbf{g}_3 \tag{1.36} $$

$$ {}^{N}\boldsymbol{\alpha}^{W_1} = \dot{u}_4\mathbf{g}_1 + (\dot{u}_5 + u_6u_7)\,\mathbf{g}_2 + (\dot{u}_6 - u_5u_7)\mathbf{g}_3 \tag{1.37} $$

$$ {}^{N}\boldsymbol{\alpha}^{W_2} = (\dot{u}_4 - u_6u_8)\mathbf{g}_1 + \dot{u}_5\mathbf{g}_2 + (\dot{u}_6 + u_4u_8)\,\mathbf{g}_3 \tag{1.38} $$

$$ {}^{N}\boldsymbol{\alpha}^{W_3} = (\dot{u}_4 + u_5u_9)\mathbf{g}_1 + (\dot{u}_5 - u_4u_9)\mathbf{g}_2 + \dot{u}_6\mathbf{g}_3 \tag{1.39} $$

Accelerations of G^*, P, Q are found by differentiating velocities in N-frame. This yields:

$$ {}^{N}\mathbf{a}^{G*} = \dot{u}_1\mathbf{g}_1 + \dot{u}_2\mathbf{g}_2 + \dot{u}_3\mathbf{g}_2 + {}^{N}\boldsymbol{\omega}^{G} \times {}^{N}\mathbf{v}^{G*} \tag{1.40} $$

$$ {}^{N}\mathbf{a}^{P} = \frac{{}^{G}d^{N}\mathbf{v}^{P}}{dt} + {}^{N}\boldsymbol{\omega}^{G} \times {}^{N}\mathbf{v}^{P} \tag{1.41} $$

$$ {}^{N}\mathbf{a}^{Q} = \frac{{}^{G}d^{N}\mathbf{v}^{Q}}{dt} + {}^{N}\boldsymbol{\omega}^{G} \times {}^{N}\mathbf{v}^{Q} \tag{1.42} $$

Table 1.2 Partial Velocity for the Dynamics of a Spacecraft with Three Reaction Wheels for Attitude Control, a Spherical Pendulum Representing Fuel Slosh, and a Nutation Damper.

r	$\mathbf{v}_r^{G*}$	$\boldsymbol{\omega}_r^{G}$	$\boldsymbol{\omega}_r^{W_1}$	$\boldsymbol{\omega}_r^{W_2}$	$\boldsymbol{\omega}_r^{W_3}$	v_r^{P}	$\mathbf{v}_r^{Q}$
1	$\mathbf{g}_1$	0	0	0	0	$\mathbf{g}_1$	$\mathbf{g}_1$
2	$\mathbf{g}_2$	0	0	0	0	$\mathbf{g}_2$	$\mathbf{g}_2$
3	$\mathbf{g}_3$	0	0	0	0	$\mathbf{g}_3$	$\mathbf{g}_3$
4	0	$\mathbf{g}_1$	$\mathbf{g}_1$	$\mathbf{g}_1$	$\mathbf{g}_1$	$-\mathbf{g}_2(z_0 - l_P c\psi_2) + \mathbf{g}_3 l_P s\psi_1 s\psi_2$	$-\mathbf{g}_2\mathbf{z_Q}$
5	0	$\mathbf{g}_2$	$\mathbf{g}_2$	$\mathbf{g}_2$	$\mathbf{g}_2$	$\mathbf{g}_1(z_0 - l_P c\psi_2) - \mathbf{g}_3 l_P c\psi_1 s\psi_2$	$\mathbf{g}_1 z_Q - \mathbf{g}_3\sigma$
6	0	$\mathbf{g}_3$	$\mathbf{g}_3$	$\mathbf{g}_3$	$\mathbf{g}_3$	$-\mathbf{g}_1 l_P s\psi_1 s\psi_2 + \mathbf{g}_2 l_P c\psi_1 c\psi_2$	$\mathbf{g}_2\sigma$
7	0	0	$\mathbf{g}_1$	0	0	0	0
8	0	0	0	$\mathbf{g}_2$	0	0	0
9	0	0	0	0	$\mathbf{g}_3$	0	0
10	0	0	0	0	0	$-\mathbf{g}_1 l_P s\psi_1 s\psi_2 + \mathbf{g}_2 l_P c\psi_1 s\psi_2$	0
11	0	0	0	0	0	$\mathbf{g}_1 l_P c\psi_1 c\psi_2 + \mathbf{g}_2 l_P s\psi_1 s\psi_2 + \mathbf{g}_3 l_P s\psi_2$	0
12	0	0	0	0	0	0	$\mathbf{g}_1$

Ayoubi *et al.* [8]. Reproduced with permission of Springer.

Negative of the generalized inertia forces for this example system are now developed as follows:

$$-F_r^* = (\mathbf{I}^G \cdot {}^N\alpha^G + {}^N\omega^G \times \mathbf{I}^G \cdot {}^N\omega^G) \cdot {}^N\omega_r^G + \sum_{i=1}^{3} (\mathbf{I}^{W_i} \cdot {}^N\alpha^{W_i} + {}^N\omega^{W_i} \times \mathbf{I}^{W_i} \cdot {}^N\omega^{W_i}) \cdot {}^N\omega_r^{W_i}$$

$$+ m_G {}^N\mathbf{a}^{G*} \cdot {}^N\mathbf{v}_r^{G*} + m_P {}^N\mathbf{a}^P \cdot {}^N\mathbf{v}_r^P + m_Q {}^N\mathbf{a}^Q \cdot {}^N\mathbf{v}_r^Q \quad (r = 1, \ldots, 12) \tag{1.43}$$

Here the quantities with boldface **I** denote the dyadic of moment of inertia about the mass center of the body concerned. Generalized active forces, due to thrust, gravity, thrust torque about G*, wheel torques, spring-damper forces from an unstretched length σ_0 on the nutation damper Q, and viscous torque on the pendulum P are:

$$F_r = \left(\mathbf{F}_{th}^{G*} + \mathbf{F}_g^{G*}\right) \cdot {}^N\mathbf{v}_r^{G*} + \mathbf{T}_{th}^G . \omega_r^G + \sum_{i=1}^{3} \mathbf{T}^{W_i} \cdot {}^G\omega_r^{W_i} + \left\{\mathbf{F}_g^Q - [k(\sigma - \sigma_0) + c\dot{\sigma}]\mathbf{g}_1\right\} \cdot {}^G\mathbf{v}_r^P$$

$$+ \left(\mathbf{F}_g^P - \mathbf{F}_v^P\right) \ldots {}^N\mathbf{v}_r^P \qquad r = 1, \ldots, 12 \tag{1.44}$$

Equations (1.43) and (1.44) are put in Kane's dynamical equations, Eq. (1.7). Finally, for an assumed body 3-2-1 rotation sequence in the angles $\theta_1, \theta_2, \theta_3$, standard kinematical equations [4] relating rotation rate to angular velocity are appended.

$$\begin{aligned}
\dot{\theta}_1 &= (u_5 s_3 + u_6 c_3)/c_2 \\
\dot{\theta}_2 &= u_5 \mathbf{c}_3 - u_6 s_3) \\
\dot{\theta}_3 &= \mathbf{s}_2 (u_5 s_3 + u_6 c_3)/c_2 + u_4
\end{aligned} \tag{1.45}$$

A detailed analysis and simulation results are in Ref. [8]. Briefly, the results of a reorientation maneuver are given in the two parts in Figure 1.4. Here the spacecraft initially spins at 5 rpm around the axis of maximum moment of inertia, $\mathbf{g}_1$ (see Figure 1.2), with zero rpm around other axes. Now we spin up the third momentum wheel along $\mathbf{g}_3$ axis as follows:

$$\begin{aligned}
{}^G\omega^{W_3} &= 0 \quad 0 \le t < 1000 \\
&= 0.9375(t - 1000) \quad 1000 \le t < 7400 \\
&= 6000 \quad 7400 \le t \le 10000
\end{aligned}$$

The result in the top curve in Figure 1.4 shows in the angular velocity components plotted that the spacecraft originally spins about the maximum moment of inertia and then gradually reorients around the minimum moment of inertia axis $\mathbf{g}_3$. In the bottom curve of Figure 1.4 we see the pitch angle of the gyrostat. For other results showing the effects of the tank location along $\mathbf{g}_3$ on the pitch angle and so on, see Ref. [8].

Equations (1.5) and (1.6) show that Kane's equations require computing accelerations and angular accelerations, and taking dot-products with partial velocities and partial angular velocities. The amount of algebra involved in doing just these two steps is not small. However,

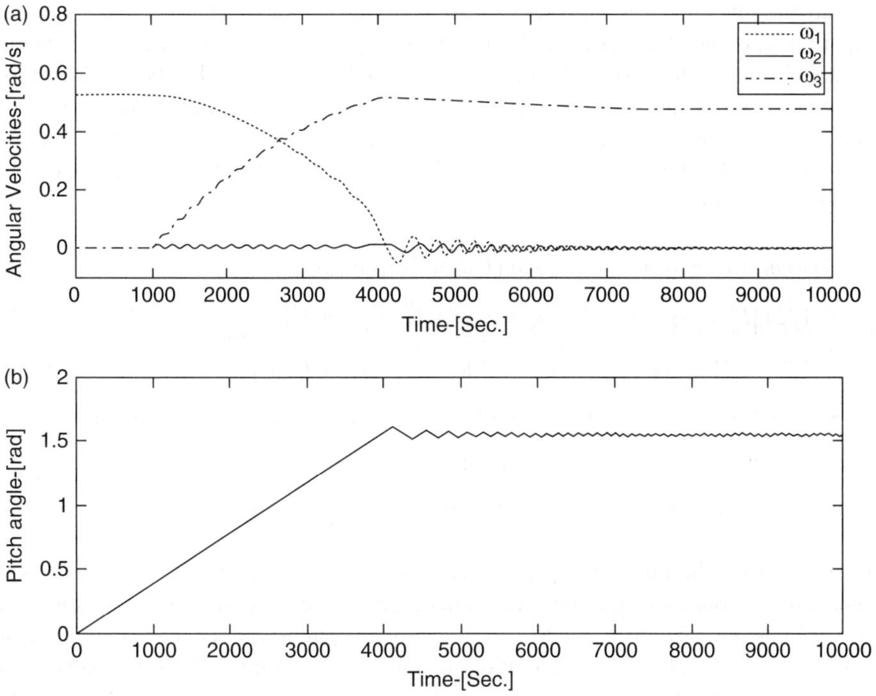

Figure 1.4 (a) Angular Velocity Components and (b) Pitch Angle in Reorientation Maneuver. Ayoubi *et al.* [8]. Reproduced with permission of Springer.

the labor is very much less in comparison to the labor required by Lagrange's method. An additional advantage of Kane's equations is the relative simplicity possible for the final equations, compared to those provided by Lagrange's or other general methods. This is made possible by the choice of motion variables – generalized speeds – allowed in Kane's method, as illustrated by Eq. (1.9). These two aspects of reduced labor and simplicity of final equations in Kane's method are in contrast to Lagrange's equations, which are reviewed next.

1.3 Comparison to Derivation of Equations of Motion by Lagrange's Method

Lagrange's equations are widely used as a method for deriving equations of motion, as illustrated in Meirovitch [9] and Likins [5]. Formulation of equations by this method consists of forming the system kinetic energy function, $K = f(q_1, \ldots, q_n, \dot{q}_1, \ldots, \dot{q}_n)$, where $q_i, i = 1, \ldots, n$ are n generalized coordinates and $\dot{q}_i$ are their time-derivatives. Lagrange's equations are given by

$$\frac{d}{dt}\frac{\partial K}{\partial \dot{q}_i} - \frac{\partial K}{\partial q_i} = Q_i - \frac{\partial V}{\partial q_i} - \frac{\partial D}{\partial \dot{q}_i}, \quad (i = 1, \ldots, n) \tag{1.46}$$

Here Q_i are the generalized active forces due to forces not derivable from a potential function or a dissipation function. Q_i are evaluated in the same way as in Kane's method if the generalized speeds are chosen as the derivatives of the generalized coordinates. Further, $V(q_1, \ldots, q_n)$ is the potential function, and $D(\dot{q}_1, \ldots, \dot{q}_n)$ is the dissipation function. Equation (1.46) is only a road map that one follows to derive equations of motion by Lagrange's method. To illustrate the process, consider the same system shown in Figure 1.2, and start with the kinetic energy function for this system, written as follows:

$$\begin{aligned}
K = {} & (1/2)m_G\left(u_1^2 + u_2^2 + u_3^2\right) + (1/2)\left(I_1 u_4^2 + I_2 u_5^2 + I_3 u_6^2\right) \\
& + (1/2)\left[I_1^W (u_4 + u_7)^2 + I_2^W (u_5 + u_8)^2 + I_3^W (u_6 + u_9)^2\right] \\
& + (1/2)m_G\{[u_1 + u_5(z_0 - l_P c q_{11}) - u_6 l_P s q_{10} s q_{11} + l_P(u_{11} c q_{10} c q_{11} - u_{10} s q_{10} s q_{11})]^2 \\
& + [u_2 + u_6 l_P c q_{10} s q_{11} - u_4(z_0 - l_P c q_{11}) + l_P(u_{10} c q_{10} s q_{11} + u_{11} s q_{10} c q_{11})]^2 \\
& + [u_3 + l_P(u_4 s q_{10} s q_{11} - u_5 c q_{10} s q_{11} + u_{11} s q_{11})]^2\} \\
& + (1/2)m_Q[(u_1 + u_5 z_Q + u_{12})^2 + (u_2 + u_6\sigma - u_4 z_Q)^2 + (u_3 - u_5\sigma)^2]
\end{aligned} \tag{1.47}$$

Here we have used the notations, $q_{10} = \psi_1, q_{11} = \psi_2$, and since Lagrange's equations work with generalized coordinates, rather than generalized speeds, we have to express $u_1, \ldots, u_{12}$ in Eq. (1.47) as: $u_i = \dot{q}_i, i = 1, \ldots, 12$. A choice of roll-pitch-yaw sequence can be made by the analyst; assuming a body 1-2-3, rotation sequence [4], the direction cosine matrix relating the body components of the mass center velocity to their inertial components, and the relation between the body angular velocity components to the Euler angle rates are given by Eqs. (1.48) and (1.45), the latter rewritten as Eq. (1.49) below:

$$\begin{bmatrix} u_1 \\ u_2 \\ u_3 \end{bmatrix} = \begin{bmatrix} cq_5 cq_6 & sq_4 sq_5 cq_6 + sq_6 cq_4 & -cq_4 sq_5 cq_6 + sq_6 sq_4 \\ -cq_5 sq_6 & -sq_4 sq_5 sq_6 + cq_6 cq_4 & cq_4 sq_5 sq_6 + cq_6 sq_4 \\ sq_5 & -sq_4 cq_5 & cq_4 cq_5 \end{bmatrix} \begin{bmatrix} \dot{q}_1 \\ \dot{q}_2 \\ \dot{q}_3 \end{bmatrix} \tag{1.48}$$

$$\begin{bmatrix} u_4 \\ u_5 \\ u_6 \end{bmatrix} = \begin{bmatrix} cq_5 cq_6 & sq_6 & 0 \\ -cq_5 sq_6 & cq_6 & 0 \\ sq_5 & 0 & 1 \end{bmatrix} \begin{bmatrix} \dot{q}_4 \\ \dot{q}_5 \\ \dot{q}_6 \end{bmatrix} \tag{1.49}$$

Here q_1, q_2, q_3 are inertial position coordinates of the mass center and q_4, q_5, q_6 are three Euler angles for attitude. Wheel angles, slosh pendulum angles, and damper location are given by:

$$\begin{aligned}
u_{6+i} &= \dot{q}_{6+i}, i = 1, 2, 3 \\
u_{9+i} &, \dot{q}_{9+i}, i = 1, 2 \\
u_{12} &= \dot{q}_{12} = \dot{\sigma}
\end{aligned} \tag{1.50}$$

Equations (1.48)–(1.50) complete the twelve kinematical equations. The generalized coordinates and their rates,$q_i, \dot{q}_i i = 1, \ldots, 12$, can now be substituted in the kinetic energy expression in terms of the generalized coordinates and their time-derivatives. The expression for K in

Eq. (1.47) resulting from these substitutions is a formidably complex function given below:

$$
\begin{aligned}
K = {} & (1/2)m_G\{[\dot{q}_1 cq_5 cq_6 + \dot{q}_2(sq_4 sq_5 cq_6 + sq_6 cq_4) + \dot{q}_3(-cq_4 sq_5 cq_6 + sq_6 sq_4)]^2 \\
& + [-\dot{q}_1 cq_5 sq_6 + \dot{q}_2(-sq_4 sq_5 sq_6 + cq_6 cq_4) + \dot{q}_3(cq_4 sq_5 sq_6 + cq_6 sq_4)]^2 \\
& + [\dot{q}_1 sq_5 - \dot{q}_2 sq_4 cq_5 + \dot{q}_3 cq_4 cq_5]^2\} + (1/2)\{I_1(\dot{q}_4 cq_5 cq_6 + \dot{q}_5 sq_6)^2 \\
& + I_2(-\dot{q}_4 cq_5 sq_6 + \dot{q}_5 cq_6)^2 + +I_3(\dot{q}_4 sq_5 + \dot{q}_6)^2\} \\
& + (1/2)\{I_1^W[(\dot{q}_4 cq_5 cq_6 + \dot{q}_5 sq_6) + \dot{q}_7]^2 + I_2^W[(-\dot{q}_4 cq_5 sq_6 + \dot{q}_5 cq_6) + \dot{q}_8]^2 \\
& + I_3^W[(\dot{q}_4 sq_5 + \dot{q}_6) + \dot{q}_9]^2\} \\
& + (1/2)m_G\{[\dot{q}_1 cq_5 cq_6 + \dot{q}_2(sq_4 sq_5 cq_6 + sq_6 cq_4) + \dot{q}_3(-cq_4 sq_5 cq_6 + sq_6 sq_4) \\
& + (-\dot{q}_4 cq_5 sq_6 + \dot{q}_5 cq_6)(z_0 - l_P cq_{11}) - (\dot{q}_4 sq_5 + \dot{q}_6) l_P sq_{10} sq_{11} \\
& + l_P(u_{11} cq_{10} cq_{11} - u_{10} sq_{10} sq_{11})]^2 \\
& + [-\dot{q}_1 cq_5 sq_6 + \dot{q}_2(-sq_4 sq_5 sq_6 + cq_6 cq_4) + \dot{q}_3(cq_4 sq_5 sq_6 + cq_6 sq_4) \\
& + (\dot{q}_4 sq_5 + \dot{q}_6) l_P cq_{10} sq_{11} - (\dot{q}_4 cq_5 cq_6 + \dot{q}_5 sq_6)(z_0 - l_P cq_{11}) \\
& + l_P(u_{10} cq_{10} sq_{11} + u_{11} sq_{10} cq_{11})]^2 + [\dot{q}_1 sq_5 - \dot{q}_2 sq_4 cq_5 + \dot{q}_3 cq_4 cq_5 \\
& + l_P(u_4 sq_{10} sq_{11} - u_5 cq_{10} sq_{11} + u_{11} sq_{11})]^2\} \\
& + (1/2)m_Q\{[\dot{q}_1 cq_5 cq_6 + \dot{q}_2(sq_4 sq_5 cq_6 + sq_6 cq_4) + \dot{q}_3(-cq_4 sq_5 cq_6 + sq_6 sq_4) \\
& + (-\dot{q}_4 cq_5 sq_6 + \dot{q}_5 cq_6) z_Q + \dot{q}_{12}]^2 \\
& + [-\dot{q}_1 cq_5 sq_6 + \dot{q}_2(-sq_4 sq_5 sq_6 + cq_6 cq_4) + \dot{q}_3(cq_4 sq_5 sq_6 + cq_6 sq_4) \\
& + (\dot{q}_4 sq_5 + \dot{q}_6)\sigma - (\dot{q}_4 cq_5 cq_6 + \dot{q}_5 sq_6) z_Q]^2 \\
& + [\dot{q}_1 sq_5 - \dot{q}_2 sq_4 cq_5 + \dot{q}_3 cq_4 cq_5 - (-\dot{q}_4 cq_5 sq_6 + \dot{q}_5 cq_6)\sigma]^2\}
\end{aligned}
\tag{1.51}
$$

Lagrange's equations, Eq. (1.46), work with generalized coordinates and require three steps for each of the n generalized coordinates: (1) forming K and differentiating it with respect to the derivatives of the generalized coordinates $\dot{q}_i$; (2) differentiating the resulting expression with respect to time; and (3) differentiating K separately with respect to the generalized coordinates q_i. Looking at Eq. (1.51) for the system in Figure 1.2, it is clear how formidable a task this can be. Not only is the labor prohibitive to an analyst; even when done by a machine using computer algebra, the resulting set of equations will take numerous pages to write. This is in contrast to the simple operations of just dot-products involved in Kane's formulation to obtain the equations. Thus, in terms of required labor alone, Kane's method is superior by far to Lagrange's method.

Another significant reason for the use of Kane's equations – as has already been seen in the example of the Gyrostat spacecraft, and will be repeatedly seen in later examples in this book – is the relative simplicity of the final equations. This is made possible by the fact that Kane uses generalized speeds rather than derivatives of generalized coordinates as the motion variables. Kane's generalized speeds can be thought of as time-derivatives of "quasi-coordinates." A form of Lagrange's equations for quasi-coordinates, also known as Boltzmann-Hamel equations, is far more complex than Eq. (1.46), and hence the advantage of using generalized speeds is completely lost there. A derivation of the Boltzmann-Hamel equations is given in Ref. [8] without details.

1.3.1 Lagrange's Equations in Quasi-Coordinates

Consider the definition of quasi-coordinates, of Eq. (1.1), in a slightly different form and its inverse relation:

$$u_j = \sum_{i=1}^{n} W_{ji}\dot{q}_i + X_j \quad (j = 1, \dots, n); \tag{1.52}$$

$$\dot{q}_k = \sum_{i=1}^{n} Z_{ki}u_i - Y_k \quad (k = 1, \dots, n) \tag{1.53}$$

The system kinetic energy can be represented in two ways, using Eq. (1.53), as follows:

$$K = K(q_1, \dots, q_n, \dot{q}_1, \dots, \dot{q}_n) = \overline{K}(q_1, \dots, q_n, u_1, \dots, u_n) \tag{1.54}$$

This leads to the derivative forms, based on Eq. (1.52), and its time-derivative:

$$\frac{\partial K}{\partial \dot{q}_i} = \sum_{j=1}^{n} \frac{\partial \overline{K}}{\partial u_j}\frac{\partial u_j}{\partial \dot{q}_i} = \sum_{j=1}^{n} \frac{\partial \overline{K}}{\partial u_j} W_{ji}; \tag{1.55}$$

$$\frac{d}{dt}\left\{\frac{\partial K}{\partial \dot{q}_i}\right\} = \sum_{j=1}^{n} \left\{\frac{\partial \overline{K}}{\partial u_j}\right\} \dot{W}_{ji} + \sum_{j=1}^{n} \frac{d}{dt}\left\{\frac{\partial \overline{K}}{\partial u_j}\right\} W_{ji} \tag{1.56}$$

Now, $\dot{W}_{ji}$ required in Eq. (1.55) can be written after recognizing the dependence of W_{ji} on the generalized coordinates:

$$\dot{W}_{ji} = \sum_{k=1}^{n} \frac{\partial W_{ji}}{\partial q_k}\dot{q}_k \tag{1.57}$$

It follows from Eqs. (1.56) and (1.57) then that:

$$\frac{d}{dt}\left\{\frac{\partial K}{\partial \dot{q}_i}\right\} = \sum_{j=1}^{n}\sum_{k=1}^{n} \left\{\frac{\partial \overline{K}}{\partial u_j}\right\} \frac{\partial W_{ji}}{\partial q_k}\dot{q}_k + \sum_{j=1}^{n} \frac{d}{dt}\left\{\frac{\partial \overline{K}}{\partial u_j}\right\} W_{ji} \tag{1.58}$$

Now, replacing from Eq. (1.53) in Eq. (1.58), one has:

$$\frac{d}{dt}\left\{\frac{\partial K}{\partial \dot{q}_i}\right\} = \sum_{j=1}^{n}\sum_{k=1}^{n} \left\{\frac{\partial \overline{K}}{\partial u_j}\right\} \frac{\partial W_{ji}}{\partial q_k}\left[\sum_{i=1}^{n} Z_{ki}u_i - Y_k\right] + \sum_{j=1}^{n} \frac{d}{dt}\left\{\frac{\partial \overline{K}}{\partial u_j}\right\} W_{ji} \tag{1.59}$$

It follows from Eqs. (1.52)–(1.54) that:

$$\frac{\partial K}{\partial q_i} = \frac{\partial \overline{K}}{\partial q_i} + \sum_{j=1}^{n} \frac{\partial \overline{K}}{\partial u_j}\frac{\partial u_j}{\partial q_i} = \frac{\partial \overline{K}}{\partial q_i} + \sum_{j=1}^{n} \frac{\partial \overline{K}}{\partial u_j}\left\langle \sum_{k=1}^{n} \frac{\partial W_{jk}}{\partial q_i}\dot{q}_k + \frac{\partial X_k}{\partial q_i} \right\rangle; \quad i = 1, \ldots, n \quad (1.60)$$

Again, replacing from Eq. (1.53) in Eq. (1.60) gives:

$$\frac{\partial K}{\partial q_i} = \frac{\partial \overline{K}}{\partial q_i} + \sum_{j=1}^{n} \frac{\partial \overline{K}}{\partial u_j}\left\langle \sum_{k=1}^{n} \frac{\partial W_{jk}}{\partial q_i}\left[\sum_{i=1}^{n} Z_{ki}u_i - Y_k\right] + \frac{\partial X_k}{\partial q_i} \right\rangle; \quad i = 1, \ldots, n \quad (1.61)$$

From Eqs. (1.59) and (1.61) we can complete the form of Boltzmann-Hamel equations:

$$\begin{aligned} &\sum_{j=1}^{n}\sum_{k=1}^{n}\left\{\frac{\partial \overline{K}}{\partial u_j}\right\}\frac{\partial W_{ji}}{\partial q_k}\left[\sum_{i=1}^{n} Z_{ki}u_i - Y_k\right] + \sum_{j=1}^{n}\frac{d}{dt}\left\{\frac{\partial \overline{K}}{\partial u_j}\right\}W_{ji} - \frac{\partial \overline{K}}{\partial q_i} \\ &- \sum_{j=1}^{n}\frac{\partial \overline{K}}{\partial u_j}\left\langle \sum_{k=1}^{n}\frac{\partial W_{jk}}{\partial q_i}\left[\sum_{i=1}^{n} Z_{ki}u_i - Y_k\right] + \frac{\partial X_k}{\partial q_i}\right\rangle = F_i \qquad i = 1, \ldots, n \end{aligned} \quad (1.62)$$

Here F_i is the generalized active force in the ith quasi-coordinate, which is formed exactly as in Kane's method using generalized speeds. Note from Eq. (1.62) that one needs to invert the coefficient matrix in the definition of the quasi-coordinates in Eq. (1.52) and compute n^3 number of non-vanishing quantities in the following term in Eq. (1.62):

$$\sum_{j=1}^{n}\sum_{k=1}^{n}\left[\left\{\frac{\partial \overline{K}}{\partial u_j}\right\}\frac{\partial W_{ji}}{\partial q_k}\right]\sum_{i=1}^{n} Z_{ki}u_i - \sum_{j=1}^{n}\frac{\partial \overline{K}}{\partial u_j}\sum_{k=1}^{n}\frac{\partial W_{jk}}{\partial q_i}\sum_{i=1}^{n} Z_{ki}u_i \qquad (i, j, k = 1, \ldots, n)$$

For the 12 degree-of-freedom system of Figure 1.1, this means forming 1,728 such terms after forming the inverse implied in Eq. (1.53). It is clear that the Boltzmann-Hamel equations, Eq. (1.62), are *prohibitively labor intensive* to use as a tool in deriving equations of motion for even rigid multibody dynamics models, as Likins reports in Ref. [5]. It seems, as has been noted in Ref. [6], that the fact that Boltzmann-Hamel equations are used at all is surprising, except perhaps to check the correctness of equations derived by other methods. As has been shown via a less complex example in Ref. [6], Kane's equations are operationally the simplest among all analytical methods available for formulating equations of motion, and with a choice of generalized speeds they produce the simplest form of the final equations, as has been indicated in the example of the gyrostat spacecraft. This latter fact greatly reduces computer simulation time.

Reader's Exercise

Consider the double pendulum problem given before, and derive Lagrange's equation in quasi-coordinates, Eq. (1.62), for it using Eq. (1.10) as the quasi-coordinates. See that one indeed forms n^3 or eight non-vanishing quantities in forming Eq. (1.62).

1.4 Kane's Method of Direct Derivation of Linearized Dynamical Equation

It is often necessary to derive linearized dynamical equations, as for example in considering the stability of an equilibrium state, or when some variables are assumed to stay small as in the case of elastic deformation of flexible bodies undergoing large overall motion. Linearized equations can, of course, be derived by first deriving the nonlinear equations and then linearizing them by discarding all nonlinear terms. A simpler method, of directly deriving the linear equations, *without deriving the nonlinear equations*, has been given by Kane in Ref. [3]. *This direct method of linearization consists of the two steps to be followed in sequence:*

1. First, form the partial velocities of all particles and mass centers of rigid bodies, and partial angular velocities of all rigid bodies from the corresponding full, *nonlinear* expressions for velocity and angular velocity, and *then* linearize these nonlinear partial velocity and partial angular velocity expressions.
2. Now linearize the velocity and angular velocity expressions, differentiate these to get accelerations and angular accelerations, and form the generalized inertia force and generalized active force expressions by dot-multiplications with the linearized partial velocity and partial angular velocity expressions, discarding all nonlinear terms.

Kane and Levinson [3] have illustrated the application of this procedure to a complex problem. The following provides an analytical basis of this process. Consider an n-dof system, with a particle P_k whose velocity in a Newtonian frame N is given by Eq. (1.2), and let its ith partial velocity vector be expanded in a Taylor series in deviations about the equilibrium state of $q_1 = \cdots = q_n = 0; u_1 = \cdots = u_n = 0$:

$$^{\mathrm{N}}\mathbf{v}_i^{P_k} = \mathbf{f}_i^k(q_1, \ldots, q_n, t) = \mathbf{f}_i^k(0, \ldots, 0, t) + \sum_{j=1}^{n} \left[\frac{\partial \mathbf{f}_i^k}{\partial q_j}\right]_0 q_j + \text{h.o.t} \tag{1.63}$$

Here the derivatives are evaluated about the zero state, $q_1 = \cdots = q_n = 0$, and h.o.t stands for higher-order terms in qs. Because of Eq. (1.2) and Eq. (1.63), the velocity of P_k in N about the equilibrium state becomes:

$$^{N}\mathbf{v}^{P_k} = \sum_{i=1}^{n} \left[\mathbf{f}_i^k(0, \ldots, 0, t) + \sum_{j=1}^{n} \left[\frac{\partial \mathbf{f}}{\partial q_j}\right]_0 q_j + \text{h.o.t}\right] u_i + {}^{N}\mathbf{v}_t^{P_k} \tag{1.64}$$

Equation (1.64) is a nonlinear expression of the velocity, because of the product of qs and us. Here the coefficient vector of u_i, $i = 1, \ldots, n$, which is nonlinear, is the ith partial velocity, and linearizing it in the generalized coordinates, qs, correctly linearized ith partial velocity of P_k is:

$$^{N}\tilde{\mathbf{v}}_i^{P_k} = \mathbf{f}_i^k(0, \ldots, 0, t) + \sum_{j=1}^{n} \left[\frac{\partial \mathbf{f}_i^k}{\partial q_j}\right]_0 q_j \tag{1.65}$$

Linearized expression for velocity of P_k in N is obtained from Eq. (1.64) by ignoring products of *u*s and *q*s, yielding:

$$^{N}\tilde{\mathbf{v}}^{P_k} = \sum_{i=1}^{n} \left[\mathbf{f}_i^k(0, \ldots, 0, t)\right] u_i + {^{N}\mathbf{v}_t^{P_k}} \tag{1.66}$$

Partial velocity of P_k with respect to the *i*th generalized speed, obtained from this linear velocity, Eq. (1.66) is:

$$^{N}\tilde{\mathbf{v}}_i^{P_k} = \mathbf{f}_i^k(0, \ldots, 0, t) \tag{1.67}$$

Comparing Eq. (1.67) with Eq. (1.65), we see that the expression for partial velocity obtained from the linearized velocity, Eq. (1.66), has already lost a term involving *q*s. This is due to an error of premature linearization, and we will see that this makes a crucial difference between correct and incorrect linear equations. A similar analysis can be done to show that the correct partial angular velocity of a rigid body is obtained only by linearizing the partial angular velocity obtained from the nonlinear angular velocity expression.

The foregoing analysis shows that the correctly linearized partial velocity / partial angular velocity already contains terms involving zeroth and first-degree terms in the *q*s. Thus, in taking dot-products of these quantities with the acceleration/angular acceleration terms – a process involving products of *q*s and *u*s – one has to remove the nonlinear terms anyway to get equations linear in *q*s and *u*s. This justifies the second step in Kane's way of directly deriving the linearized equations.

This process of deriving linearized equations will be demonstrated here by means of a simple example shown in Figure 1.5, where a massless rod of length L is attached to a rotating

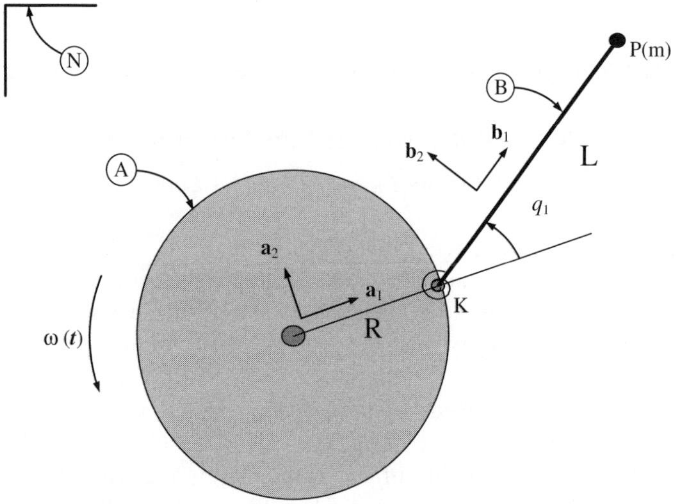

Figure 1.5 Elastically Restrained Bar Connected to a Table with Prescribed Rotation.

table by a torsion spring, with a particle P of lumped mass m, at the end of the rod. The table has a prescribed angular velocity $\omega(t)$, and the orientation of the rod is given by the angle $q_1(t)$.

We introduce a generalized speed,

$$u_1 = \dot{q}_1 \tag{1.68}$$

In terms of the basis vectors $\mathbf{a}_1, \mathbf{a}_2$ fixed in the table, the nonlinear expression for velocity of the particle P located by the position vector $L[\cos q_1 \mathbf{a}_1 + \sin q_1 \mathbf{a}_2]$, is given by:

$$^{\mathrm{N}}\mathbf{v}^P = \omega(t)R\mathbf{a}_2 + [\omega(t) + u_1]L[\cos q_1 \mathbf{a}_2 - \sin q_1 \mathbf{a}_1] \tag{1.69}$$

which may be rewritten in the format of Eq. (1.63) as:

$$^{\mathrm{N}}\mathbf{v}^P = L[\cos q_1 \mathbf{a}_2 - \sin q_1 \mathbf{a}_1]u_1 + \omega(t)\{[R + L\cos q_1]\mathbf{a}_2 - \sin q_1 \mathbf{a}_1]\} \tag{1.70}$$

Using the notations used above, the nonlinear partial velocity of P becomes:

$$\mathbf{f}_1^P = L[\cos q_1 \mathbf{a}_2 - \sin q_1 \mathbf{a}_1] \tag{1.71}$$

Linearizing this nonlinear partial velocity, following Eq. (1.65), we get the correctly linearized partial velocity:

$$^{N}\tilde{\mathbf{v}}_1^P = \mathbf{f}_1^P(0) + \left[\frac{\partial \mathbf{f}_1^P}{\partial q_1}\right]_0 q_1 = L[\mathbf{a}_2 - q_1 \mathbf{a}_1] \tag{1.72}$$

Now linearize the velocity, Eq. (1.69), to get:

$$^{\mathrm{N}}\tilde{\mathbf{v}}^P = -\omega q_1 L\mathbf{a}_1 + [\omega(R + L) + Lu_1]\mathbf{a}_2 \tag{1.73}$$

The linearized acceleration of P in N is obtained by differentiating Eq. (1.73) in N:

$$\begin{aligned} ^{\mathrm{N}}\tilde{\mathbf{a}}^P &= -[\dot{\omega} q_1 + \omega u_1]L\mathbf{a}_1 + [\dot{\omega}(R + L) + L\dot{u}_1]\mathbf{a}_2 \\ &\quad + \omega\{-\omega q_1 L\mathbf{a}_2 - [\omega(R + L) + Lu_1)]\mathbf{a}_2\} \end{aligned} \tag{1.74}$$

Finally, dot-multiplying the linearized partial velocity, Eq. (1.72), with mass times the acceleration, Eq. (1.74), linearized in q_1, and removing resulting nonlinear terms gives the linearized generalized inertia force. The linearized generalized active force being simply $-kq_1$, Kane's way of directly deriving linearized dynamical equations gives for this system:

$$\dot{u}_1 = -[k/(mL^2) + (R/L)\omega^2]q_1 - (1 + R/L)\dot{\omega} \tag{1.75}$$

Equation (1.75) shows that the natural stiffness, $k/(mL^2)q_1$, is augmented by $(R/L)\omega^2 q_1$. This is clearly identifiable as centrifugal stiffening, where the effective stiffness increases with the rotational speed of the base. It can be checked by the method of Section 1.2 that the full, nonlinear equation of the system in Figure 1.5 is:

$$\dot{u}_1 = -k/(mL^2)q_1 - (R/L)\omega^2 \sin q_1 - [1 + (R/L)\cos q_1]\dot{\omega} \tag{1.76}$$

Linearization of this equation in q_1 would indeed give Eq. (1.75), verifying that Eq. (1.75) is the correctly linearized equation for this system.

The above approach of directly deriving linearized dynamical equations, without deriving the full, nonlinear equations, is a special feature of Kane's method that has no analog in Lagrange's method. The method has been extended to special cases of continua like beams and plates in Refs. [10, 11]. The direct linearization method, however, is not extensible to general elastic continua, because it is extremely difficult to derive the velocity expressions for nonlinear deformation in general continua. In search for a general method that does work for arbitrary continua, we will presently discuss, rather paradoxically, a procedure of premature linearization that actually produces incorrect equations but can, however, be redeemed to get the correct linear equations, with a posteriori addition of load-dependent geometric stiffness due to inertia loads [12].

1.5 Prematurely Linearized Equations and a Posteriori Correction by *ad hoc* Addition of Geometric Stiffness due to Inertia Loads

In directly deriving the linear equation without deriving the nonlinear equations, premature linearization occurs when the partial velocities of particles and mass centers of rigid bodies, as well as partial angular velocities of rigid bodies, are obtained from the linearized velocities and angular velocities. Ref. [3] hints at this without going into details. We will first show that premature linearization leads to incorrect equations, with loss of structural stiffness with motion, but going beyond, we show how to compensate for the error in stiffness by adding load-dependent geometric stiffness. Again, the ideas are best explained by a simple example. So we will consider the same example as in Section 1.4. We repeat the expression for linearized velocity of P in N here from Eq. (1.73): ${}^{\mathrm{N}}\tilde{\mathbf{v}}^P = -\omega q_1 L\mathbf{a}_1 + [\omega(R+L) + Lu_1]\mathbf{a}_2$.

Prematurely linearized partial velocity with respect to u_1 from this linearized velocity is:

$${}^{N}\hat{\mathbf{v}}_1^P = L\mathbf{a}_2 \tag{1.77}$$

Note that we have already lost a term in Eq. (1.77) from the partial velocity expression, Eq. (1.72). Dot-multiplying this partial velocity expression with mass times the linearized acceleration from Eq. (1.74) gives the prematurely linearized generalized inertia force. The linearized generalized active force being simply $-kq_1$, linearized equations for the system with premature linearization can be written after rearranging terms as:

$$\dot{u}_1 = -[k/\mathrm{m}\mathrm{L}^2 - \omega^2]q_1 - (R/L + 1)\dot{\omega} \tag{1.78}$$

It shows that the effective natural frequency of the system shown in Figure 1.6 decreases with any rotational speed of the base! Obviously this is physically wrong, as centrifugal force would make the rod in Figure 1.5 more inclined to stay straight along $\mathbf{a}_1$ and hence it would be stiffer to deflect, and so Eq. (1.78) is incorrect. The error in loss of stiffness can be traced to the incorrect partial velocity in Eq. (1.77), since multiplication with linearized acceleration is common to both approaches.

When is the error due to premature linearization tolerable? It can be seen from Eq. (1.78) that when the rotation rate of the frame, ω, is far smaller than the natural frequency of the structure – that is, $\omega^2 \ll [k/(\mathrm{mL}^2)]$ – the error committed is small. Thus, in such situations of relatively very small rotation rates, the indigenous error due to premature linearization may be acceptable. Certain simplifications are then acceptable, and we will illustrate that in Chapter 6.

It may be noted in passing that error due to using linearized velocity also affects other methods, such as Lagrange's equation. Thus, computing the kinetic energy for this example problem based on the linearized velocity given by Eq. (1.72), one obtains:

$$K = \frac{m}{2}[\omega^2 q_1^2 L^2 + \omega^2 (R+L)^2 + L^2 \dot{q}_1^2 + 2\omega(R+L)L\dot{q}_1] \tag{1.79}$$

Upon substitution in Lagrange's equation with generalized force due to the torsional spring,

$$\frac{d}{dt}\frac{\partial K}{\partial \dot{q}_1} - \frac{\partial K}{\partial q_1} = -kq_1 \tag{1.80}$$

the reader can check that one gets the same incorrect equation as Eq. (1.78).

Ironically, for most general elastic bodies undergoing large overall motion, where vibration modes are used to describe small deformation, one is innately doing premature linearization, since vibration modes arise from a linear theory. Hence, it is of interest to see if there is a way to redeem the incorrect equation, a posteriori, with some appropriate augmentation of stiffness. In fact, this is possible by appealing to the notion of geometric stiffness due to loads [12]. To illustrate this, we continue with the same example problem.

Cook [13] has shown that for a bar or rod type of a structure, geometric stiffness due to an axial load F_a is expressed as follows:

$$\begin{Bmatrix} F_1 \\ F_2 \end{Bmatrix} = \begin{Bmatrix} \dfrac{\partial}{\partial \delta_1} \\ \dfrac{\partial}{\partial \delta_2} \end{Bmatrix} K_G; \quad K_G = \frac{1}{2}\frac{F_a}{L}[\delta_1 \ \delta_2] \begin{bmatrix} 1 & -1 \\ -1 & 1 \end{bmatrix} \begin{Bmatrix} \delta_1 \\ \delta_2 \end{Bmatrix} \tag{1.81}$$

Here δ_1, δ_2 are transverse end displacements of the bar under axial inertia load and F_a. For the rod in Figure 1.6, end displacements and load are: $\delta_1 = 0, \delta_2 = Lq_1; F_a = m\omega^2(R+L)$. Equation (1.81) gives:

$$K_G = 0.5m\omega^2(R+L)L(q_1)^2 \tag{1.82}$$

Generalized active force due to potential energy in geometric stiffness is defined as:

$$F_G = -\frac{dK_G}{dq_1} = -m\omega^2(R + L)Lq_1 \tag{1.83}$$

Adding this term, a posteriori, to the right-hand side of the prematurely linearized equation, Eq. (1.78), produces:

$$mL^2\dot{u}_1 = -[k + mRL\omega^2]q_1 - mL(R + L)\dot{\omega} \tag{1.84}$$

Equation (1.84) is precisely the same as Eq. (1.75) obtained by direct linearization in the previous section. In conclusion, we have shown here by a simple example that the error in loss of motion-induced stiffness due to premature linearization can be rectified, a posteriori, by adding geometric stiffness due to the relevant inertia load. This idea is developed more fully for an arbitrary flexible body in Ref. [12].

1.6 Kane's Equations with Undetermined Multipliers for Constrained Motion

For systems with constraints on the motion, the constraints may be of the (holonomic) form,

$$\phi_i(q_1, \ldots, q_n) = 0, \quad i = 1, \ldots, m \tag{1.85}$$

or in the non-holonomic, that is, non-integrable form stated in terms of generalized speeds [3]:

$$\sum_{j=1}^{n} a_{ij}u_j + b_i = 0, \quad i = 1, \ldots, m \tag{1.86}$$

Kane [1–3] eliminates m of the generalized speeds from Eq. (1.86) to work with (n-m) independent generalized speeds, developing non-holonomic velocities and angular velocities, non-holonomic partial velocities and partial angular velocities, and associated generalized active and inertia forces. See Ref. [3] for details. If the constraints are "workless" – that is, their satisfaction involves no work of the constraint forces – the method produces a reduced set of differential equations of motion and is computationally efficient, though the labor involved in solving Eq. (1.86) should not be overlooked. However, if the constraint forces are also of interest, then a modified version of Kane's method, called *Kane's method with undetermined multipliers* [14], may well have an advantage. The method is described as follows.

The velocity and angular velocity constraints can always be written as m constraint equations in the n generalized speeds:

$$\sum_{j=1}^{n} \frac{\partial \mathbf{v}^i}{\partial u_j} u_j + \mathbf{v}_t^i = 0, \quad i = 1, \ldots., m \tag{1.87}$$

Here $\mathbf{v}^i$ is a vector that stands symbolically for either the ith particle velocity or the ith body frame angular velocity in the Newtonian frame N. Now Eq. (1.87) can be differentiated in N with respect to time to get the corresponding acceleration form of the constraint equations, including the acceleration remainder terms, that absorb all terms *not* involving $\dot{u}_j$:

$$\sum_{j=1}^{n} \frac{\partial \mathbf{v}^i}{\partial u_j} \dot{u}_j + \mathbf{a}_t^i = 0, \quad i = 1, \ldots., m \tag{1.88}$$

Now if certain forces and torques are at play to maintain the m number of constraints, given by Eq. (1.86), then if the constraints are relaxed, these forces will do work. When the m number of as yet undetermined force measure numbers $\lambda_1, \ldots., \lambda_m$, representing scalar values of forces acting at the points or torques on frames, is introduced, the generalized active forces can be evaluated due to forces/torques in the augmented set of generalized speeds, in the usual manner. Since constraint forces do no work when the constraints are active, the ideal constraint force is assumed to be in the direction perpendicular to the partial velocity of the point constrained, and the right-hand side term in Eq. (1.89), representing generalized force due to constraint forces, will be as follows when the constraints are relaxed,

$$F_i^C = \sum_{j=1}^{m} \frac{\partial \mathbf{v}^j}{\partial u_i} \cdot \lambda_j \left(\frac{\partial \mathbf{v}^j}{\partial u_i} \middle/ \left| \frac{\partial \mathbf{v}^j}{\partial u_i} \right| \right) \qquad i = 1, \ldots, n; j = 1, \ldots, m \tag{1.89}$$

where $\mathbf{v}^j$ is the velocity of the point (or angular velocity of the frame) j. Equation (1.88) can be written in the matrix form shown below, where A is an (mxn) matrix and B and U are column matrices with m elements:

$$A\dot{U} + B = 0 \tag{1.90}$$

Similarly, the matrix form of Eq. (1.89), the generalized forces due to constraints, is written in terms of the A-matrix and an $(m \times 1)$ matrix of so-called undetermined multipliers, Λ:

$$F^c = A^t \Lambda \tag{1.91}$$

1.7 Summary of the Equations of Motion with Undetermined Multipliers for Constraints

Equations of motion with constraint equations in the augmented set of generalized speeds are:

$$M\dot{U} = C + F + A^t \Lambda \tag{1.92}$$

$$A\dot{U} + B = 0 \tag{1.93}$$

From these equations, the constraint force measures and the equations of motion are determined as:

$$\Lambda = -(AM^{-1}A^t)^{-1}[AM^{-1}(C+F)+B] \tag{1.94}$$

$$\dot{U} = M^{-1}(C+F+A^t\Lambda) \tag{1.95}$$

This concludes the formulation of the equations of motion with constraints.

1.8 A Simple Application

Consider the slider crank shown in Figure 1.6. It is actually a single-dof system, with the constraint that the vertical displacement of the slider is zero. To maintain this constraint, there is a vertical force R applied by the slider. If we want to determine this force R, we relax the constraint and augment the degrees of freedom of the two-link system by one, as shown at the lower part of the figure.

The constraint equation follows from the bottom part of the figure to match the actual system:

$$r\sin q_1 + l\sin(q_1+q_2) = 0 \tag{1.96a}$$

Differentiating the above equation with respect to time twice yields:

$$\begin{aligned}&\left[r\cos q_1 + l\cos(q_1+q_2) \quad l\cos(q_1+q_2)\right]\begin{Bmatrix}\ddot{q}_1\\ \ddot{q}_2\end{Bmatrix}\\ &+\{-r(\dot{q}_1)^2\sin q_1 - l(\dot{q}_1+\dot{q}_2)^2\sin(q_1+q_2)\} = 0\end{aligned} \tag{1.96b}$$

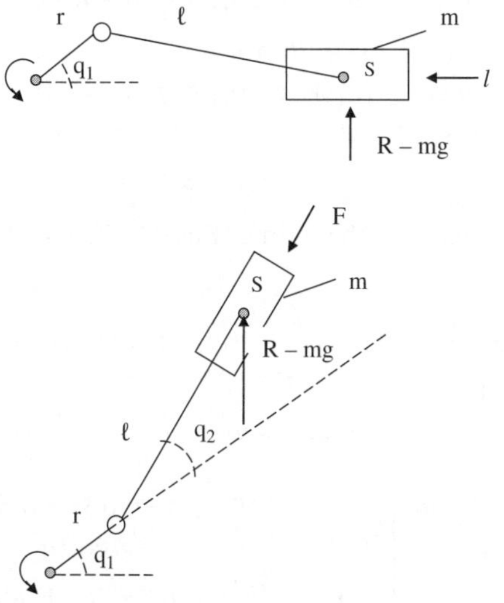

Figure 1.6 Slider Crank, Actual and Unconstrained Versions.

Defining the augmented generalized speeds, $U = [\dot{q}_1; \dot{q}_2]$, this conforms to Eq. (1.93) by:

$$A = \begin{bmatrix} A_{11} & A_{12} \end{bmatrix}; A_{11} = r\cos q_1 + l\cos(q_1 + q_2); A_{12} = l\cos(q_1 + q_2);$$
$$B = \{-r(\dot{q}_1)^2 \sin q_1 - l(\dot{q}_1 + \dot{q}_2)^2 \sin(q_1 + q_2)\} \tag{1.97}$$

Using Kane's method for the augmented set of generalized speeds for this system, the equations of motion of the slider under external force F_e and gravity and reactive load R are written by expanding the terms in Eq. (1.92). For simplicity we assume that the rods are massless, and only the slider has mass m. This yields:

$$\begin{bmatrix} M_{11} & M_{12} & A_{11} \\ M_{12} & M_{22} & A_{12} \\ A_{11} & A_{12} & 0 \end{bmatrix} \begin{Bmatrix} \ddot{q}_1 \\ \ddot{q}_2 \\ -R \end{Bmatrix} + \begin{Bmatrix} C_1 \\ C_2 \\ 0 \end{Bmatrix} = \begin{Bmatrix} F_1 nr \\ F_{2nr} \\ B \end{Bmatrix} \tag{1.98}$$

Here F_{1nr}, F_{2nr} are parts of the generalized force not coming from reactions, but rather due to gravity and any other external forces, and:

$$\begin{aligned}
M_{11} &= m\left[r^2 + l^2 + rl\cos q_2\right];\\
M_{12} &= m[rl\cos q_2 + l^2];\\
M_{22} &= ml^2;\\
C_1 &= m(\dot{q}_2^2 + 2\dot{q}_1\dot{q}_2)rl\sin q_2;\\
C_2 &= -mrl\,\dot{q}_1^2 \sin q_2;\\
F_1 &= -F_e r\sin q_2 - mg[r\cos q_1 + l\cos(q_1 + q_2)] + R[r\cos q_1 + l\cos(q_1 + q_2)]\\
F_2 &= -mgl\cos(q_1 + q_2) + Rl\cos(q_1 + q_2)\\
F_{1nr} &= -F_e r\sin q_2 - mg[r\cos q_1 + l\cos(q_1 + q_2)]\\
F_{2nr} &= -mgl\cos(q_1 + q_2)
\end{aligned} \tag{1.99}$$

These equations can be combined in the form of Eqs. (1.92) and (1.93) as:

$$\begin{bmatrix} m\left[r^2 + l^2 + rl\cos q_2\right] & m[rl\cos q_2 + l^2] & [r\cos q_1 + l\cos(q_1 + q_2)] \\ m[rl\cos q_2 + l^2] & ml^2 & [l\cos(q_1 + q_2)] \\ [r\cos q_1 + l\cos(q_1 + q_2)] & [l\cos(q_1 + q_2)] & 0 \end{bmatrix} \begin{Bmatrix} \ddot{q}_1 \\ \ddot{q}_1 \\ -R \end{Bmatrix}$$
$$= \begin{Bmatrix} -m(\dot{q}_2^2 + 2\dot{q}_1\dot{q}_2)rl\sin q_2 \\ mrl\dot{q}_1^2 \sin q_2 \\ 0 \end{Bmatrix} + \begin{Bmatrix} -F_e r\sin q_2 - mg[r\cos q_1 + l\cos(q_1 + q_2)] \\ -mgl\cos(q_1 + q_2) \\ -r(\dot{q}_1^2)\sin q_1 - l(\dot{q}_1 + \dot{q}_2)^2 \sin(q_1 + q_2) \end{Bmatrix} \tag{1.100}$$

Appendix 1.A Guidelines for Choosing Efficient Motion Variables in Kane's Method

1.A.1 [Contributed by Prof. Paul Mitiguy of Stanford University]

The guidelines for choosing efficient motion variables for Kane's method are given in Ref. [15]. The underlying rationale is to choose variables that produce the simplest possible expressions for inertial angular velocities of rigid bodies (which figure prominently in expressing other important quantities such as velocities, accelerations, partial angular velocities, partial velocities, angular accelerations, inertia torques, and kinetic energy). This is illustrated with an example.

Figure 1.A.1 shows the space shuttle with its manipulator arm, a system made up of the shuttle, body *A* in the figure, with a revolving body *B*, and the manipulator links *C, D*, and *E*. Bodies *B, C*, and *D* are mounted on revolute joints, and *E* has a ball-and-socket joint. While this is really a 12 dof system, in the force-free condition, the constraint of conservation of linear momentum reduces by 3 degrees of freedom, yielding a 9 dof system. In the following description of the configuration of the system, in terms of the various vector bases we use bold notation instead of the overbar shown for them in the figure.

Body A rotates freely in inertial frame N, and its orientation in N can be described by three successive right-handed rotations characterized by $q_1\mathbf{a}_x, q_2\mathbf{a}_y, q_3\mathbf{a}_z$; body B is connected at the center of mass of A at a revolute joint and turns through the angle $q_4\mathbf{b}_y$; body C is connected at the center of mass of A at a revolute joint and turns through the angle $q_5\mathbf{c}_z$; body D is connected to body C at point P by a revolute joint and turns through the angle $q_6\mathbf{d}_z$; body *E* is connected to body D at a spherical joint subjected to relative right-hand rotations successively about $q_7\mathbf{e}_x, q_8\mathbf{e}_y, q_9\mathbf{e}_z$. In the table that follows we show three sets of possible choices for generalized speeds, dubbed as simple, customary, and efficient, for this problem. Figure 1.A.2 lists the kinematical equations corresponding to the customary and efficient choices for generalized speeds. Figure 1.A.3 shows the dynamical equations corresponding to the generalized speed u_2, for the customary choice and the efficient choice, with $c_4 = \cos q_4, s_{56} = \sin(q_5 + q_6), t_4 = \tan(q_4)$, and so on. This shows how the equations become simpler with efficient choice of generalized speeds, and illustrates how the choice of motion variables makes a vast difference in the final equations.

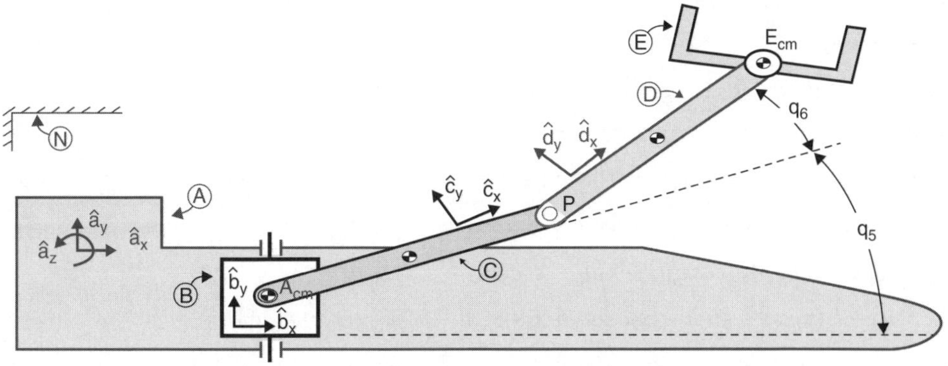

Figure 1.A.1 The Shuttle with Its Manipulator Arm.

Simple	Customary	Efficient
$u_1 = \dot{q}_1$	$u_1 = {}^N\boldsymbol{\omega}^A \cdot \mathbf{a}_x$	$u_1 = {}^N\boldsymbol{\omega}^A \cdot \mathbf{a}_x$
$u_2 = \dot{q}_2$	$u_2 = {}^N\boldsymbol{\omega}^A \cdot \mathbf{a}_y$	$u_2 = {}^N\boldsymbol{\omega}^A \cdot \mathbf{a}_y$
$u_3 = \dot{q}_3$	$u_3 = {}^N\boldsymbol{\omega}^A \cdot \mathbf{a}_z$	$u_3 = {}^N\boldsymbol{\omega}^A \cdot \mathbf{a}_z$
$u_4 = \dot{q}_4$	$u_4 = \dot{q}_4$	$u_4 = {}^N\boldsymbol{\omega}^B \cdot \mathbf{b}_y$
$u_5 = \dot{q}_5$	$u_5 = \dot{q}_5$	$u_5 = {}^N\boldsymbol{\omega}^C \cdot \mathbf{c}_z$
$u_6 = \dot{q}_6$	$u_6 = \dot{q}_6$	$u_6 = {}^N\boldsymbol{\omega}^D \cdot \mathbf{d}_z$
$u_7 = \dot{q}_7$	$u_7 = {}^D\boldsymbol{\omega}^E \cdot \mathbf{e}_x$	$u_7 = {}^N\boldsymbol{\omega}^E \cdot \mathbf{e}_x$
$u_8 = \dot{q}_8$	$u_8 = {}^D\boldsymbol{\omega}^E \cdot \mathbf{e}_x$	$u_8 = {}^N\boldsymbol{\omega}^E \cdot \mathbf{e}_x$
$u_9 = \dot{q}_9$	$u_9 = {}^D\boldsymbol{\omega}^E \cdot \mathbf{e}_z$	$u_9 = {}^N\boldsymbol{\omega}^E \cdot \mathbf{e}_z$

Customary

$$\dot{q}_1 = (c_3u_1 - s_3u_2)/c_2$$
$$\dot{q}_2 = s_3u_1 + c_3u_2$$
$$\dot{q}_3 = u_3 + t_2(s_3u_2 - c_3u_1)$$
$$\dot{q}_{3+i} = u_{3+i} \quad i = 1, 2, 3$$
$$\dot{q}_7 = (c_9u_7 - s_9u_8)/c_8$$
$$\dot{q}_8 = s_9u_7 + c_9u_8$$
$$\dot{q}_9 = u_9 + t_8(s_9u_8 - c_9u_7)$$

Efficient

$$\dot{q}_1 = (c_3u_1 - s_3u_2)/c_2$$
$$\dot{q}_2 = s_3u_1 + c_3u_2$$
$$\dot{q}_3 = u_3 + t_2(s_3u_2 - c_3u_1)$$
$$\dot{q}_4 = u_4 - u_2$$
$$\dot{q}_5 = u_5 - (s_4u_1 + c_3u_3)$$
$$\dot{q}_6 = u_6 - u_5$$
$$\dot{q}_7 = (c_7s_8u_6 + c_9u_7 - s_9u_8)/c_8 + (s_4c_{56} - s_7t_8s_{56})u_3 + c_4(s_7t_8s_{56} - c_{56})u_1 - (s_7t_8c_{56} + s_{56})u_2$$
$$\dot{q}_8 = s_9u_7 + c_9u_8 + c_4c_7s_{56}u_1 - s_7u_6 - c_7c_{56}u_4 - s_4c_7s_{56}u_3$$
$$\dot{q}_9 = u_9 + t_8(s_9u_8 - c_9u_7) + (s_7c_{56}u_4 + s_4s_7s_{56}u_3 - c_7u_6 - s_7c_4s_{56}u_1)/c_8$$

Figure 1.A.2 Definition and Kinematical Equations for Customary and Efficient Generalized Speeds.

Customary

$$
\begin{aligned}
&\dot{u}_2 = (I_1^B - I_3^B)(s_4u_1+c_4u_3)(s_4u_3-c_4u_1)+c_5(I_1^C(u_5+s_4u_1+c_4u_3)(s_4c_5u_3-s_5u_2-s_5u_4-c_4c_5u_1)-I_3^C(u_5+\\
&s_4u_1+c_4u_3)(s_4c_5u_3-s_5u_2-s_5u_4-c_4c_5u_1)-I_2^C(s_5u_4(s_4u_1+c_4u_3)-u_5(s_5(u_2+u_4)-c_5(s_4u_3-c_4u_1))))+\\
&c_{56}(I_1^D(u_5+u_6+s_4u_1+c_4u_3)(s_4c_{56}u_3-s_{56}u_2-s_{56}u_4-c_4c_{56}u_1)-I_3^D(u_5+u_6+s_4u_1+c_4u_3)(s_4c_{56}u_3-\\
&s_{56}u_2-s_{56}u_4-c_4c_{56}u_1)-I_2^D(s_6(c_5u_4(s_4u_1+c_4u_3)-u_5(c_5(u_2+u_4)+s_5(s_4u_3-c_4u_1)))+c_6(s_5u_4(s_4u_1+\\
&c_4u_3)-u_5(s_5(u_2+u_4)-c_5(s_4u_3-c_4u_1)))-u_6(s_6(c_5u_2+c_5u_4+s_4s_5u_3-s_5c_4u_1)-c_6(s_4c_5u_3-s_5u_2-s_5u_4-\\
&c_4c_5u_1))))+(s_9c_8s_{56}-c_{56}(c_7c_9-s_7s_8s_9))(I_1^E(u_9+c_7c_8u_5+c_7c_8u_6+(s_8s_{56}-s_7c_8c_{56})u_2+(s_8s_{56}-s_7c_8c_{56})u_4+\\
&(s_4c_7c_8+s_8c_4c_{56}+s_7c_4c_8s_{56})u_1+(c_4c_7c_8-s_4s_8c_{56}-s_4s_7c_8s_{56})u_3)(u_7+(s_7s_9-s_8c_7c_9)u_5+(s_7s_9-s_8c_7c_9)u_6+\\
&(c_8c_9s_{56}+c_{56}(s_9c_7+s_7s_8c_9))u_2+(c_8c_9s_{56}+c_{56}(s_9c_7+s_7s_8c_9))u_4+(c_4c_8c_9c_{56}+s_4(s_7s_9-s_8c_7c_9)-c_4s_{56}(s_9c_7+\\
&s_7s_8c_9))u_1-(s_4c_8c_9c_{56}-c_4(s_7s_9-s_8c_7c_9)-s_4s_{56}(s_9c_7+s_7s_8c_9))u_3)-I_3^E(u_9+c_7c_8u_5+c_7c_8u_6+(s_8s_{56}-\\
&s_7c_8c_{56})u_2+(s_8s_{56}-s_7c_8c_{56})u_4+(s_4c_7c_8+s_8c_4c_{56}+s_7c_4c_8s_{56})u_1+(c_4c_7c_8-s_4s_8c_{56}-s_4s_7c_8s_{56})u_3)(u_7+\\
&(s_7s_9-s_8c_7c_9)u_5+(s_7s_9-s_8c_7c_9)u_6+(c_8c_9s_{56}+c_{56}(s_9c_7+s_7s_8c_9))u_2+(c_8c_9s_{56}+c_{56}(s_9c_7+s_7s_8c_9))u_4+\\
&(c_4c_8c_9c_{56}+s_4(s_7s_9-s_8c_7c_9)-c_4s_{56}(s_9c_7+s_7s_8c_9))u_1-(s_4c_8c_9c_{56}-c_4(s_7s_9-s_8c_7c_9)-s_4s_{56}(s_9c_7+\\
&s_7s_8c_9))u_3)-I_2^E((s_7c_9+s_8s_9c_7)u_4(s_4u_3-c_4u_1)-u_7(c_7c_8(u_5+u_6+s_4u_1+c_4u_3)-s_8(s_4c_{56}u_3-s_{56}u_2-s_{56}u_4-\\
&c_4c_{56}u_1)-s_7c_8(c_{56}u_2+c_{56}u_4+s_4s_{56}u_3-c_4s_{56}u_1))-u_9(c_8c_9(s_4c_{56}u_3-s_{56}u_2-s_{56}u_4-c_4c_{56}u_1)-(s_7s_9-\\
&s_8c_7c_9)(u_5+u_6+s_4u_1+c_4u_3)-(s_9c_7+s_7s_8c_9)(c_{56}u_2+c_{56}u_4+s_4s_{56}u_3-c_4s_{56}u_1))-s_9c_8(c_6(c_5u_4(s_4u_1+c_4u_3)-\\
&u_5(c_5(u_2+u_4)+s_5(s_4u_3-c_4u_1)))-s_6(s_5u_4(s_4u_1+c_4u_3)-u_5(s_5(u_2+u_4)-c_5(s_4u_3-c_4u_1)))-u_6(c_6(c_5u_2+\\
&c_5u_4+s_4s_5u_3-s_5c_4u_1)+s_6(s_4c_5u_3-s_5u_2-s_5u_4-c_4c_5u_1)))-(c_7c_9-s_7s_8s_9)(s_6(c_5u_4(s_4u_1+c_4u_3)-\\
&u_5(c_5(u_2+u_4)+s_5(s_4u_3-c_4u_1)))+c_6(s_5u_4(s_4u_1+c_4u_3)-u_5(s_5(u_2+u_4)-c_5(s_4u_3-c_4u_1)))-u_6(s_6(c_5u_2+\\
&c_5u_4+s_4s_5u_3-s_5c_4u_1)-c_6(s_4c_5u_3-s_5u_2-s_5u_4-c_4c_5u_1)))))+T_y^A - \quad \bullet\bullet\bullet\ \text{9 MORE PAGES}
\end{aligned}
$$

Efficient

$$\dot{u}_2 = [\,T_y^A - T^{A/B} - (I_1^A - I_3^A)\,u_1u_3\,]\ /\ I_2^A$$

Figure 1.A.3 Sample Dynamical Equations in u_2 with Customary and Efficient Generalized Speeds for the Shuttle Manipulator System, taken from Ref. [15].

Problem Set 1

1.1 Consider the double pendulum example worked out in Figure 1.1. Derive the equations of motion of the system for large angles q_1, q_2, in two ways:

(a) Using Lagrange's equations in the generalized coordinates q_1, q_2.

(b) Using Kane's equations, with the choice of generalized speeds, $u_1 = \dot{q}_1; u_2 = \dot{q}_2$. Note that the dynamical equations are now more complex than those worked out in Example 1. This exemplifies the simplicity of equations obtainable with a proper choice of generalized speeds.

1.2 Consider the same double pendulum, but placed horizontally, being attached to the rotating disk of Figure 1.6, with torsional springs of stiffness k between the disk and first link and also between the two links. Derive the nonlinear dynamical equations by Kane's method.

1.3 Derive the equations of motion of the double pendulum, linearized in q_1, q_2, by Kane's method of direct linearization, and verify that they agree with the equations obtainable by linearizing the nonlinear equations.

References

[1] Kane, T.R. and Wang, C.F. (1965) On the derivation of equations of motion. *Journal of the Society of Industrial and Applied Mechanics*, **13**(2), 487–492.

[2] Kane, T.R. (1968) *Dynamics*, Holt, Rinehart and Winston, Inc.

[3] Kane, T.R. and Levinson, D.A. (1985) *Dynamics*, McGraw-Hill.

[4] Kane, T.R., Likins, P.W., and Levinson, D.A. (1983) *Spacecraft Dynamics*, McGraw-Hill.

[5] Likins, P.W. (1974) Analytical Dynamics and Nonrigid Spacecraft Simulation. Technical Report 32-1593, Jet Propulsion Laboratory, Pasadena, CA, July 15.

[6] Kane, T.R. and Levinson, D.A. (1980) Formulation of equations of motion for complex spacecraft. *Journal of Guidance, Control, and Dynamics*, **3**(2), 99–112.

[7] Levinson, D.A. and Kane, T.R. (1990) AUTOLEV-A new approach on multibody dynamics, in *Multibody Systems Handbook* (ed. W. Schiehlen), Springer-Verlag, pp. 81–102.

[8] Ayoubi, M.A., Goodarzi, F.A., and Banerjee, A.K (2011) Attitude motion of a spinning spacecraft with fuel sloshing and nutation damping. *Journal of the Astronautical Sciences*, **58**(4), 551–568.

[9] Meirovitch, L., (1970), *Methods of Analytical Dynamics*, McGraw-Hill Book.

[10] Kane, T.R., Ryan, R.R., and Banerjee, A.K. (1987) Dynamics of a cantilever beam attached to a moving base. *Journal of Guidance, Control, and Dynamics*, **10**(2), 139–151.

[11] Banerjee, A.K. and Kane, T.R. (1989) Dynamics of a plate in large overall motion. *Journal of Applied Mechanics*, **56**(6), 887–892.

[12] Banerjee, A.K. and Dickens, J.M. (1990) Dynamics of an arbitrary flexible body in large rotation and translation. *Journal of Guidance, Control, and Dynamics*, **13**(2), 221–227.

[13] Cook, R.D. (1974), *Concepts and Applications of Finite Element Analysis*, John Wiley & Sons, Inc.

[14] Wang, J.T. and Huston, E.L. (1987) Kane's equations with undetermined multipliers – Application to constrained multibody systems. *Journal of Applied Mechanics*, **54**, 424–429.

[15] Mitiguy, P.C. and Kane, T.R. (1996) Motion variables leading to efficient equations of motion. *International Journal of Robotics Research*, **15**(5), 522–532.

2

Deployment, Station-Keeping, and Retrieval of a Flexible Tether Connecting a Satellite to the Shuttle

In this chapter we consider an application of Kane's equations of motion, without any form of linearization, to an interesting problem, namely the dynamics of the shuttle-tethered satellite.

The objective of the shuttle-tethered satellite project was to study the earth from a low altitude where a satellite could not stay by itself because of high atmospheric drag, and the idea was to tow the satellite by a cable attached to the more stable carrying vehicle – the space shuttle. Dynamics of the system was originally considered in Refs. [1–4]. This chapter considers the same problem in further detail based on the work described in Refs. [5–8], starting with (a) deployment of a tethered satellite from the orbiting shuttle, where the satellite is basically pulled by the earth's gravity, with the tether considered elastic and taking up a curved shape; (b) retrieval of the satellite where the elastic tether is reeled in by a motor and can become slack, thus needing a thruster to apply tension; and (c) the simpler problem of station-keeping itself. As previous investigators found, all three processes require control for effectiveness. One basic assumption is that a control system for the shuttle maintains its desired orbit in the presence of disturbance due to mass going out or coming in, as represented by deployment and retrieval of the tether and the satellite. Station-keeping of the satellite after its deployment offers an interesting control problem to keep the tether swing angles small for good earth observation; it has been treated in Ref. [8] and is considered here. Matters of impact dynamics during ejection of a nose cap with a parachute package [10] and formation flying of tethered satellites will be discussed briefly in the appendix to this chapter. We will first treat the deployment problem.

Flexible Multibody Dynamics: Efficient Formulations and Applications, First Edition. Arun K. Banerjee.

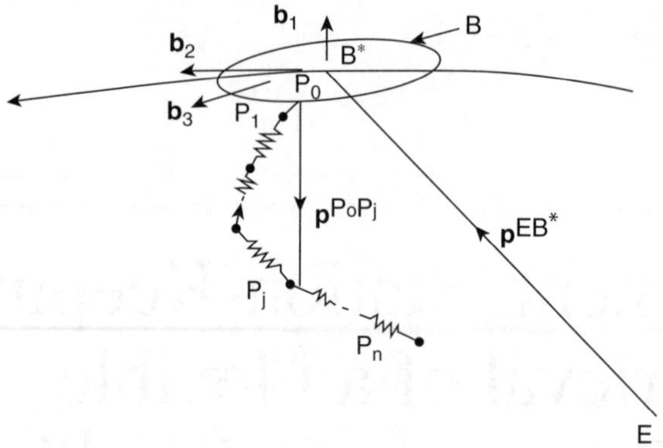

Figure 2.1 Discrete Model of a Tethered Satellite Being Deployed out of the Orbiter. Banerjee [6]. Reproduced with permission of the American Institute of Aeronautics and Astronautics, Inc.

2.1 Equations of Motion of a Tethered Satellite Deployment from the Space Shuttle

A tether is an elastic cable that can be modeled either as an elastic continuum or as a discrete system of particles connected by springs. The exposition given here for deployment is based on Refs. [5, 6] where a discrete tether model is used to allow possibly large deviations of the tether from a straight line. We will use a tether continuum model for retrieval. Consider the model from Ref. [6] of a tethered satellite shown in Figure 2.1, where E is the center of the earth, B is the shuttle orbiter, and the tethered satellite system is modeled with a series of spring-connected lumped mass points numbered as $P_1, \ldots, P_n$, where P_n is the satellite at the end of the tether.

The (shuttle) orbiter is modeled as a rigid body with orthogonal B-basis vectors, $\mathbf{b}_1, \mathbf{b}_2, \mathbf{b}_3$, and mass center at B^*, tether lumped mass points are measured from the tether exit point P_0 on the orbiter, and E is the point mass representing the earth. Generalized speeds, the basic motion variables in Kane's method, are chosen in Ref. [6] as:

$$
\begin{aligned}
u_i &= {}^N\boldsymbol{\omega}^B \cdot \mathbf{b}_i \quad (i = 1, 2, 3) \\
u_{3+i} &= {}^N\mathbf{v}^{B^*} \cdot \mathbf{b}_i \quad (i = 1, 2, 3) \\
u_{3+3j+i} &= {}^N\mathbf{v}^{Pj} \cdot \mathbf{b}_i \quad (i = 1, 2, 3;\ j = 1, \ldots, n)
\end{aligned}
\tag{2.1}
$$

Here ${}^N\boldsymbol{\omega}^B, {}^N\mathbf{v}^{B^*}, {}^N\mathbf{v}^{\mathbf{P}_\mathbf{j}}$ represent, respectively, the vectors of angular velocity of B in a Newtonian reference frame N, the velocity of the mass center of B^* in N, and the velocity of the jth lumped mass P_j. The configuration variables for the system are defined in terms of the following generalized coordinates (see Figure 2.1), rather than by inertial coordinates measured from E, to avoid differencing two nearly close large numbers when measured from E:

$$
\begin{aligned}
q_{3j-3+i} &= \mathbf{p}^{P_oP_j} \cdot \mathbf{b}_i \quad (i = 1, 2, 3;\ j = 1, \ldots, n) \\
q_{3n+i} &= \mathbf{p}^{EB^*} \cdot \mathbf{b}_i \quad (i = 1, 2, 3)
\end{aligned}
\tag{2.2}
$$

Here $\mathbf{p}^{\mathbf{P_0P_j}}, \mathbf{p}^{\mathbf{EB^*}}$ are, respectively, the position vectors from P_0 to P_j, $(j = 1, \ldots, n)$ and from E to B^*. The tether exits the shuttle from the point P_0 with its location components in the B-basis:

$$p_i = \mathbf{p}^{B^*P_0}.\mathbf{b}_i \quad (i = 1, 2, 3) \tag{2.3}$$

2.1.1 Kinematical Equations

The angular velocity of B, the velocity of B^*, and the velocity of P_j, $(j = 1, \ldots n)$ in N follow from Eq. (2.1), and are given by the three equations below:

$$\begin{aligned} {}^N\boldsymbol{\omega}^B &= \sum_{i=1}^{3} u_i\mathbf{b}_i \\ {}^N\mathbf{v}^{B^*} &= \sum_{i=1}^{3} u_{3+i}\mathbf{b}_i \\ {}^N\mathbf{v}^{P_j} &= \sum_{i=1}^{3} u_{3+3j+i}\mathbf{b}_i \end{aligned} \tag{2.4}$$

Note that these angular velocity and velocity expressions would have been far more complicated had one used $u_j = \dot{q}_j$, $j = 1, \ldots, 6 + 3n$, required when deriving equations of motion using Lagrange's method. Substituting Eqs. (2.3) and (2.4) in the equation for the velocity of P_j in N,

$${}^N\mathbf{v}^{P_j} = {}^N\mathbf{v}^{B^*} + {}^N\boldsymbol{\omega}^B x \left[\mathbf{p}^{B^*P_0} + \sum_{i=1}^{3} q_{3j-3+i}\mathbf{b}_i\right] + \sum_{i=1}^{3} \dot{q}_{3j-3+i}\mathbf{b}_i, \quad (j = 1, \ldots n) \tag{2.5}$$

leads to the equations:

$$\begin{aligned} \dot{q}_{3j-2} &= u_{4+3j} - (u_4 + u_2 z_{3j} - u_3 z_{2j}) \quad (j = 1, \ldots, n) \\ \dot{q}_{3j-1} &= u_{5+3j} - (u_5 + u_3 z_{1j} - u_1 z_{3j}) \quad (j = 1, \ldots, n) \\ \dot{q}_{3j} &= u_{6+3j} - (u_6 + u_1 z_{2j} - u_2 z_{1j}) \quad (j = 1, \ldots, n) \end{aligned} \tag{2.6}$$

Here

$$z_{ij} = p_i + q_{3j-3+i} \quad (i = 1, 2, 3;\ j = 1, \ldots, n)$$

Equating the derivative of the position vector from E to B^* with the velocity of B^* in N yields:

$$\begin{aligned} \dot{q}_{3n+1} &= u_4 - (u_2 q_{3n+3} - u_3 u_{3n+2}) \\ \dot{q}_{3n+2} &= u_5 - (u_3 q_{3n+1} - u_1 u_{3n+3}) \\ \dot{q}_{3n+3} &= u_6 - (u_1 q_{3n+2} - u_2 u_{3n+1}) \end{aligned} \tag{2.7}$$

This completes the kinematical equations.

2.1.2 Dynamical Equations

Equation (2.4) can be used to calculate the angular acceleration of B, the accelerations of B^* and P_j ($j = 1, \ldots, n$). The generalized inertia forces can then be written as the following set of equations, using Kane's method reviewed in Chapter 1:

$$\begin{aligned}
F_1^* &= -I_1\dot{u}_1 + (I_2 - I_3)u_2u_3 \\
F_2^* &= -I_2\dot{u}_2 + (I_3 - I_1)u_3u_1 \\
F_3^* &= -I_3\dot{u}_3 + (I_1 - I_2)u_1u_2 \\
F_4^* &= -m_B(\dot{u}_4 + u_2u_6 - u_3u_5) \\
F_5^* &= -m_B(\dot{u}_5 + u_3u_4 - u_1u_6) \\
F_6^* &= -m_B(\dot{u}_6 + u_1u_5 - u_2u_4)
\end{aligned} \tag{2.8}$$

$$\begin{aligned}
F_{4+3j}^* &= -m_j(\dot{u}_{3j+4} + u_2u_{3j+6} - u_3u_{3j+5}) \quad (j = 1, \ldots, n) \\
F_{5+3j}^* &= -m_j(\dot{u}_{3j+5} + u_3u_{3j+4} - u_1u_{3j+6}) \quad (j = 1, \ldots, n) \\
F_{6+3j}^* &= -m_j(\dot{u}_{3j+6} + u_1u_{3j+5} - u_2u_{3j+4}) \quad (j = 1, \ldots, n)
\end{aligned} \tag{2.9}$$

Here I_1, I_2, I_3 are, respectively, the centroidal principal moments of inertia along axes that define the $\mathbf{b}_1, \mathbf{b}_2, \mathbf{b}_3$ basis vectors, and m_B, m_j are the mass of B and of P_j. It is assumed here that the mass and moment of inertia of the body B do not change to any significant level due to tether mass deployed or retrieved.

Generalized active forces due to gravity (and neglecting gravitational moments) can also be expressed as follows using Kane's method, as reviewed in Chapter 1:

$$\begin{aligned}
F_i^g &= 0 && (i = 1, 2, 3) \\
F_{3+i}^g &= -\mu\, m_B q_{3n+i} \Bigg/ \left(\sum_{j=1}^{3} q_{3n+j}^2\right)^{3/2} && (i = 1, 2, 3) \\
z_{3+i,j} &= q_{3n+i} + z_{i,j} && (i = 1, 2, 3;\ j = 1, \ldots, n) \\
z_{7,j} &= \mu\, m_j \Bigg/ \left(\sum_{j=1}^{3} z_{3+i}^2\right)^{3/2} && (i = 1, 2, n) \\
F_{3+3j+i}^g &= -z_{7,j}z_{3+i,j} && (i = 1, 2, 3;\ j = 1, \ldots, n)
\end{aligned} \tag{2.10}$$

Here μ is the universal gravitational constant multiplied by the mass of the Earth. Generalized active forces due to the springs, in Figure 2.1, representing tether elasticity are formed assuming that the tether cannot sustain compression, and has extensional stiffness k per segment of the tether. Defining intermediate variables,

$$\begin{aligned}
z_{7+i,j} &= q_{3j-3+i} - q_{3j-6+i} && (i = 1, 2, 3;\ j = 2, \ldots, n) \\
z_{11,j} &= \left(\sum_{j=1}^{3} z_{7+i,j}^2\right)^{1/2} && (j = 2, \ldots, n) \\
z_{12j} &= z_{11,j} - L/n && (j = 2, \ldots, n)
\end{aligned} \tag{2.11}$$

or

$$\begin{aligned} z_{12,j} &= 0 \quad \text{if} \quad z_{12,j} < 0 \quad (z = 2, \dots, n) \\ z_{13,j} &= k_j z_{12,j} / z_{11,j} \qquad\quad (z = 2, \dots, n) \end{aligned} \tag{2.12}$$

one gets the expressions for the generalized active forces due the springs as follows:

$$\begin{aligned} F_i^s &= 0 && (i = 1, \dots, 9; n = 1) \\ F_i^s &= 0 && (i = 1, \dots, 6) \\ F_{6+i}^s &= z_{13,2} z_{7+i,2} && (i = 1, 2, 3; n \geq 2) \\ F_{9+i}^s &= -F_{6+i}^s && (i = 1, 2, 3; n = 2) \\ F_{3+3j+i}^s &= -z_{13,j} z_{7+i,j} + z_{13,j+1} z_{7+i,j+1} && (i = 1, 2, 3;\ j = 2, \dots, n-1) \end{aligned} \tag{2.13}$$

Force due to the thrust-type interaction needed for continuous deployment (pushing out material) is computed by defining the intermediate variable τ:

$$\tau = T \bigg/ \left(\sum_{i=1}^{3} q_i^2 \right)^{1/2} \tag{2.14}$$

Here T is the tether tension at P_1, q_1, q_2, q_3 being the coordinates of P_1 with respect to the origin P_0 (see Eq. 2.2), in the B-basis. Then the generalized force due to this action-reaction thrust on P_0, P_1 is:

$$F_i^t = \left[\tau \sum_{i=1}^{3} q_i \mathbf{b}_i \right] \cdot (\mathbf{v}_i^{P_0} - \mathbf{v}_i^{P_1}) \tag{2.15}$$

Here $\mathbf{v}_i^{P_0}, \mathbf{v}_i^{P_1}$ are the partial velocity vectors of the points P_0 and P_1 (see Figure 2.1) with respect to the ith generalized speed that are evaluated from the expression:

$$\mathbf{v}^{P_1} - \mathbf{v}^{P_0} = (u_2 q_3 - u_3 q_2 + \dot{q}_1)\, \mathbf{b}_1 + (u_3 q_1 - u_1 q_3 + \dot{q}_2)\, \mathbf{b}_2 + (u_1 q_2 - u_2 q_1 + \dot{q}_3)\, \mathbf{b}_3 \tag{2.16}$$

By using Eq. (2.6) for $j = 1$, in Eq. (2.16) one finds:

$$\begin{aligned} \mathbf{v}^{P_1} - \mathbf{v}^{P_0} &= (u_7 - u_4 - u_2 p_3 + u_3 p_2)\, \mathbf{b}_1 \\ &\quad + (u_8 - u_5 - u_3 p_1 + u_1 p_3)\, \mathbf{b}_2 + (u_9 - u_6 - u_1 p_2 + u_2 p_1)\, \mathbf{b}_3 \end{aligned} \tag{2.17}$$

Now the partial velocity vectors needed in Eq. (2.15) can be obtained by inspection of Eq. (2.17), and the generalized forces expressions due to deployment or pushing out the tether, given in Eq. (2.15), can be evaluated with the following sequence of expressions:

$$\begin{aligned} y_1 &= p_2 q_3 - p_3 q_2 \\ y_2 &= p_3 q_1 - p_1 q_3 \\ y_3 &= p_1 q_2 - p_2 q_1 \\ F_i^t &= y_i && (i = 1, 2, 3) \\ F_{3+i}^t &= q_i && (i = 1, 2, 3) \\ F_{6+i}^t &= -q_i && (i = 1, 2, 3) \end{aligned} \tag{2.18}$$

Kane's dynamical equations for tether deployment can now be written as:

$$F_i^* + F_i^g + F_i^t + F_i^c = 0 \quad (i = 1, \ldots, 6 + n) \tag{2.19}$$

Here F_i^c is a generalized force due to motion constraint implied by having to follow a deployment law, which is a statement of how the commanded length of the tether changes as a prescribed function of time, $l(t)$. Note that we have control over how the distance between P_1, and P_0 changes with deployment. This can be stated as a constraint condition on the "error" in following the command, defined by:

$$y_4 = l(t) - L(n-1)/n \tag{2.20a}$$

$$e = \sum_{i=1}^{3} q_i^2 - (y_4)^2 = 0 \tag{2.20b}$$

Here L is the total length of the tether, unstretched. Now Eqs. (2.19) and (2.20b) constitute a system of differential-algebraic equations. We choose to solve them as a system of ordinary differential equations, by differentiating Eq. (2.20b) twice:

$$\begin{aligned} \dot{e} &= 2\sum_{i=1}^{3} \{q_i \dot{q}_i - \dot{l} y_4\} = 0 \\ \ddot{e} &= 2\left\{\sum_{i=1}^{3} (q_i \ddot{q}_i + \dot{q}_i^2) - \dot{l}^2 - \ddot{l} y_4\right\} = 0 \end{aligned} \tag{2.21}$$

Equation (2.19) and the second derivative equation in Eq. (2.21) can be solved together. However, due to unavoidable approximations in numerical integration, constraint satisfaction will drift with time. To remedy this, Ref. [7] invokes Baumgarte's constraint stabilization procedure [12], which consists in writing the constraint equation with $\ddot{\mathbf{e}}, \dot{\mathbf{e}}, \mathbf{e}$ as follows:

$$\ddot{e} + k_d \dot{e} + k_p e = 0 \quad (k_d > 0.\ k_p > 0) \tag{2.22}$$

Use of the equations of motion in Eq. (2.6) for the expression for $\ddot{e}$ in Eq. (2.21) renders Eq. (2.22) into the matrix differential equation:

$$\begin{aligned} H\dot{U}_1 &= G \\ \text{where} \quad H &= [y_1,\ y_2,\ y_3,\ q_1,\ q_2,\ q_3 - q_1 - q_2 - q_3] \\ U_1 &= [u_1,\ u_2,\ u_3,\ u_4,\ u_5,\ u_6,\ u_7,\ q_8,\ u_9]^T \\ G &= -q_1(u_2,\ q_3 - u_3,\ q_2) - q_2(u_3 q_1 - u_1 q_3) \\ &\quad -q_3(u_1,\ q_2 - u_2,\ q_1) - \dot{l}^2 + \sum_{i=1}^{3} \dot{q}_i^2 - \ddot{l} y_4 - k_d \dot{e} - k_p e \end{aligned} \tag{2.23}$$

Looking back on the dynamical equations (Eq. 2.19), those can be rewritten in matrix form and interpreted as Kane's equations with undetermined multipliers [11].

$$\begin{aligned} M_1\dot{U}_1 &= -C_1 + F_1 + H^T\tau \\ M_2\dot{U}_2 &= -C_2 + F_2 \\ U_2 &= [u_{10}, \ldots, u_{6+3n}]^T \quad (n > 2) \end{aligned} \tag{2.24}$$

Here $M_1, M_2, C_1, C_2, F_1, F_2$ are defined implicitly by the negative of the generalized inertia force and active force, with

$$\begin{aligned} -F^* &= \begin{bmatrix} M_1 & 0 \\ 0 & M_2 \end{bmatrix} \begin{Bmatrix} \dot{U}_1 \\ \dot{U}_2 \end{Bmatrix} + \begin{Bmatrix} C_1 \\ C_2 \end{Bmatrix} \\ \begin{Bmatrix} F_1 \\ F_2 \end{Bmatrix} &= F^g + F^s \end{aligned} \tag{2.25}$$

Note that F^*, F^g, F^s represent column matrices forming generalized forces due to inertia, gravity, and spring terms in Eqs. (2.9), (2.10), and (2.13). Equations (2.8) and (2.9) show that M_1, M_2 are diagonal matrices. Equations (2.23) and (2.24) permit the elimination of τ in Eq. (2.24) and the explicit development of the dynamical equations.

$$\begin{aligned} y_5 &= \sum_{i=1}^{3} y_i^2/I_1 + \sum_{i=1}^{3} q_i^2(m_B + m_1)/(m_B m_1) \\ \tau &= \left\langle G - \sum_{i=1}^{3} [y_i\{[F_1(i) - C_1(i)]/I_i - q_i[F_1(3+i) - C_1(3+i)]/m_B \right. \\ &\quad \left. + q_i[F_1(6+i) - C_1(6+i)]/m_1\} \right\rangle / y_5 \\ \dot{u}_i &= [F_1(i) - C_1(i) + y_i\tau]/I_i \qquad (i = 1, 2, 3) \\ \dot{u}_{3+i} &= [F_1(3+i) - C_1(3+i) + q_i\tau]/m_B \qquad (i = 1, 2, 3) \\ \dot{u}_{3+3j+i} &= [F_2(3j-6+i) - C_2(3j-6+i)]/m_j \quad (i = 1, 2, 3; \quad j = 2, \ldots, n) \end{aligned} \tag{2.26}$$

Here $F_1(i), F_2(i), C_1(i), C_2(i)$ refer to the ith element in the column matrices F_1, F_2, C_1, C_2, respectively. Note that the derivatives of the generalized speeds in Eq. (2.26) are uncoupled. This fact is primarily responsible for the speed of the numerical simulations.

2.1.3 Simulation Results

Equations (2.6) and (2.26) were numerically integrated for the following system parameters: satellite orbit altitude 600 nautical miles, $m_B = 272$ kg, $I_1 = 41$ kg-m^2, $I_2 = I_3 = 542$ kg-m^2, tether attachment offset $p_1 = 1$ m, payload mass 45 kg, tether length $L = 1$ km, tether segment stiffness 373 N/m, constraint gains $k_v = 5.0$, $k_p = 6.32$; constraint gains in Eq. (2.22) were chosen for good error frequency and damping. The tether mass was divided into five lumped

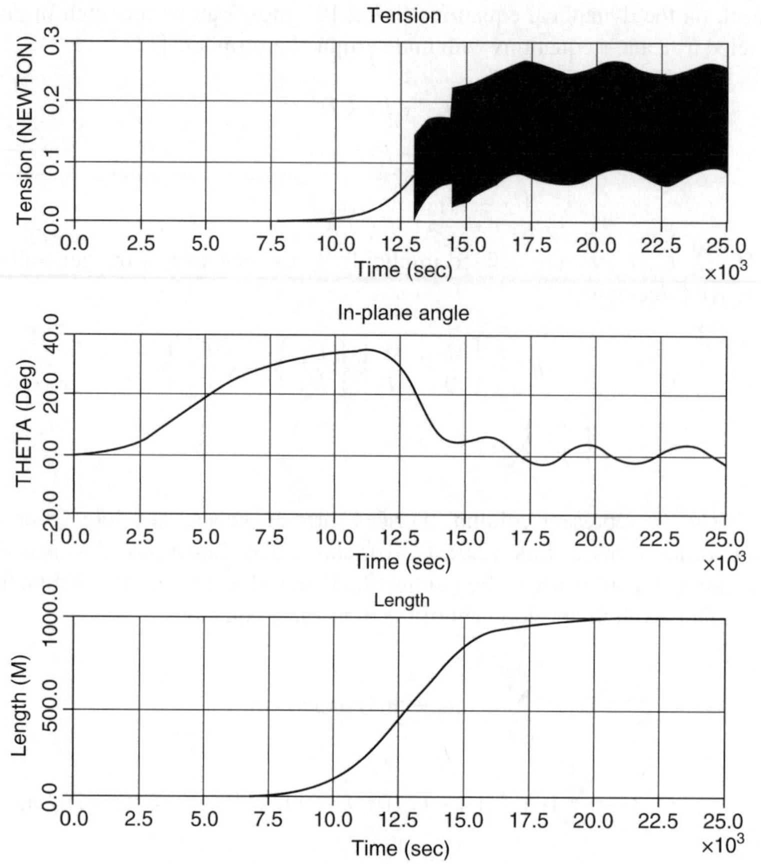

Figure 2.2 Numerical Simulation Results. Banerjee [6]. Reproduced with permission of the American Institute of Aeronautics and Astronautics, Inc.

masses from four tether segments when completely deployed, so as to reduce the computational burden while keeping representative complexity. The method of updating the dynamical states by momentum conservation as the number of particles increases by one during deployment followed Kallaghan *et al.* in Ref. [1]. Results [6] given in Figure 2.2 show the plots for a deployed length rate control law proposed by Rheinfurth in Ref. [2], which is as follows:

$$\begin{aligned} \frac{dl}{dt} &= \alpha l, & 0 \le l \le l_1 \\ &= \alpha l_1 & l_1 \le l \le l_2 \\ &= \alpha[l_1 - (l - l_2)] & l_2 \le l \le L \end{aligned} \tag{2.27}$$

where $l_1 = 250$ m; $l_2 = 750$ m.

Figure 2.2 shows tether tension at P_1 in Fig. 2.1; tether orbit in-plane angle for the line of sight to the payload, and tether instantaneous deployed length for a value of $\alpha = 0.7$ in Eq. (2.27). Note the snatch load action as the deployment rates are changed at l_1, l_2.

The sharp increase in tension in the plot as the number of particle masses deployed increases by one appears to be a reflection of the modeling of the lumped mass deployment process. The dark band in the figure is due to plotting at 0.5 sec interval the tether tension arising from longitudinal oscillations at 1.36 Hz, added on the fluctuations due to in-plane libration period of the tether at 3750 sec. This in-plane angle, shown as "THETA," being positive implies that the tether leads the orbiting body during deployment, and it has been reported in Ref. [1]. Additional simulations show that the tether becomes slack for a higher rate of deployment, such as $\alpha = 0.8$ in Eq. (2.27). Finally, note that the above tether deployment model can be used for station-keeping by setting $\dot{l} = 0$. Station-keeping is basically a control problem, and is presented in Section 2.3.

2.2 Thruster-Augmented Retrieval of a Tethered Satellite to the Orbiting Shuttle

Tethered satellite retrieval is basically an unstable process, as seen from the consideration of conservation of angular momentum in the absence of external moments. Thus if $\omega_i, \omega_f, l_i, l_f$ are initial and final angular velocity and initial and final lengths, respectively, for a rigid massless tether connecting the shuttle to a satellite of mass m, $ml_i^2\omega_i = ml_f^2\omega_f$, yielding $\omega_f = \omega_i(l_i/l_f)^2$, meaning that as the final length is smaller compared to the initial length, the final angular velocity of a rigid tether-satellite system is larger than the initial angular velocity, implying in effect that the tether will tend to "wrap up around the shuttle" in a large angle. The problem is exacerbated by aerodynamics, which bows out a non-rigid tether and causes it to stretch, and can even make the tether become slack. Originally tethered satellite retrieval was studied in Ref. [1] using a tension control law in terms of commanded length, actual length, and its rate. References [2, 3] accounted for tether extensibility and transverse motion flexibility and used an augmented tension control law in feeding back in-plane and out-of-plane tether angles and their rates for improved performance. However, this comes at the cost of high simulation time; indeed, it appears that the cost of simulating retrieval of a tether less than 5 km long is prohibitive, which means that information of crucial importance, as the tether gets shorter, cannot be obtained from such a model. First mode transverse vibration frequency of a both-ends-fixed string being $(\pi/L)\sqrt{T/\rho}$, where T is the tension, ρ is mass per unit length, and L is the length, frequency increases with shorter length. In this book, we exclude transverse flexibility and retain extensibility, hoping to make up for what we lose in model fidelity by *usefulness* in use with a tension control law. Tension control has its own drawback in that tension becomes very small for short tether lengths. To augment tension, we propose using a tether-aligned thruster mounted on the satellite connected to the tether, and also thrusters capable of exerting forces transverse to a tensioned tether, to stabilize tether retrieval and speed up the retrieval process. The following exposition is based on the work in Ref. [7].

2.2.1 *Dynamical Equations*

Consider Figure 2.3, which is a schematic of an orbiter-mounted smooth drum D of radius r and axial moment of inertia J, an extensible tether T of mass ρ per unit length and having a total length L, and a particle S of mass m, representing the satellite. The orbiter (not shown) is presumed to be Earth-pointing and moving in an Earth-centered polar circular orbit of radius R.

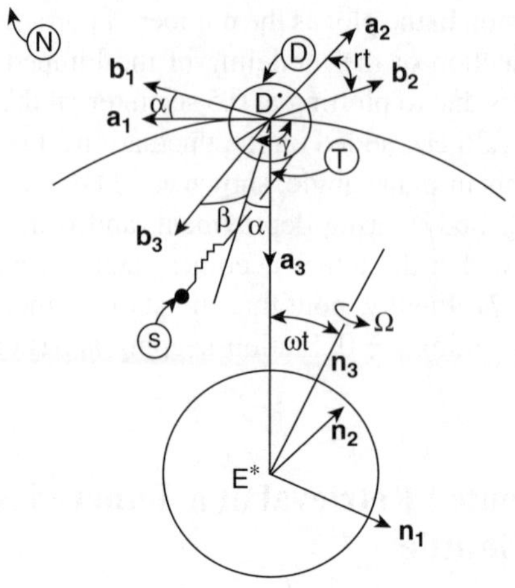

Figure 2.3 Representation of a Continuum Model Tether Connecting an Earth-Orbiting Satellite. Banerjee and Kane [7]. Reproduced with permission of the American Institute of Aeronautics and Astronautics, Inc.

To describe the orientation of a tether frame, we first introduce a dextral set of mutually perpendicular unit vectors $\mathbf{a}_1, \mathbf{a}_2, \mathbf{a}_3$, with $\mathbf{a}_1$ pointing in the direction of motion of D^*, the center of D, and $\mathbf{a}_3$ pointing from D^* to E^*, the center of the Earth. Next, we let $\mathbf{b}_1, \mathbf{b}_2, \mathbf{b}_3$ form a similar set of unit vectors, align these with $\mathbf{a}_1, \mathbf{a}_2, \mathbf{a}_3$, respectively, and then subject them successively to a rotation of amount α about a line parallel to $\mathbf{a}_2$, and a rotation of amount β about a line parallel to $\mathbf{b}_2$. Finally, we take the tether aligned to be parallel to $\mathbf{b}_3$. Thus α and β are, respectively, the in-plane and out-of-plane tether angle when the tether remains straight, which defines a tether frame. The two tether frame angles are shown in Figure 2.3 Drum rotations are characterized by the angle θ between a line parallel to $\mathbf{a}_1$ and a line fixed in D. The line joining D^* to E^* is presumed to move in a Newtonian reference frame N, with a constant orbital rate, ω; this amounts to assuming that the center of mass of the shuttle-tether-satellite system moves in a circular orbit. Letting $\mathbf{n}_1, \mathbf{n}_2, \mathbf{n}_3$ be mutually perpendicular unit vectors fixed in the inertial frame N, with $\mathbf{n}_3$ parallel to Earth's polar axis, we can thus set the angle between the polar axis and line E^*D^* equal to ωt. Finally, the angular speed of Earth in N is denoted by Ω, a quantity needed in evaluation of the aerodynamic drag force.

Let ν be the number of extensional modes of tether vibration to be taken into account; denoting the associated modal coordinates $q_1, \ldots, q_\nu$, one can introduce $\nu + 3$ generalized speeds as:

$$
\begin{aligned}
u_1 &= r\dot{\theta} \\
u_{1+i} &= \dot{q}_i \quad (i = 1, \ldots, \nu) \\
u_{\nu+2} &= \dot{\alpha} \\
u_{\nu+3} &= \dot{\beta}
\end{aligned}
\tag{2.28}
$$

Now $\nu + 3$ differential equations then can be formulated by Kane's method, considering generalized active forces due to gravity, aerodynamic drag due to a rotating Earth, tether elasticity, control torque action on the tether drum D, and thruster action on the satellite S. The key steps in the derivation of the equations of motion follow.

The generalized inertia force, F_i^* corresponding to the ith generalized speed is given by

$$F_i^* = -\left[{}^N\boldsymbol{\omega}_i^D \cdot (\mathbf{J}.{}^N\boldsymbol{\alpha}^D) + \int_0^{L/r} {}^N\mathbf{v}_i^{P'} \cdot {}^N\mathbf{a}^{P'} \rho r d\psi + \int_{L-r\theta}^{l} {}^N\mathbf{v}_i^P \cdot \mathbf{a}^P \rho dx + {}^N\mathbf{v}_i^S \cdot m\mathbf{a}^S\right],$$

$$(i = 1, \ldots, \nu + 3) \tag{2.29}$$

In Eq. (2.29) ${}^N\boldsymbol{\omega}_i^D, {}^N\mathbf{v}_i^{P'}, {}^N\mathbf{v}_i^P, {}^N\mathbf{v}_i^S$ are, respectively, the ith partial angular velocity of D, partial velocity of a particle P' of the tether on D, partial velocity of a particle P on the deployed part of the tether, and partial velocity of the particle satellite S. Quantities ${}^N\boldsymbol{\alpha}^D, {}^N a^{P'}, {}^N a^P, {}^N a^S$ are, respectively, the angular acceleration of D and the accelerations of P', P, and S. The required partial velocities and accelerations are all derivable from the corresponding velocity / angular velocity expressions. The inertial angular velocity of the drum D in N is the orbit angular velocity plus relative angular velocity of drum:

$${}^N\boldsymbol{\omega}^D = (\omega - r\dot{\theta})\,\mathbf{a}_2 \tag{2.30}$$

The velocity of the point P', which is the unstretched location of a generic point P for which the tether wrap angle is ψ and the tangent vector $\boldsymbol{\tau}^{P'}$ is:

$${}^N\mathbf{v}^{P'} = \omega R\mathbf{a}_1 + \left(\frac{u_1}{r} - \omega\right) r\boldsymbol{\tau}^{P'} + \sum_{i=1}^{\nu} \phi_i \left[\frac{(\psi - \theta)r}{L}\right] u_{1+i}\boldsymbol{\tau}^{P'} \tag{2.31}$$

The velocity of the particle P, which, when the tether is unstretched, is at a distance x from the point at which T is attached to D, is given in Eq. (2.32), where the stretched location is given by Eq. (2.33):

$${}^N\mathbf{v}^P = \omega R\mathbf{a}_1 + \left[(u_{2+\nu} - \omega)\,\mathbf{a}_2 + u_{3+\nu}\mathbf{b}_1\right] \times y\mathbf{b}_3 + \left[u_1 + \sum_{i=1}^{\nu} \phi_i \left(\frac{x}{L}\right) u_{1+i}\right] \mathbf{b}_3 \tag{2.32}$$

$$y = x + \sum_{i=1}^{\nu} \phi_i \left(\frac{x}{L}\right) q_i(t) - (L - r\theta) \tag{2.33}$$

${}^N\mathbf{v}^S$, velocity of the satellite, is obtained from Eq. (2.32) by setting $x = L$ in Eqs. (2.32) and (2.33). The ith generalized active force F_i', reflecting contributions from the control torque $T_c\mathbf{a}_2$ on drum D, the gravitational forces on the tether and satellite, and the thrust force $\mathbf{F}$ on the satellite, is given by

$$F_i' = \boldsymbol{\omega}_i^D \cdot (T_c\mathbf{a}_2) + \int_\theta^{L/r} \mathbf{v}_i^{P'} \cdot (\rho_T g_0 \mathbf{a}_3) r d\psi + \int_{L-r\theta}^{L} \mathbf{v}_i^P \cdot \mathbf{f}_g(x) dx$$
$$+ \mathbf{v}_i^S \cdot [m\mathbf{f}_g(L) + \mathbf{F}] \quad (i = 1, \ldots \nu + 3) \tag{2.34}$$

$$\mathbf{f}_g(x) = \rho_T g_0 [(1 + 3(y/R)\cos\alpha\cos\beta]\mathbf{a}_3 - (y/R)\,\mathbf{b}_3 \tag{2.35}$$

and g_0 is the acceleration due to Earth's gravity at the orbit, and ρ_T is the linear mass density of the tether. The contribution to the generalized forces associated with rotating Earth aerodynamics is given by:

$$F_i'' = -0.5\, C_d f \int_{L-r\theta}^{L} \mathbf{v}_\mathbf{i}^\mathbf{P} \cdot \mathbf{v}_w \rho_T dx - 0.5\, C_D \rho_S A_S \mathbf{v}_\mathbf{i}^\mathbf{P} . \left|\mathbf{v}_w\right| \mathbf{v}_w \quad (i = 1, \ldots, \nu + 3) \tag{2.36}$$

Here

$$\mathbf{v}_w = \omega R \mathbf{a}_1 + \Omega R \sin \omega t\, \mathbf{a}_2 \tag{2.37}$$

and the notations C_d, C_D stand for aerodynamic drag coefficients for tether and the satellite, f is a factor expressed in the details that follow, A_s is satellite cross-sectional area, ρ_s is the satellite mass density, and ρ_T, A_T are the tether linear mass density and cross-sectional area. The generalized force due to tether longitudinal visco-elasticity, evaluated as shown in Ref. [4], is obtained by referring to Eqs. (2.32) and (2.33) as:

$$\begin{aligned} F_i''' &= -A_T \sum_{i=1}^{\nu} (\text{Eq}_i + Fu_{1+i}) \int_{L-r\theta}^{L} \frac{d}{dx}\left[\phi_i\left(\frac{x}{L}\right)\right] \mathbf{b}_3 \cdot \frac{d}{dx}\left({}^{N}\mathbf{v}_i^{P}\right) dx \\ &\qquad (i = 1, \ldots, \nu + 3) \\ &\quad - A_T \sum_{i=1}^{\nu} (\text{Eq}_i + Fu_{1+i}) \frac{1}{r} \int_0^{L/r} \frac{d}{d\psi}\left[\phi_i\left\{(\psi-\theta)\frac{r}{L}\right\} \tau^{P'} \mathbf{b}_3\right] \cdot \frac{d}{d\psi}\left({}^{N}\mathbf{v}_i^{P'}\right) d\psi \end{aligned} \tag{2.38}$$

Substitution from Eq. (2.2) and Eqs. (2.7)–(2.11) into Kane's dynamical equations written as

$$-F_i^* = F_i' + F_i'' + F_i''' \quad (i = 1, \ldots, \nu + 3) \tag{2.39}$$

one arrives at the equations of motion, with all notations explained following these equations.

$$\begin{aligned} &\left(m + \rho L + \frac{J}{r^2}\right) \dot{u}_1 + \sum_{i=1}^{\nu} \left[m \sin \gamma_i + \frac{\rho L}{\gamma_i}(1 - \cos \gamma_i)\right] \dot{u}_{1+i} \\ &= I_1 \left[u_{\nu+3}^2 + (u_{\nu+2} - \omega)^2 \cos^2 \beta + \omega^2 (3 \cos^2 \alpha \cos^2 \beta - 1)\right] \\ &\quad - 0.5\, R(\omega \sin \alpha \cos \beta - \Omega \sin \beta \sin \omega t) \left[f\, C_d \int_{L-r\theta}^{L} \rho_T dx\right] + \frac{T_c}{r} \end{aligned} \tag{2.40}$$

$$\begin{aligned} &\dot{u}_1 \left[m \sin \gamma_i + \frac{\rho L}{\gamma_i}(1 - \cos \gamma_i)\right] + \sum_{j=1}^{\nu} \dot{u}_{1+j} \left\{m \sin \gamma_i \sin \gamma_j + \frac{\rho L}{2}\left[\frac{\sin(\gamma_i - \gamma_j)}{\gamma_i - \gamma_j} - \frac{\sin(\gamma_i + \gamma_j)}{\gamma_i + \gamma_j}\right]\right\} \\ &= I_{3+i} \left[u_{\nu+3}^2 + (u_{\nu+2} - \omega)^2 \cos^2 \beta + \omega^2 (3 \cos^2 \alpha \cos^2 \beta)\right] \\ &\quad - 0.5 R(\omega \sin \alpha \cos \beta - \Omega \sin \beta \sin \omega t) \left[f\, C_d A_T \int_{L-r\theta}^{L} \rho_T \phi_i\left(\frac{x}{L}\right) dx + h\phi_i(1)\right] \\ &\quad - \sum_{j=1}^{\nu} (\text{Eq}_j + Fu_{1+j}) \frac{\gamma_i \gamma_j}{2L} \left[\frac{\sin(\gamma_i - \gamma_j)}{\gamma_i - \gamma_j} - \frac{\sin(\gamma_i + \gamma_j)}{\gamma_i + \gamma_j}\right] + \tau \phi_i(1) \quad (i = 1, \ldots, \nu) \end{aligned} \tag{2.41}$$

$$\dot{u}_{\nu+2} = 2(u_{\nu+2} - \omega)u_{\nu+3}\tan\beta - 3\omega^2 \sin\alpha\cos\alpha - 2\frac{I_3}{I_2}(u_{\nu+3} - \omega)$$
$$- \frac{\omega R\cos\alpha}{2I_2\cos\beta}\left\{f\,C_d \int_{L-r\theta}^{L} \rho_T \left[x + \sum_{i=1}^{\nu}\phi_i\left(\frac{x}{L}\right)q_i - L + r\theta\right]dx + h\left[r\theta + \sum_{i=1}^{\nu}\phi_i(1)q_i\right]\right\} + F_\alpha \tag{2.42}$$

$$\dot{u}_{\nu+3} = -[(u_{\nu+2} - \omega)^2 + 3\omega^2\cos^2\alpha]\sin\beta\cos\beta - 2\frac{I_3}{I_2}u_{\nu+3} + \frac{R(\omega\sin\alpha\sin\beta + \Omega\sin\omega t\cos\beta}{2I_2}$$
$$\times\left\{f\,C_d \int_{L-r\theta}^{L} \rho_T A_T \left[x + \sum_{i=1}^{\nu}\phi_i\left(\frac{x}{L}\right)q_i - L + r\theta\right]dx + h\left[r\theta + \sum_{i=1}^{\nu}\phi_i(1)q_i\right]\right\} + F_\beta \tag{2.43}$$

In the above equations of motion, F_α, F_β are generalized forces due to in-plane and out-of-plane thrust, defined respectively by referring Figure 2.3, with F_α in orbit in-plane and F_β out-of-plane along $\mathbf{b}_3$. The modal integrals $I_1, \ldots, I_{3+i}, (i = 1, \ldots, \nu)$, and f, h are shown below:

$$I_1 = \int_{L-r\theta}^{L} \rho_T \left[x + \sum_{i=1}^{\nu}\phi_i\left(\frac{x}{L}\right)q_i - L + r\theta\right]dx + m\left[r\theta + \sum_{i=1}^{\nu}\phi_i(1)q_i\right] \tag{2.44}$$

$$I_2 = \int_{L-r\theta}^{L} \rho_T \left[x + \sum_{i=1}^{\nu}\phi_i\left(\frac{x}{L}\right)q_i - L + r\theta\right]^2 dx + m\left[r\theta + \sum_{i=1}^{\nu}\phi_i(1)q_i\right]^2 \tag{2.45}$$

$$I_3 = \int_{L-r\theta}^{L} \rho_T \left[x + \sum_{i=1}^{\nu}\phi_i\left(\frac{x}{L}\right)q_i - L + r\theta\right]\left[u_1 + \sum_{j=1}^{\nu}\phi_i\left(\frac{x}{L}\right)u_{j+1}\right]dx$$
$$+ m\left[r\theta + \sum_{i=1}^{\nu}\phi_i(1)q_i\right]\left[u_1 + \sum_{j=1}^{\nu}\phi_i\left(\frac{x}{L}\right)u_{j+1}\right] \tag{2.46}$$

$$I_{3+\nu} = \int_{L-r\theta}^{L} \rho_T \left[x + \sum_{j=1}^{\nu}\phi_j\left(\frac{x}{L}\right)q_j - L + r\theta\right]\phi_i\left(\frac{x}{L}\right)dx + m\left[r\theta + \sum_{i=1}^{\nu}\phi_i(1)q_i\right]\phi_i(1)$$
$$(i = 1, \ldots, \nu) \tag{2.47}$$

$$f = \left\{\frac{4A_T R^2}{\pi}\left[\omega^2(1 - \sin^2\alpha\cos^2\beta) + \Omega^2\cos^2\beta\sin^2\omega t + 2\omega\Omega\sin\alpha\sin\beta\cos\beta\sin\omega t\right]\right\}^{0.5} \tag{2.48}$$

$$h = C_d\rho_s A_s R\left\{\omega^2 + \Omega^2\sin^2\omega t\right\}^{0.5} \tag{2.49}$$

In all of the above tether extensional vibrations are described as in Ref. [5] in terms of modal functions

$$\phi_i\left(\frac{x}{L}\right) = \sin\left(\frac{\gamma_i x}{L}\right) \quad (i = 1, \dots, \nu) \tag{2.50}$$

where the vibration frequencies satisfy the characteristic equations:

$$\gamma_i L - \tan(\gamma_i L) = \rho L/m \quad (i = 1, \dots, \nu) \tag{2.51}$$

In addition to the thrust control, tension control used by previous investigators is also required. To this end we use a form of a control law proposed in Ref. [2], and a command

$$\begin{aligned} \frac{T_c}{r} &= -3\left(m + \frac{\rho r\theta}{2}\right)\omega^2 r\theta + \left(m + \rho L + \frac{J}{r^2}\right) \\ &\quad \times \left\{4\omega^2(L_c - r\theta) - 4\omega u_1 - L_c\omega[u_{\nu+2} + u_{\nu+3} + \omega(\alpha + \beta)]\right\} \end{aligned} \tag{2.52}$$

with commanded tether length for retrieval given by:

$$L_c = L_0 \exp(-\varepsilon t) \tag{2.53}$$

In Eq. (2.53), describing commanded length of the tether, L_0, ε are, respectively, initial length and a retrieval rate control parameter. In the results shown in Figures 2.4–2.8, $\varepsilon = 1.27$, and in Figures 2.9–2.12, $\varepsilon = 2.00$. As regards thrust control, in-plane thruster, represented by F_α, is turned on when α first reaches (−0.35) rad and continues thereafter. For out-of-plane control represented by F_β in Eq. (2.43), thruster is fired whenever β first reaches $|\beta| > 0.35$ rad and continues thereafter. The threshold values for the angles are set to reduce the burden on the thrusters, and thrusters go into action only after tension control has removed some of the

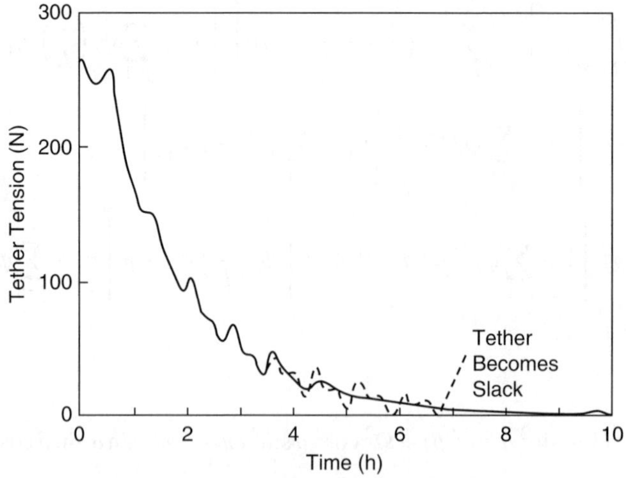

Figure 2.4 Tether Tension with and without Thruster Augmentation, vs. Time (dashed line – tension control only; solid line – tension control + thrust). Banerjee and Kane [7]. Reproduced with permission of the American Institute of Aeronautics and Astronautics, Inc.

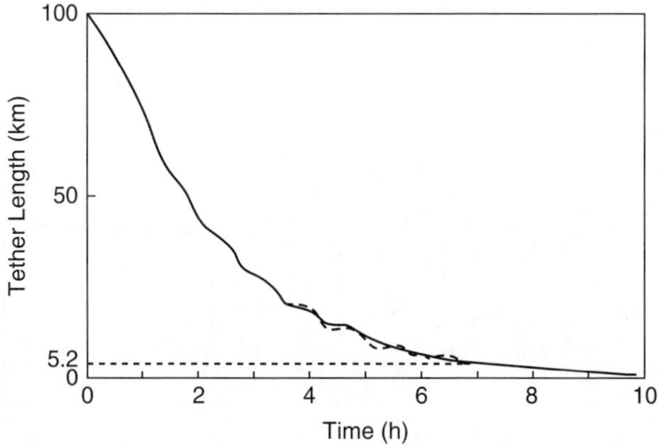

Figure 2.5 Tether Length with and without Thruster, and Tension Control (tether slack at 6.7 hrs). Banerjee and Kane [7]. Reproduced with permission of the American Institute of Aeronautics and Astronautics, Inc.

energy in the system. However, relying only on tension control and delaying the thruster onset time too long can make recovery impossible. The threshold value of ±0.35 rad represents a compromise that highlights the necessity of doing numerous simulations, as was done here. Expressions for orbit in-plane and out-of plane thrusters are chosen as follows:

$$
\begin{aligned}
F_\alpha &= 2(\delta - \varsigma\omega\sqrt{3})u_{\nu+2} \\
F_\beta &= 2(\delta - 2\varsigma\omega)u_{\nu+2}
\end{aligned}
\tag{2.54}
$$

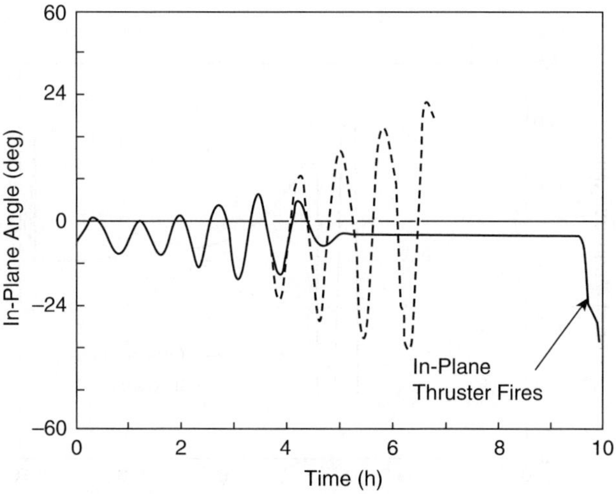

Figure 2.6 Tether In-Plane Angle with and without Thrust Augmentation of Tension Control. Banerjee and Kane [7]. Reproduced with permission of the American Institute of Aeronautics and Astronautics, Inc.

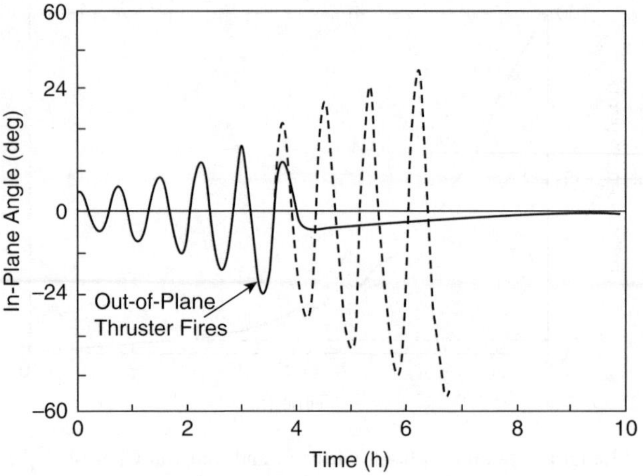

Figure 2.7 Tether Out-of-Plane Angle with and without Thrust Augmentation of Tension Control. Banerjee and Kane [7]. Reproduced with permission of the American Institute of Aeronautics and Astronautics, Inc.

with $\varsigma = 10.0$ and

$$\delta = (m + 0.5\rho r\theta)r\theta u_1 \Big/ \left\{ mr^2\theta^2 + \frac{\rho L}{3}L^2 \left[l - \left(1 - \frac{r\theta}{L}\right)^3 \right] - \rho r\theta L^2 \left(1 - \frac{r\theta}{L}\right) \right\} \tag{2.55}$$

Again, Eqs. (2.54), (2.55) were arrived at after several numerical experiments. The rationale underlying such functional forms is to attempt to annul the $\dot{L}$ terms normally associated with

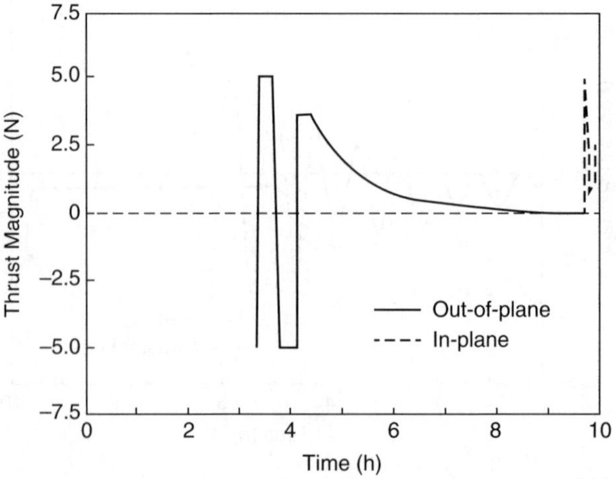

Figure 2.8 In-Plane and Out-of-Plane Thrust Histories for Figures 2.6 and 2.7, vs. Time. Banerjee and Kane [7]. Reproduced with permission of the American Institute of Aeronautics and Astronautics, Inc.

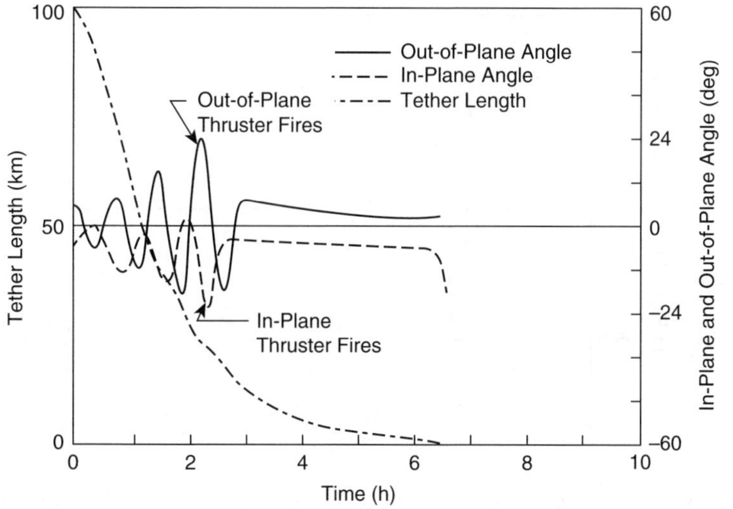

Figure 2.9 Fast Retrieval with Thruster-Aided Tension Control and Tension Control Alone (dashed). Banerjee and Kane [7]. Reproduced with permission of the American Institute of Aeronautics and Astronautics, Inc.

negative damping (coefficients of $\dot{L}$, which is negative for retrieval) in the dynamical equations. Equations (2.54) are expressions for generalized active force due to thrust; the corresponding physical thrust has to be of the form

$$\mathbf{F} = F_1\mathbf{b}_1 + F_2\mathbf{b}_2 \tag{2.56}$$

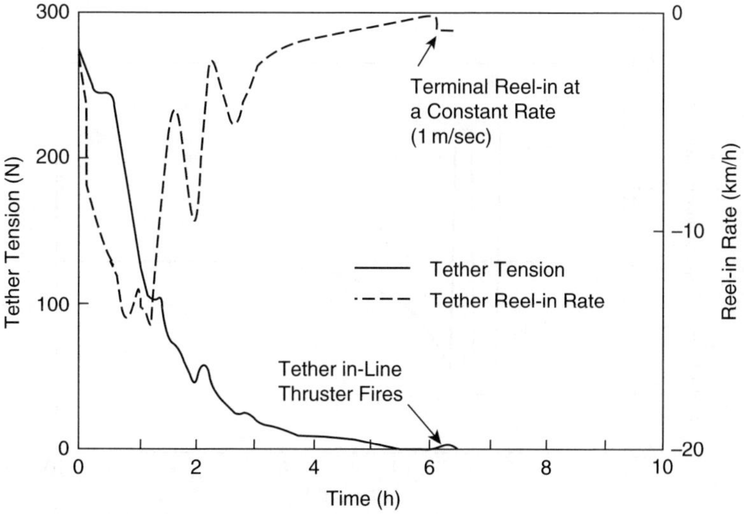

Figure 2.10 Fast Retrieval Tether Tension and Reel-In Rate, vs. Time. Banerjee and Kane [7]. Reproduced with permission of the American Institute of Aeronautics and Astronautics, Inc.

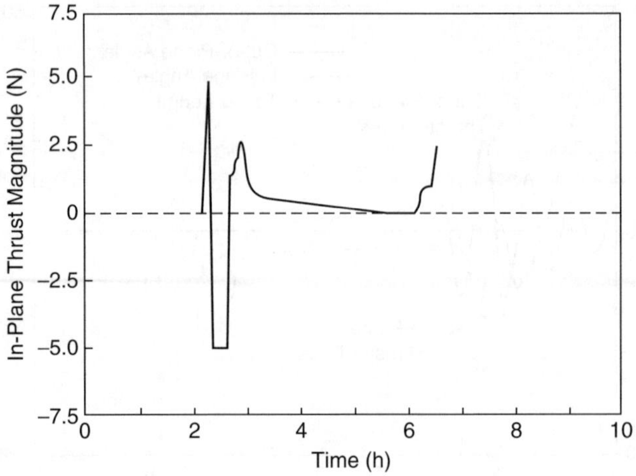

Figure 2.11 In-Plane Thruster-Firing Time History for Fast Retrieval, vs. Time. Banerjee and Kane [7]. Reproduced with permission of the American Institute of Aeronautics and Astronautics, Inc.

which in turn requires the following, with I_2 given by Eq. (2.45):

$$F_1 = F_\alpha I_2 \cos\beta/(r\theta) \tag{2.57}$$

$$F_2 = -F_\beta I_2/(r\theta) \tag{2.58}$$

Finally, a key terminal step in control of tethered satellite retrieval is to turn on a tether in-line thruster of magnitude 2 N away from the orbiter and reel in the tether rapidly, while relying on in-plane thrust to prevent a tether wrap-up. A thrust level of 2 N is chosen because

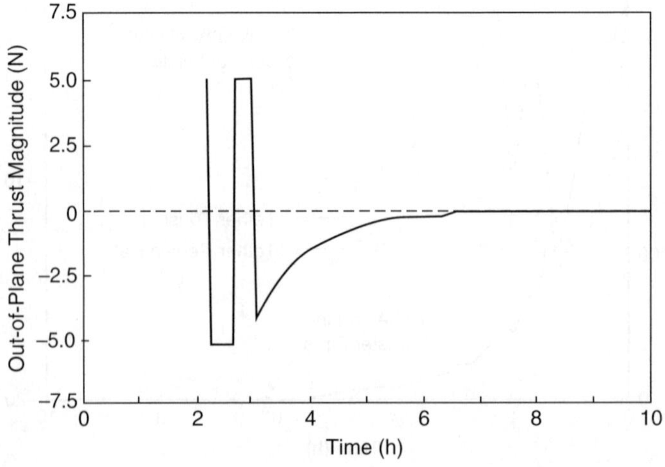

Figure 2.12 Out-of-Plane Thruster Time History for Fast Retrieval, vs. Time. Banerjee and Kane [7]. Reproduced with permission of the American Institute of Aeronautics and Astronautics, Inc.

hardware tests reveal that the same level of thrust is needed during initial deployment of the satellite merely to overcome the friction in the tether pulley system. With tether tension maintained in this way, the final strategy is then to reel in the satellite at a high rate, such as 1 m/s used in the simulations, while relying on the tether in-plane thruster to prevent a tether wrap-up around the shuttle.

2.2.2 *Simulation Results*

Figure 2.4 shows results for tethered satellite retrieval for tether tension with only tension control law (dashed line) and tension control together with tether in-plane and out-of-plane control (solid line). The figure shows that the tether becomes slack after some time, with tension control alone, while thrust augmentation keeps the tether taut. Figure 2.5 shows the corresponding tether lengths.

Figures 2.6 and 2.7 show the in-plane and out-of-plane angles with respect to the local vertical with and without thruster augmentation of tension control. Again, the dashed line is for tension control alone. Thrust vs. time histories associated with Figures 2.4–2.7 are shown in Figure 2.8. Figures 2.9 and 2.10 show results for tether length and tension for fast retrieval; again, dashed lines mean tension control acting alone. Corresponding thruster firing histories for in-plane and out-of-plane fast retrieval are shown in Figures 2.11 and 2.12.

2.2.3 *Conclusion*

The effectiveness of both in-plane and out-of-plane thrust augmentation control, compared to tension control acting alone, has been shown. It has also been shown that thrust augmentation can cut down retrieval time, whereas tension control for fast retrieval can cause a tether wrap-up on the shuttle. Retrieval is basically an unstable process, and thrust control laws can stabilize the system, by annulling the negative damping terms inherent with negative rates of length change describing retrieval.

2.3 Dynamics and Control of Station-Keeping of the Shuttle-Tethered Satellite

The purpose of having a shuttle-borne tethered satellite is, of course, to do station-keeping, with a small satellite to look at the earth from an altitude where the satellite cannot stay by itself due to atmospheric drag. The analysis presented here is based on Ref. [8]. Consider the spherical pendulum shown in Figure 2.13. Let it represent a rigid model of a tether of length L attached to an orbiting body going in a circular orbit with angular speed ω. Rotations through an orbit in-plane angle θ followed by an out-of-plane angle ϕ describe the orientation of the tether. To keep the angles θ and ϕ small, we consider a yo-yo pumping control of the tether.

For prescribed deployment rate $\dot{L}$ the equations of motion of a particle as a lumped end-mass on a massless rigid tether, acted on by gravity gradient force and tether tension, can be written from Newton's law, as follows:

$$\ddot{\theta} = -2(\omega + \dot{\theta})[(\dot{L}/L) + \dot{\phi}\tan\phi] - 3\omega^2 \sin\theta\cos\theta \tag{2.59}$$

$$\ddot{\phi} = -2(\dot{L}/L)\dot{\phi} - (\omega + \dot{\theta})^2 \sin\phi\cos\phi - 3\omega^2\cos^2\theta\sin\phi\cos\phi \tag{2.60}$$

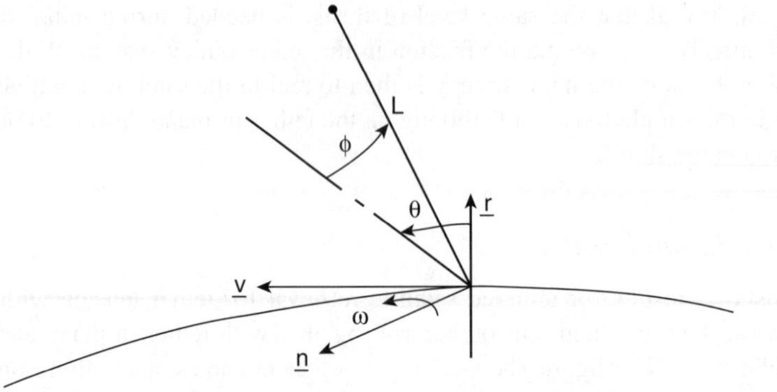

Figure 2.13 Spherical Pendulum Representing a Rigid Tether Attached to an Orbiting Body. Davis and Banerjee [8]. Reproduced with permission of the American Institute of Aeronautics and Astronautics, Inc.

For the tether to always remain taut, it is required that the tension T be always positive, where the tension is evaluated by projecting all forces along the tether, yielding:

$$T = m\{L[\dot{\phi}^2 + (\omega + \dot{\theta})^2 \cos^2 \phi + \omega^2(3\cos^2 \theta \cos^2 \phi - 1)] - \ddot{L}\} \tag{2.61}$$

This places a constraint on the maximum value of $\ddot{L}$, which from Eq. (2.61) requires for small angles and angular rates $\dot{\theta} < \omega, \dot{\phi} < \omega$, that:

$$\ddot{L} < 3\omega^2 L \tag{2.62}$$

The station-keeping objective is to keep θ and ϕ small. For this Eqs. (2.59), (2.62) reduce to

$$\ddot{\theta} = -2(\dot{L}/L)(\omega + \dot{\theta}) - 3\omega^2\theta \tag{2.63}$$

and

$$\ddot{\phi} = 2(\dot{L}/L)\dot{\phi} - 4\omega^2\phi \tag{2.64}$$

Now yo-yo length control of the tether is visualized as changing the rate of length, $\dot{L}$, to effect reductions in θ and ϕ. Equation (2.63) and (2.64) show that changing L L changes θ more than ϕ, due to the presence of the term involving ω. As Eq. (2.64) indicates, when ϕ reaches a maximum, $\dot{\phi}$ is zero, and tether can be deployed or retracted without building up ϕ.

On the other hand when roll angle (or ϕ) is zero, $\dot{\phi}$ is maximum, and a change in tether length at this time has the maximum effect on roll momentum and energy. Since lengthening the tether removes energy by decreasing roll rate, a roll-damping effect can be accomplished by lengthening the tether as it passes through the local vertical and then shortening the tether the same amount at the maximum roll angle. This action described is similar to a child's pumping of a swing, and is shown as the out-of-plane control strategy in Figure 2.14. The situation for in-plane control is similar to a spinning ice-skater. This describes the overall strategy for station-keeping.

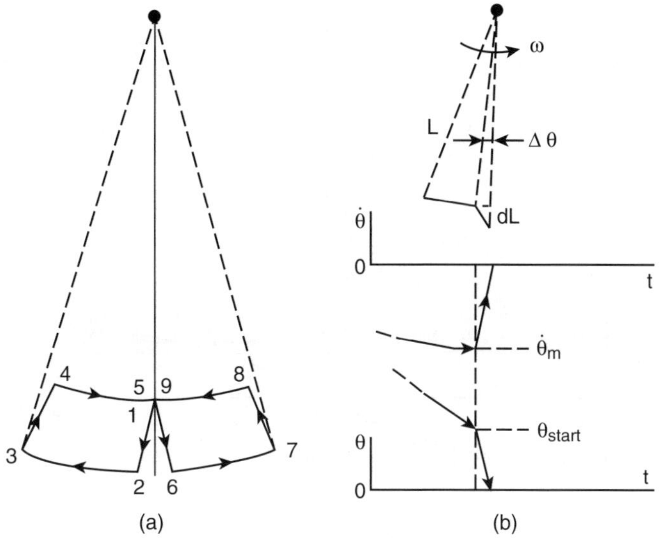

Figure 2.14 (a) Out-of-plane and (b) In-plane Yo-Yo Control Strategy Scenarios. Davis and Banerjee [8]. Reproduced with permission of the American Institute of Aeronautics and Astronautics, Inc.

If the tether length is changed by an appropriate amount at the time the pitch angle is zero, pitch libration can be completely eliminated. The magnitude and direction of the increment in tether length is found by equating pitch angular momentum before and after length increment,

$$mL^2(\omega + \dot{\theta}) = (mL^2 + 2mLdL)\omega \tag{2.65}$$

which leads to the result:

$$2\omega dL = L\dot{\theta} \tag{2.66}$$

The sign and magnitude of $\dot{\theta}$ determines the sign and magnitude of dL. This length control situation is described in Figure 2.15a, while typical results for combined in-plane and out-of plane control by length adjustment, with digital implementation, are shown in Figure 2.15b. Detailed explanations are given in Ref. [8].

Appendix 2.A Sliding Impact of a Nose Cap with a Package of Parachute Used for Recovery of a Booster Launching Satellites

This section deals with a problem associated with the recovery of boosters launching the shuttle, for possible reuse described in Ref. [10]. Figure 2.A.1 shows the container nose cap for the parachute package coming off the nose cone of a shuttle solid rocket booster (SRB), before the parachute opens to decelerate the descent of the SRB for its recovery on the ocean surface.

Restricting ourselves to the analysis of a planar impact, with generalized coordinates q_r, $r = 1, 2, 3$ representing the centroidal coordinates of the nose cap and its orientation, x, y, θ,

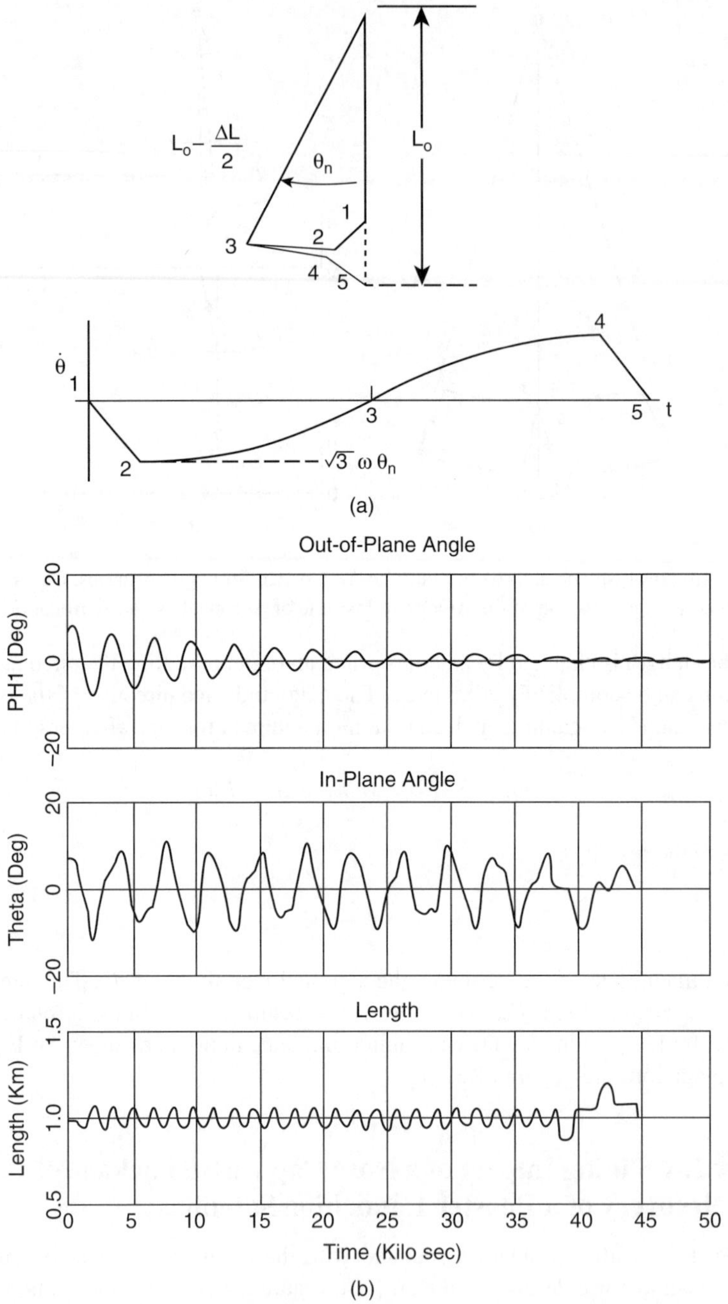

Figure 2.15 (a) Yo-Yo Tether Length Command Control; (b) Overall Station-Keeping Control Results. Davis and Banerjee [8]. Reproduced with permission of the American Institute of Aeronautics and Astronautics, Inc.

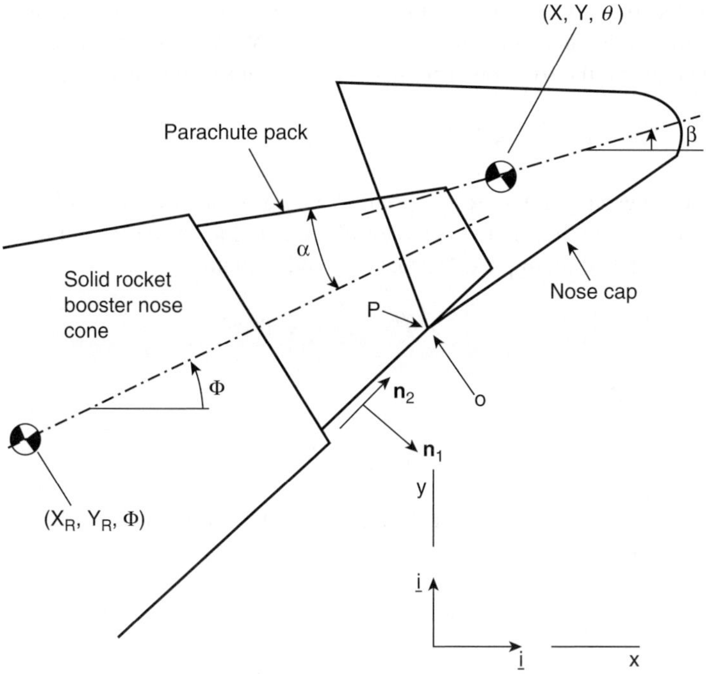

Figure 2.A.1 Nose Cap of Parachute Package Coming Off a Solid Rocket Booster. Banerjee and Coppey [10]. Reproduced with permission of the American Institute of Aeronautics and Astronautics, Inc.

respectively, the linear and angular impulses for impact at point Q are obtained from Kane's definition [9] of a generalized impulse,

$$\begin{aligned} \mathbf{I}_r &= {}^N\mathbf{v}_r^Q.(N_1\mathbf{n}_1 + N_2\mathbf{n}_2) \quad && r = 1,2,3 \\ N_i &= \int_{t_1}^{t_2} F_i dt && i = 1,2 \end{aligned} \tag{2.67}$$

where ${}^N\mathbf{v}_r^Q$ are the rth partial velocities of Q, (see Figure 2.A.1), N_1, N_2 are the magnitudes of normal impulse and the tangential impulse, $\mathbf{n}_1, \mathbf{n}_2$ are the normal and tangential unit vectors, and t_1, t_2 are two instants of time infinitesimally separated. The basic impulse-momentum relationship governs the difference in generalized momentum at instants infinitesimally after and before impact,

$$I_r = p_r(t_+) - p_r(t_-) \quad r = 1,2,3 \tag{2.68}$$

where for mass m and centroidal moment of inertia I, for planar impact

$$p_r = \frac{\partial K}{\partial \dot{q}_r} = \frac{\partial}{\partial \dot{q}_r}\left[0.5\,\mathrm{m}\left(\dot{q}_1^2 + \dot{q}_2^2\right) + 0.5\,I\dot{q}_3^2\right] \quad r = 1,2,3 \tag{2.69}$$

Treating the effect of the impact on the massive SRB as negligible, one can look at Eq. (2.68) as three equations in the five unknowns, $\dot{q}_r, r = 1, 2, 3; N_1, N_2$. Two standard assumptions are now made to complete the solution. The first is the law of restitution, expressed as

$$^N\mathbf{v}^{Q/P}(t_+) \cdot \mathbf{n}_1 = -e^N\mathbf{v}^{Q/P}(t_-) \cdot \mathbf{n}_1; \qquad ^N\mathbf{v}^{Q/P} = {^N\mathbf{v}^Q} - {^N\mathbf{v}^P} \tag{2.70}$$

where P in the superscript refers to the point P on the parachute pack in Figure 2.A.1. The second assumption is on the friction behavior of the colliding surfaces, and results in either a normal rebound or a sliding impact. Normal rebound with no slip means

$$^N\mathbf{v}^{Q/P}(t_+) \cdot \mathbf{n}_2 = 0 \tag{2.71}$$

and is possible only if

$$|N_2| < \mu N_1 \tag{2.72}$$

where μ is the coefficient of friction. Sliding impact occurs when $|N_2| = \mu N_1$, which is interpreted as either

$$N_2 = \mu N_1 \tag{2.73}$$

or

$$N_2 = -\mu N_1 \tag{2.74}$$

It is seen from Figure 2.A.1 that the condition $N_2 > 0$ indicates that the nose cap slips backward on the cone, and $N_2 < 0$ indicates the cap is sliding forward. Both cases are covered by the condition that the tangential component of separation velocity opposes the sense of the tangential impulse:

$$N_2[^N\mathbf{v}^{Q/P}(t_+) \cdot \mathbf{n}_2] < 0 \tag{2.75}$$

Analysis of an actual impact is started using Eqs. (2.68), (2.70), and (2.71). If the test in Eq. (2.72) fails, Eq. (2.71) is replaced by Eq. (2.73) and the test (2.75) applied. If this too fails, the only possible solution is given by Eqs. (2.68), (2.70), and (2.74).

Impact detection is made by integrating the free-flight condition of the two colliding bodies from a non-impact configuration to a situation where the interpenetration of their boundaries is indicated. Noting the penetration depth d of the point Q normal to the surface, a time interval, $\Delta t = t - t_-$ is calculated from the knowledge of relative velocity and relative acceleration:

$$d = -{^N\mathbf{v}^{Q/Q'}} \cdot \mathbf{n}_1 \Delta t - 0.5{^N\mathbf{a}^{Q/Q'}} \cdot \mathbf{n}_1 \Delta t^2 \tag{2.76}$$

The generalized coordinates and their rates are then computed from Δt as follows:

$$q_r(t_-) = q_r(t) - \dot{q}_r(t)\Delta t - 0.5\ddot{q}_r(t)\Delta t^2; \quad \dot{q}_r(t_-) = \dot{q}_r(t) - \ddot{q}_r(t)\Delta t \tag{2.77}$$

Successive impact configurations for a nose cap with respect to a parachute package with slip forward, slip backward and no slip are shown in Fig. 2.A.2.

Appendix 2.B Formation Flying with Multiple Tethered Satellites

Tethered satellite dynamics has reached a level of maturity where an entire book [13] has now been written. We conclude this chapter by pointing to a very interesting body of work on formation flying in space, with tethers deployed from a central body, deploying four or more subsatellites. The principal reference is Ref. [14]. We explain the idea of formation flying below.

Two examples of formation flying, as shown in Figure 2.B.1, are considered in Ref. [14], with N satellites connected to a main hub, the $(N + 1)$ satellite, with the closed hub version forming an outer loop. Each particle satellite has 3 dof, because all tethers are considered stretchable. Another possible scenario is a seven-body double-pyramid configuration shown in Figure 2.B.2.

With the motion of the primary body prescribed, gravitational potential energy of the system is obtained by expanding the standard denominator term of gravitational potential in a binomial series and neglecting terms beyond the third power,

$$V_g = GM_e \sum_{i=1}^{N+1} \frac{m_i}{R_p} - \frac{GM_e}{R_P^2} \sum_{i=1}^{N+1} m_i(\mathbf{O}_1 \cdot \mathbf{r}_i) + \frac{GM_e}{2R_P^3} m_i \left[\mathbf{r}_i \cdot \mathbf{r}_i - 3(\mathbf{O}_1 \cdot \mathbf{r}_i)^2\right] \tag{2.78}$$

where $\mathbf{O}_1$ is an orbit radius unit vector going from the earth center to the primary satellite, and $\mathbf{r}_i$ is the three-dimensional position vector from the primary to the ith satellite. Potential energy stored in the springs is given in terms of the spring stiffness k_i and instantaneous and nominal length l_i, l_{ni} for the ith tether (see Figure 2.B.3a) as:

$$V_e = \sum_{i=1}^{N} k_i(l_i - l_{ni})^2 \tag{2.79}$$

Generalized force for damping in ith tether is stated via the Rayleigh dissipation function:

$$Q_{d,j} = \frac{\partial}{\partial \dot{q}_j} \left[0.5 \sum_{i=1}^{N} c_i(\dot{l})_i^2\right] \tag{2.80}$$

Finally, Lagrange's equations of motion, used in Ref. [14], is written as:

$$\frac{d}{dt}\frac{\partial K}{\partial \dot{q}_j} - \frac{\partial K}{\partial q_j} = -\frac{\partial}{\partial q_j}(V_g + V_e) - Q_{d,j}, \quad (j = 1, \ldots, n) \tag{2.81}$$

Ref. [14] gives explicit equations of motion, based on Eqs. (2.78)–(2.81), for the case when the parent body is in a circular orbit. Figure 2.B.3a is shown in the sequel, where α_i, β_i are the angles to the projection of the ith tether on the y-z plane and the angle from this projection to the actual tether.

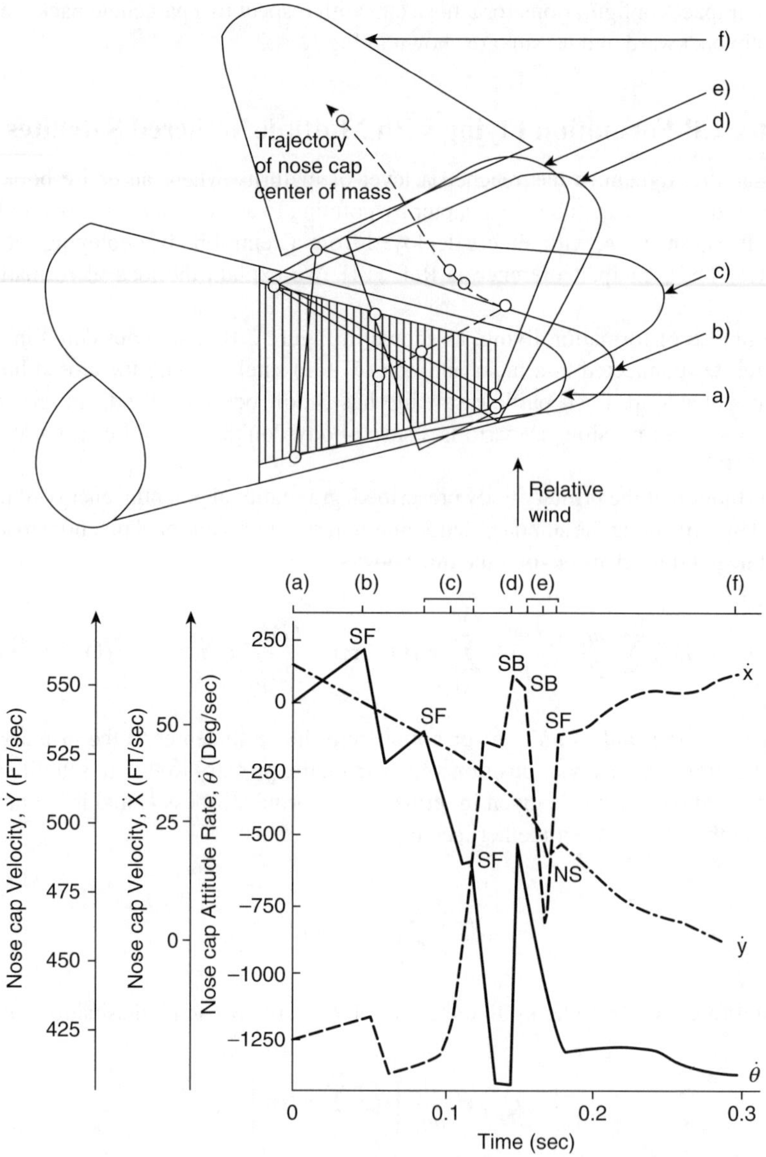

Figure 2.A.2 Successive Impact Configurations and Nose-Cap X- and Y-Velocity and Attitude; Configurations *a, b, c, d, e, f* of the Nose-Cap with Respect to Parachute Shown; Notations: SF = Slip Forward; SB = Slip Backward; NS = No Slip. Banerjee and Coppey [10]. Reproduced with permission of the American Institute of Aeronautics and Astronautics, Inc.

Figure 2.B.3b shows the time response of one of the alignment angles for bodies 1 and 3 in the closed hub-and-spoke four-body system of Figure 2.B.1. Figure 2.B.4 shows the other angle needed for describing tether alignment, for bodies 1–4 in the seven-body pyramid configuration of Figure 2.B.2.

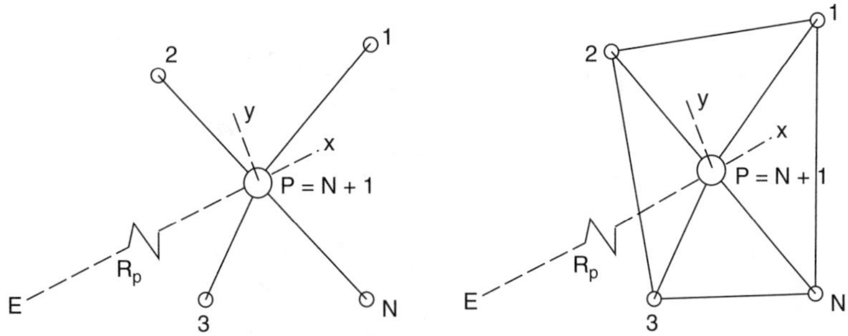

Figure 2.B.1 Open Hub-and-Spoke and Closed Hub-and-Spoke Formation Flying. Pizarro-Chong and Misra [14]. Reproduced with permission of Elsevier.

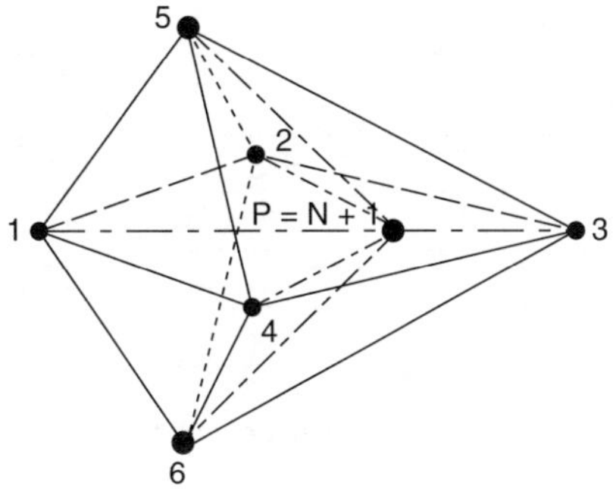

Figure 2.B.2 A Seven-Body Double-Pyramid Configuration. Pizarro-Chong and Misra [14]. Reproduced with permission of Elsevier.

The stability of the configurations is investigated by simulation in Ref. [14]. It was noted that, for planar formations, the closed configuration is stable, but there is no guarantee that the open one will be stable. For the three-dimensional case, only the double-pyramid configuration was found to be stable.

Appendix 2.C Orbit Boosting of Tethered Satellite Systems by Electrodynamic Forces

Use of the tether, in a tethered satellite system, as a current-carrying conductor moving in the earth's magnetic field to generate a propulsive force has been investigated by several authors, an example being [15]. The process is executed by turning the current in the tether for a time to generate the force-time impulse needed for orbit boosting. A mathematical basis of the idea, and the means to modulate the current pulse shaped to suppress librations of the tethered satellite system, as well as to steady down the orbit, are given later [16].

Basically, the tether is discretized into several rigid segments, as shown in Section 2.1, and forces due to gravity and aerodynamic drag are computed as before. A schematic of the electrodynamic tethered satellite system is shown in Figure 2.C.1. The electrodynamic force

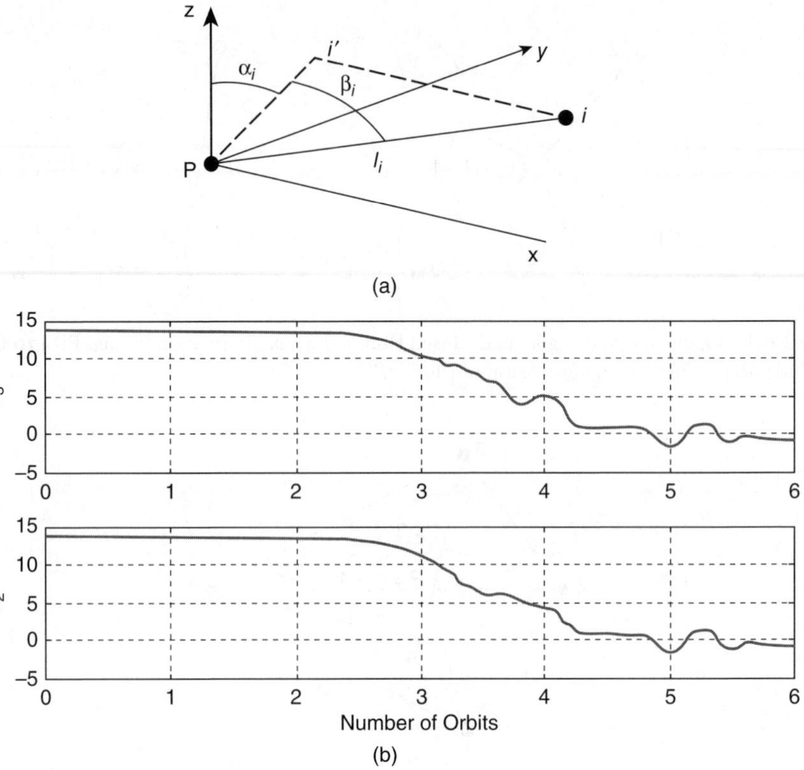

Figure 2.B.3 (a) In-Plane Angles $\boldsymbol{\alpha}_1, \boldsymbol{\alpha}_3$ for Bodies; (b) In-plane Angle Rate $\dot{\alpha}_1, \dot{\alpha}_3$ for Bodies 1 and 3 in Closed Hub-and-Spoke Four-Bbody System. Pizarro-Chong and Misra [14]. Reproduced with permission of Elsevier.

on a tether of length L moving in the earth's magnetic field vector $\mathbf{B}$ with a current I flowing along the segment direction $\mathbf{u}_j$ is given by:

$$\mathbf{F}_e^{P_j} = IL\mathbf{u}_j \times \mathbf{B} \tag{2.82}$$

The magnetic field of the earth can be evaluated at the midpoint R_j of the tether segment as due to a magnetic dipole, and written as:

$$\begin{aligned} B &= \mu_e R_e^3 [\cos \lambda \mathbf{u}_n + 2 \sin \lambda \mathbf{u}_r] / (\mathbf{p}^{ER_j} \cdot \mathbf{p}^{ER_j})^{1.5} \\ \sin \boldsymbol{\lambda} &= \mathbf{u}_m \cdot \mathbf{u}_r \\ \mathbf{u}_n &= \mathbf{u}_r \times (\mathbf{u}_r \times \mathbf{u}_m) / \left| \mathbf{u}_r \times \mathbf{u}_m \right| \\ \mathbf{u}_m &= \sin \boldsymbol{\delta} (\cos \boldsymbol{\omega}_e t \mathbf{n}_1 + \sin \boldsymbol{\omega}_e t \mathbf{n}_2) + \cos \boldsymbol{\delta} \mathbf{n}_3 \end{aligned} \tag{2.83}$$

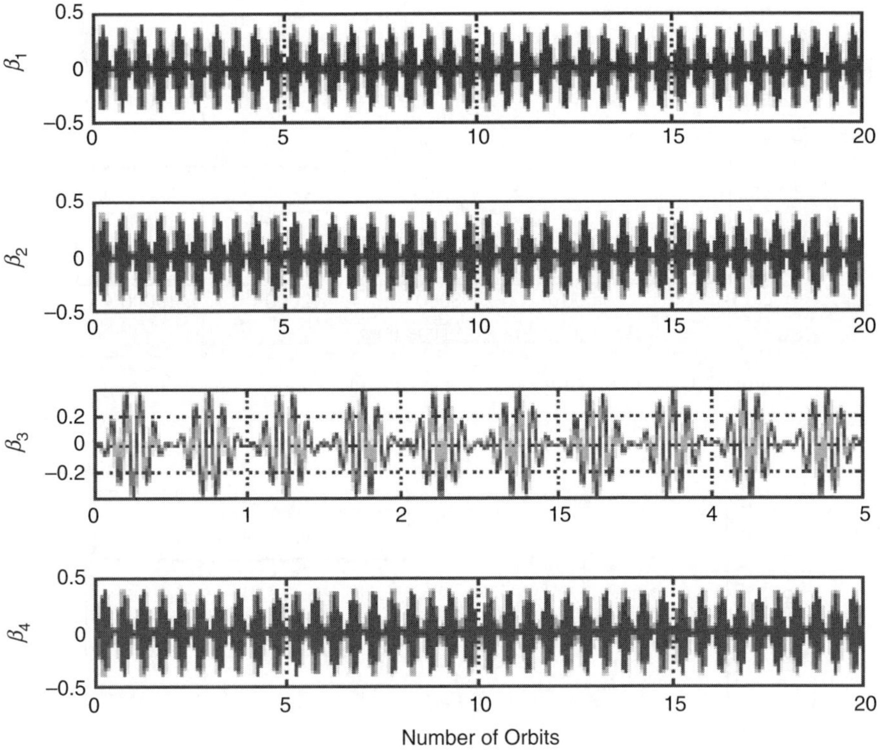

Figure 2.B.4 Out-of-Plane Angle β History for Bodies 1–4 in Seven-Body Double Pyramid of Figure 2.19. Pizarro-Chong and Misra [14]. Reproduced with permission of Elsevier.

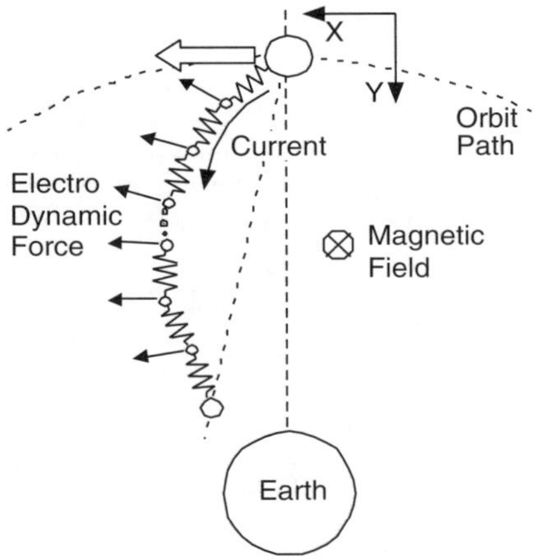

Figure 2.C.1 Schematic of the Tethered Satellite Model.

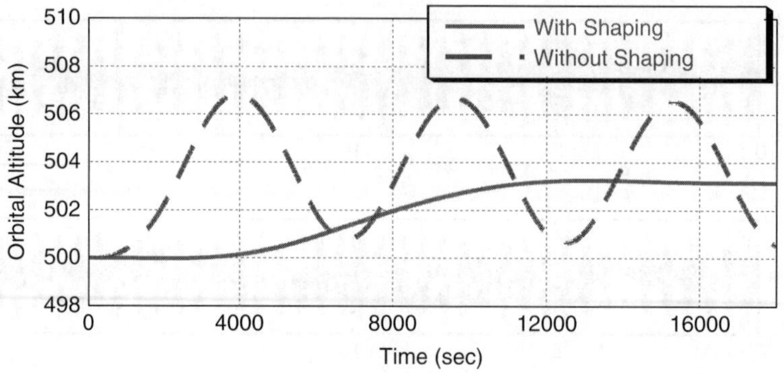

Figure 2.C.2 Orbit Altitude vs. Time with and without Current Shaping.

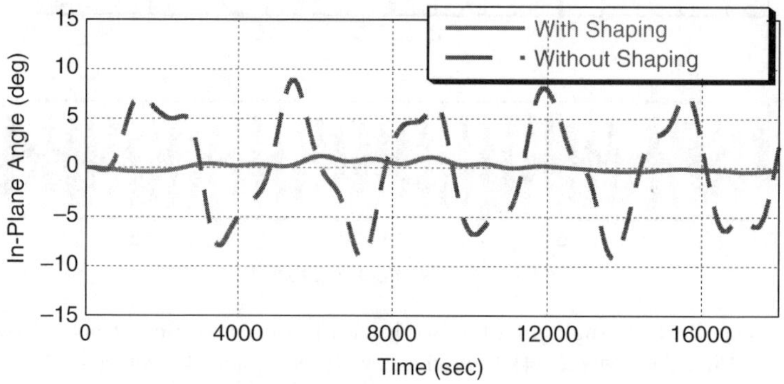

Figure 2.C.3 Tether in-Plane Angle with and without Current Shaping.

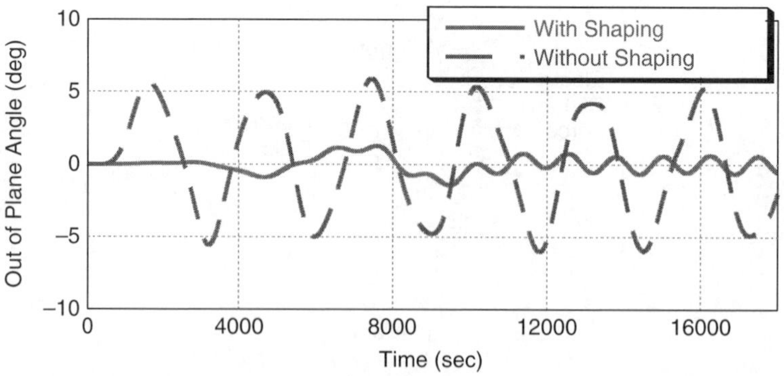

Figure 2.C.4 Tether Out-of-Plane Angle with and without Current Shaping.

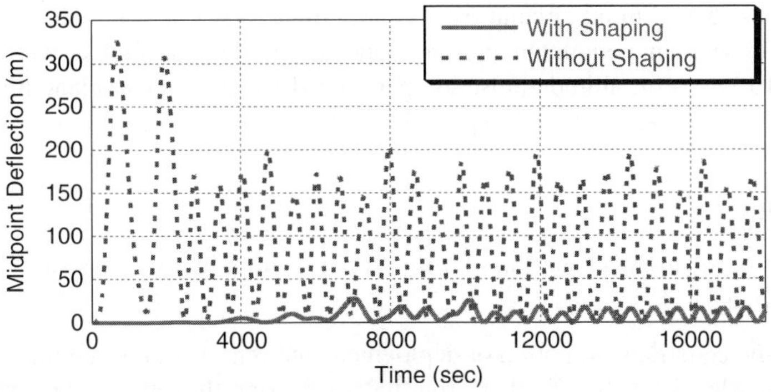

Figure 2.C.5 Tether Mid-Point Deflection in the Discretized Tether Model.

Here $\boldsymbol{\mu}_e, \mathbf{u}_r, \mathbf{u}_m$ are, respectively, the earth's magnetic field constant, 3.5e-5 Wb/m^2, the unit vector from the earth center E to R_j, and the unit vector along the earth's magnetic axis; $\boldsymbol{\delta}$ is the angle between the earth's magnetic axis and its spin axis. Once the electrodynamic force is calculated using Eqs. (2.82) and (2.83), generalized active forces can be evaluated by dot-multiplying the force vector given by Eq. (2.82) with the partial velocity of R_j for all j, and added to the equations of motion.

The application of tether electrodynamic force for orbit boosting is considered here. For higher-energy orbits, the system mass center in Figure 2.1, the required change in angular momentum is equated to the time-integral of the component of the electrodynamic force in the orbit tangent vector $\mathbf{b}_2$ (see Figure 2.C.1). Approximating this force to be constant, the duration of the pulse to raise the orbit from R_1 to R_2 for a total system mass M at R^* would be:

$$\begin{aligned} \Delta t &= [\sqrt{\mu R_2} - \sqrt{\mu R_1}]/[F(R_1 + R_2)/(2M)] \\ F &= \mathbf{F}^{R^*} \cdot \mathbf{b}_2 \end{aligned} \tag{2.84}$$

In the following, we show results for imparting a pulse of force level F for a duration Δt without pre-shaping, and pre-shaping by convolution of the pulse with a three-impulse sequence [16] for each mode of vibrations in the tethered model of Figure 2.C.1. The following data were used for generating the results in Figures 2.C.2–2.C.6: mass of satellite 300 kg; tether mass

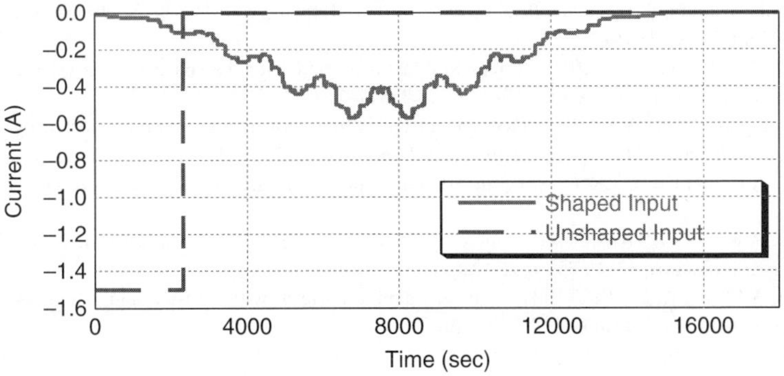

Figure 2.C.6 Shaping of Rectangular Pulse with Three-Impulse Sequence for Vibration Suppression.

20 kg; end mass 5 kg; initial circular earth orbit altitude 500 km; total tether length 20 km; length of tether segment conducting current, upper 10 km; length of tether with aerodynamic drag, lower 10 km; orbit altitude pulsation period 5817 sec; tether in-plane libration period 3398 sec; tether out-of-plane libration period 2828 sec; tether string vibration period 652 sec.

Problem Set 2

2.1 Derive the equations of motion of a rigid tether with a lumped mass as in Eqs. (2.59)–(2.61) for a 3 dof system.

2.2 Derive the equations of motion of deployment and retrieval of a satellite tethered to the space shuttle going on a circular orbit, by considering the tether to be a rigid massless rod with the only mass being that of the satellite, that is, with the tether instantaneous length, in-plane and out-of-plane angles, as 3 dof.

2.3 Derive the equations of motion for one main satellite deploying four subsatellites in the orbital plane of the main satellite, as in the closed hub-and-spoke configuration. Refer to Ref. [14] for the details of the derivation.

2.4 Consider the station-keeping problem of a 20-km-long tether of mass 1 kg/km connected to a small 200 kg satellite and a large spacecraft orbiting the earth with an orbit radius of 6800 km. Examine small longitudinal elastic oscillations of the tether, assuming that the transverse vibrations are negligible. Derive linearized dynamical equations of the system by Kane's method, and calculate the lowest four longitudinal natural frequencies of the system from the associated eigenvalue problem.

2.5 Consider the same problem as above and assume that the tether has a steady longitudinal strain but no longitudinal oscillations. Assuming that the tether executes small transverse vibrations, develop a dynamics model to describe this motion, and obtain the first four transverse vibration frequencies.

References

[1] Colombo, G., Gaposchkin, E.M., Grossi, M.D., and Weiffenbach, G.C. (1974) Shuttle-borne Skyhook, A New Tool for Low-Orbital Attitude Research. Smithsonian Astrophysical Observatory Reports in Geoastronomy, No. 1.

[2] Baker, W.P., Dunkin, J.A., Galaboff, Z.J., *et al.* (1976) Tethered Subsatellite Study. Marshall Space Flight Center, NASA TM X-73314.

[3] Misra, A.K. and Modi, V.J. (1979) A general dynamical model for the space shuttle based tethered satellite system. *Advances in the Astronautical Sciences*, **40**, 537–557.

[4] Misra, A.K. and Modi, V.J. (1982) Deployment and retrieval of a subsatellite connected by a tether to the space-shuttle. *Journal of Guidance, Control, and Dynamics*, **5**, 278–285.

[5] Banerjee, A.K. and Kane, T.R. (1982) Tether deployment dynamics. *Journal of the Astronautical Sciences*, **30**, 347–366.

[6] Banerjee, A.K. (1990) Dynamics of tethered payloads with deployment rate control. *Journal of Guidance, Control, and Dynamics*, **13**(4), 759–762.

[7] Banerjee, A.K. and Kane, T.R. (1984) Tethered satellite retrieval with thruster augmented control. *Journal of Guidance, Control, and Dynamics*, **7**(1), 45–50.

[8] Davis, W.R. and Banerjee, A.K. (1990) Libration damping of a tethered satellite by Yo-Yo control with angle measurement. *Journal of Guidance, Control, and Dynamics*, **13**(2), 370–374.

[9] Kane, T.R. and Levinson, D.A. (1985) *Dynamics: Theory and Application*, McGraw-Hill, pp. 225–241.

[10] Banerjee, A.K. and Coppey, J.M. (1975) Post-ejection impact of the space shuttle booster nose cap. *Journal of Spacecraft and Rockets*, **12**(10), 632–633.

[11] Wang, J.T. and Huston, E.L. (1987) Kane's equations with undetermined multipliers – application to constrained multibody systems. *Journal of Applied Mechanics*, **54**, 424–429.

[12] Baumgarte, J. (1972) Stabilization of constraints and integrals of motion. *Computer Methods in Applied Mechanics and Engineering*, **1**, 1–16.

[13] Alpatov, A.P., Beletsky, V.V., Dranovskii, V.I., *et al.* (2010) *Dynamics of Tethered Space Systems*, CRC Press, Taylor and Francis Group, London-New York, pp. 1–241.

[14] Pizarro-Chong, A. and Misra, A.K. (2008) Dynamics of multi-tethered satellite formations containing a parent body. *Acta Astronautica*, **63**, 1188–1202.

[15] Vas, I.E., Kelley, T.J., and Scarl, E.A. (2000) Space station reboost with electrodynamic tethers. *Journal of Spacecraft and Rockets*, **37**(2), 154–164.

[16] Banerjee, A., Singhose, W., and Blackburn, D. Orbit Boosting of an Electrodynamic Tethered Satellite with Input-Shaped Current. Unpublished Report.

3

Kane's Method of Linearization Applied to the Dynamics of a Beam in Large Overall Motion

This chapter is based on a paper [1] that deals with the issue of linearization with respect to modal coordinates used with vibration mode shapes, for a beam executing small vibrations in a reference frame that is itself undergoing large motion. Study of the behavior of flexible bodies attached to moving supports has been vigorously pursued for over sixty years in connection with a number of diverse disciplines, such as machine design, robotics, aircraft dynamics, and spacecraft dynamics. In particular, beams attached to moving bases have received attention in hundreds of technical papers dealing variously with elastic linkages, rotating machinery, robotic manipulator arms, aircraft propellers, helicopter rotor blades, and flexible satellites. Indeed, the existing literature is so voluminous as to preclude comprehensive review even within the confines of a chapter. A 1974 review by Modi [2] of the literature on rigid bodies with flexible appendages contains more than 200 references, and we will add to it with relatively recent pertinent publications as the discussion proceeds. Presently, we will get down to the task of deriving equations of motion for large overall motion of beams.

3.1 Nonlinear Beam Kinematics with Neutral Axis Stretch, Shear, and Torsion

The system to be described, shown in Figure 3.1, consists of a cantilever beam B built into a rigid body A whose motion in a Newtonian frame N is prescribed as a function of time. The beam is characterized by material properties, $E(x)$, $G(x)$, $\rho(x)$ and cross-sectional properties $A_0(x)$, $I_2(x)$, $I_3(x)$, $\alpha_2(x)$, $\alpha_3(x)$, $\kappa(x)$, $\Gamma(x)$, $e_2(x)$, $e_3(x)$ defined as follows. Let x be the distance from a point O, located at the root of B, to the plane of a generic cross section of B, when B is undeformed. Then $E(x)$, $G(x)$, $\rho(x)$ are the modulus of elasticity, the shear modulus, and the mass per unit length of the beam, respectively, as functions of x. The area of the cross section located at a distance x from O is denoted as $A(x)$, and the Saint Venant torsion factor

Flexible Multibody Dynamics: Efficient Formulations and Applications, First Edition. Arun K. Banerjee.

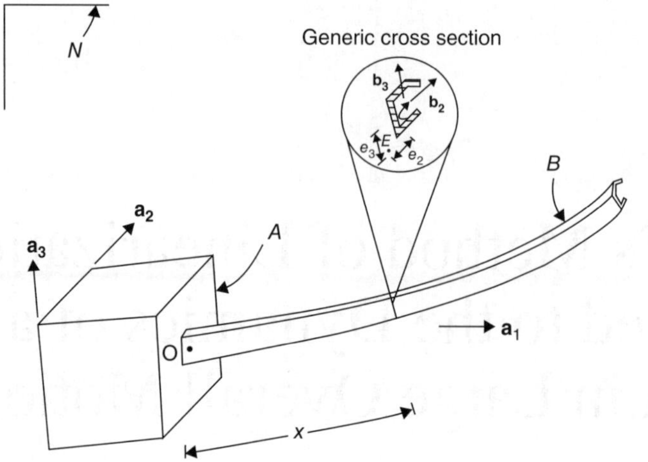

Figure 3.1 Beam Attached to a Moving Rigid Base. Kane *et al.* [1]. Reproduced with permission of the American Institute of Aeronautics and Astronautics, Inc.

and warping factor are denoted by $\kappa(x), \Gamma(x)$. Symbols $I_2(x)$, $I_3(x)$, $\alpha_2(x)$, $\alpha_3(x)$, $e_2(x)$, $e_3(x)$ can be defined by introducing a dextral set of unit vectors $\mathbf{b}_1, \mathbf{b}_2, \mathbf{b}_3$, fixed in the plane of the cross section located at a distance x from point O, and oriented such that $\mathbf{b}_1$ is parallel to the centroidal axis of B, while $\mathbf{b}_2$ and $\mathbf{b}_3$ are parallel to central principal axes of the cross section. Additionally, introduce a similar set of unit vectors $\mathbf{a}_1, \mathbf{a}_2, \mathbf{a}_3$, fixed in A and parallel respectively to $\mathbf{b}_1, \mathbf{b}_2, \mathbf{b}_3$, when B is undeformed. The symbols $I_2(x)$, $I_3(x)$ represent, respectively, the central principal second moments of inertia of the cross section about $\mathbf{b}_2, \mathbf{b}_3$, respectively, and $\alpha_2(x)$, $\alpha_3(x)$ are the shear correction factors; $e_2(x)$, $e_3(x)$ are components of the eccentricity vector from the elastic center to the centroid of the area.

The goal of the following analysis is to produce equations governing the extension, transverse bending deflection, shear, and torsion of the beam introduced in the previous section. The motion of A in N is characterized by the six scalars $v_1, v_2, v_3, \omega_1, \omega_2, \omega_3$, defined as follows:

$$
\begin{aligned}
v_i &= {}^N\mathbf{v}^{\mathrm{O}}.\mathrm{a}_i \qquad (i = 1, 2, 3) \\
\omega_i &= {}^N\boldsymbol{\omega}^{\mathrm{A}}.\mathrm{a}_i \qquad (i = 1, 2, 3)
\end{aligned}
\tag{3.1}
$$

Equation (3.1) implies that the velocity of O in N and the angular velocity of A in N in Figure 3.2 are:

$$
\begin{aligned}
{}^N\mathbf{v}^O &= v_1\mathbf{a}_1 + v_2\mathbf{a}_2 + v_3\mathbf{a}_3 \\
{}^N\boldsymbol{\omega}^A &= \omega_1\mathbf{a}_1 + \omega_2\mathbf{a}_2 + \omega_3\mathbf{a}_3
\end{aligned}
\tag{3.2}
$$

To describe the motion of B in N, introduce a rigid differential element dB of B, with centroid C and elastic center E located at a distance x from point O when B is undeformed. The differential element dB can be brought into a general orientation in A by aligning $\mathbf{b}_1, \mathbf{b}_2, \mathbf{b}_3$ with $\mathbf{a}_1, \mathbf{a}_2, \mathbf{a}_3$, respectively, and subjecting dB to successive dextral rotations in A of amounts θ_1, θ_2, and θ_3 , about lines parallel to $\mathbf{a}_1, \mathbf{a}_2, \mathbf{a}_3$, in turn. Similarly, point E of dB can be brought

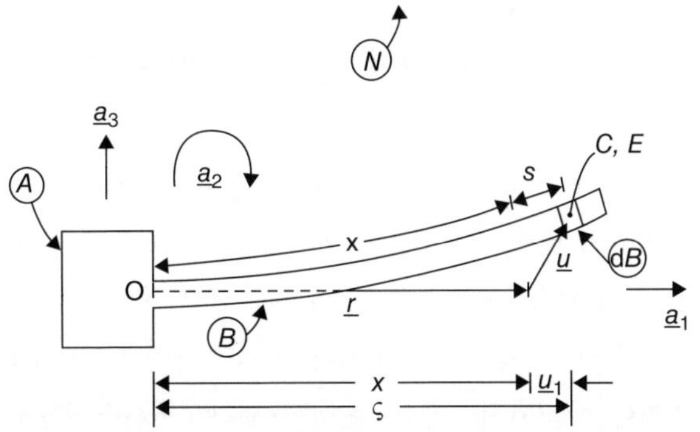

Figure 3.2 Deformed Beam with Differential Element. Kane *et al.* [1]. Reproduced with permission of the American Institute of Aeronautics and Astronautics, Inc.

into a general position by subjecting E to successive translations $u_1\mathbf{a}_1$, $u_2\mathbf{a}_2$, $u_3\mathbf{a}_3$, which allows us to define two vectors,

$$\begin{aligned}\mathbf{u} &= u_1\mathbf{a}_1 + u_2\mathbf{a}_2 + u_3\mathbf{a}_3 \\ \mathbf{r} &= x\mathbf{a}_1\end{aligned} \tag{3.3}$$

so that the position vector from O to E is:

$$\mathbf{p}^{OE} = \mathbf{r} + \mathbf{u} \tag{3.4}$$

The position vector from E to C is given by $\mathbf{p}^{EC} = e_2\mathbf{b}_2 + e_3\mathbf{b}_3$, so that

$$\mathbf{p}^{OC} = (x + u_1)\mathbf{a}_1 + u_2\mathbf{a}_2 + u_3\mathbf{a}_3 + e_2\mathbf{b}_2 + e_3\mathbf{b}_3 \tag{3.5}$$

Orientation of dB in A is described by the space 3 orientation matrix $C^{dB/A}$ such that its elements are Ref. [4]:

$$\begin{aligned}C_{ij}^{dB/A} &= \mathbf{a}_i \cdot \mathbf{b}_j \quad i,j = 1,2,3 \\ C^{dB/A} &= \begin{bmatrix} c_2c_3 & s_1s_2c_3 - s_3c_1 & c_1s_2c_3 + s_3s_1 \\ c_2s_3 & s_1s_2s_3 + c_3c_1 & c_1s_2s_3 - c_3s_1 \\ -s_2 & s_1c_2 & c_1c_2 \end{bmatrix}\end{aligned} \tag{3.6}$$

Here $s_i = \sin\theta_i$, $c_i = \cos\theta_i, i = 1, 2, 3$. The velocity of the centroid C of the cross section in N is given by

$$^N\mathbf{v}^C = {}^N\mathbf{v}^O + {}^N\boldsymbol{\omega}^A \times \mathbf{p}^{OC} + {}^A\mathbf{v}^C \tag{3.7}$$

where ${}^{A}\mathbf{v}^{C}$, the velocity of C in A can be found by differentiating $\mathbf{p}^{OC}$ with respect to time in frame A. Substitution of the resulting expression and using Eqs. (3.2), (3.5) in Eq. (3.7) yields:

$$
\begin{aligned}
{}^{N}\mathbf{v}^{C} = \mathbf{a}_1\{& v_1 + \omega_2[u_3 + e_2 s_1 c_2 + e_3 c_1 c_2] - \omega_3[u_2 + e_2(s_1 s_2 s_3 + c_3 c_1) \\
& + e_3(c_1 s_2 s_3 - c_3 s_1)] + \dot{u}_1 + e_2[(c_1 s_2 c_3 + s_3 s_1)\dot{\theta}_1 + s_1 c_2 c_3 \dot{\theta}_2 - (s_1 s_2 s_3 + c_3 c_1)\dot{\theta}_3] \\
& + e_3[(-s_1 s_2 c_3 + s_3 c_1)\dot{\theta}_1 + c_1 c_2 c_3 \dot{\theta}_2 - (c_1 s_2 s_3 - c_3 s_1)\dot{\theta}_3]\} \\
& + \mathbf{a}_2\{v_2 + \omega_3[x + u_1 - e_2(-s_1 s_2 c_3 + s_3 c_1) + e_3(c_1 s_2 c_3 + s_3 s_1)] + \dot{u}_2 \\
& - \omega_1\left[u_3 + e_2 s_1 c_2 + e_3 c_1 c_2\right] + e_2[(c_1 s_2 s_3 - c_3 s_1)\dot{\theta}_1 + s_1 c_2 s_3 \dot{\theta}_2 + (s_1 s_2 c_3 - s_3 c_1)\dot{\theta}_3] \\
& + e_3[-(s_1 s_2 s_3 + c_3 c_1)\dot{\theta}_1 + c_1 c_2 s_3 \dot{\theta}_2 + (c_1 s_2 c_3 + s_3 s_1)\dot{\theta}_3]\} \\
& + \mathbf{a}_3\{v_3 + \omega_1\left[u_2 + e_2(s_1 s_2 s_3 + c_3 c_1) + e_3(c_1 s_2 s_3 - c_3 s_1)\right] - e_3[s_1 c_2 \dot{\theta}_1 + c_1 s_2 \dot{\theta}_2] \\
& - \omega_2[x + u_1 - e_2(-s_1 s_2 c_3 + s_3 c_1) + e_3(c_1 s_2 c_3 + s_3 s_1)] + \dot{u}_3 + e_2[c_1 c_2 \dot{\theta}_1 - s_1 s_2 \dot{\theta}_2]\}
\end{aligned}
\tag{3.8}
$$

The previous rotation sequence $\theta_1\ \theta_2\ \theta_3$, consistent with the direction cosine matrix of Eq. (3.6), gives the angular velocity expression for *dB* in *A*, as shown in Ref. [4]:

$$
{}^{A}\boldsymbol{\omega}^{dB} = (\dot{\theta}_1 - \dot{\theta}_3 s_2)\mathbf{b}_1 + (\dot{\theta}_2 c_1 + \dot{\theta}_3 s_1 c_2)\mathbf{b}_2 + (-\dot{\theta}_2 s_1 + \dot{\theta}_3 c_1 c_2)\mathbf{b}_3 \tag{3.9}
$$

The angular velocity of *dB* in *N* is found by the angular velocity addition theorem as:

$$
\begin{aligned}
{}^{N}\boldsymbol{\omega}^{dB} &= {}^{N}\boldsymbol{\omega}^{A} + {}^{A}\boldsymbol{\omega}^{dB} \\
&= \mathbf{b}_1[\omega_1 c_2 c_3 + \omega_2 c_2 s_3 - \omega_3 s_2 + \dot{\theta}_1 - \dot{\theta}_3 s_2] \\
&\quad + \mathbf{b}_2[\omega_1(s_1 s_2 c_3 - s_3 c_1) + \omega_2(s_1 s_2 s_3 + c_3 c_1) + (\omega_3 + \dot{\theta}_3)s_1 c_2 + \dot{\theta}_2 c_1] \\
&\quad + \mathbf{b}_3[\omega_1(c_1 s_2 c_3 + s_3 c_1) + \omega_2(c_1 s_2 s_3 - c_3 s_1) + (\omega_3 + \dot{\theta}_3)c_1 c_2 - \dot{\theta}_2 s_1]
\end{aligned}
\tag{3.10}
$$

The variables $\dot{u}_1, \dot{u}_2, \dot{u}_3, \dot{\theta}_1, \dot{\theta}_2, \dot{\theta}_3$ in Eqs. (3.8) and (3.10) have to be expressed in terms of system generalized coordinates. To that end, we introduce the variable $\varsigma = \mathbf{p}^{OE} \cdot \mathbf{a}_1$ (see Figure 3.2), and let $s(x,t)$ denote the stretch in the beam along the elastic axis; then the distance along the deformed elastic axis from O to E, which point is at a distance x from O when B is undeformed, is

$$
x + s(x,t) = \int_0^{\varsigma} \left\{ 1 + \left[\frac{\partial u_2(\sigma,t)}{\partial \sigma}\right]^2 + \left[\frac{\partial u_3(\sigma,t)}{\partial \sigma}\right]^2 \right\}^{0.5} d\sigma \tag{3.11}
$$

where σ is a dummy variable along $\mathbf{a}_1$. Denoting for convenience

$$
J(\sigma,t) \equiv 1 + \left[\frac{\partial u_2(\sigma,t)}{\partial \sigma}\right]^2 + \left[\frac{\partial u_3(\sigma,t)}{\partial \sigma}\right]^2 \tag{3.12}
$$

and differentiating under the integral sign in Eq. (3.11) with respect to time, one obtains

$$\dot{s}(x,t) = 0.5\int_0^{\zeta} [J(\sigma,t)]^{-0.5}\,\frac{\partial J(\sigma,t)}{\partial t}\,d\sigma + \dot{\zeta}\,[J(\zeta,t)]^{0.5} \tag{3.13}$$

which produces the solution for the stretch rate:

$$\dot{\zeta} = \frac{-0.5\int_0^{\zeta}[J(\sigma,t)]^{-0.5}[\partial J(\sigma,t)/\partial t]d\sigma + \dot{s}(x,t)}{[J(\zeta,t)]^{0.5}} \tag{3.14}$$

From Eqs. (3.5)–(3.7) it is apparent that: $\varsigma = x + u_1$. Therefore,

$$\dot{u}_1 = \dot{\varsigma} \tag{3.15}$$

because x is the unstretched length. At this stage we express the stretch and the transverse displacements and the rotations, in terms of a number of assumed vibration modes:

$$\begin{aligned} s(x,t) &= \sum_{i=1}^{\nu}\phi_{1i}(x)q_i(t) \\ u_j(x,t) &= \sum_{i=1}^{\nu}\phi_{ji}(x)q_i(t) \qquad (j=2,3) \\ \theta_j(x,t) &= \sum_{i=1}^{\nu}\phi_{j+3,i}(x)q_i(t) \qquad (j=1,2,3) \end{aligned} \tag{3.16}$$

Here q_i, $i = 1, \ldots, \nu$ are the generalized coordinates, and ν is the number of modes. Based on Eq. (3.16), one can differentiate Eq. (3.12) with respect to t to form

$$\frac{\partial J(\sigma,t)}{\partial t} = 2\sum_{i=1}^{\nu}\sum_{j=1}^{\nu}\left[\phi'_{2i}(\sigma)\phi'_{2j}(\sigma) + \phi^{p}_{3i}rime(\sigma)\phi'_{3j}(\sigma)\right]\dot{q}_i(t)\dot{q}_j(t) \tag{3.17}$$

In Eq. (3.17) a prime denotes differentiation with respect to the dummy variable σ. Putting this relation in Eqs. (3.14), (3.15), $\dot{s} = \dot{u}_1$ yields Eq. (3.18), with the other derivatives as in (3.19):

$$\begin{aligned} \dot{u}_1 = [J(\zeta,t)]^{-0.5}\sum_{i=1}^{\nu}\Bigg\{\phi_{1i}(x) - \sum_{j=1}^{\nu}q_j(t)\int_0^{\zeta}[J(\sigma,t)]^{-0.5} \\ \left[\phi'_{2i}(\sigma)\phi'_{2j}(\sigma) + \phi'_{3i}(\sigma)\phi'_{3j}(\sigma)\right]d\sigma\Bigg\}\,\dot{q}_i(t) \end{aligned} \tag{3.18}$$

$$\begin{aligned} \dot{u}_j(x,t) &= \sum_{i=1}^{\nu}\phi_{ji}(x)\dot{q}_i(t) \qquad (j=2,3) \\ \dot{\theta}_j(x,t) &= \sum_{i=1}^{\nu}\phi_{j+3,i}(x)\dot{q}_i(t) \qquad (j=1,2,3) \end{aligned} \tag{3.19}$$

Substitution of $\dot{u}_1, \dot{u}_2, \dot{u}_3, \dot{\theta}_1, \dot{\theta}_2, \dot{\theta}_3$ in Eq. (3.8) results in the velocity expression, ${}^N\mathbf{v}^C$ (Eq. 3.20), and the angular velocity ${}^N\boldsymbol{\omega}^{dB}$ (Eq. 3.21), shown below. Nonlinear partial velocities, needed for correctly deriving linearized equations, follow from these as the coefficient of the generalized speeds $u_i, i = 1, \ldots, \nu$ in these equations, and are given in Eqs. (3.22) and (3.23).

$$
\begin{aligned}
{}^N\mathbf{v}^C = {} & \mathbf{a}_1 \langle\, v_1 + \omega_2(u_3 + e_2 s_1 c_2 + e_3 c_1 c_2) - \omega_3[u_2 + e_2(s_1 s_2 s_3 + c_3 c_1) + e_3(c_1 s_2 s_3 - c_3 s_1)] \\
& + [J(\zeta,t)]^{-1/2} \sum_{i=1}^{\nu} \left\{ \varphi_{1i}(x) - \sum_{j=1}^{\nu} q_j(t) \int_0^{\zeta} [J(\sigma,t)]^{-1/2} [\varphi'_{2i}(\sigma)\varphi'_{2j}(\sigma) \right. \\
& \left. + \varphi'_{3i}(\sigma)\varphi'_{3j}(\sigma)] d\sigma \right\} \dot{q}_i(t) \\
& + e_2 \sum_{i=1}^{\nu} \{[c_1 s_2 c_3 + s_3 s_1]\varphi_{4i}(x) + s_1 c_2 c_3 \varphi_{5i}(x) + [-c_1 s_2 c_3 + c_3 s_1]\varphi_{6i}(x)\}\, \dot{q}_i(t) \\
& + e_3 \sum_{i=1}^{\nu} \{[-s_1 s_2 c_3 + s_3 c_1]\varphi_{4i}(x) + c_1 c_2 c_3 \varphi_{5i}(x) + [-c_1 s_2 s_3 + c_3 s_1]\varphi_{6i}(x)\}\dot{q}_i(t) \,\rangle \\
& + \mathbf{a}_2 \langle\, v_2 + \omega_3(x + u_1 + e_2(s_1 s_2 c_3 - s_3 c_1) + e_3(c_1 s_2 c_3 + s_3 s_1) \\
& - \omega_1(u_3 + e_2 s_1 c_2 + e_3 c_1 c_2) \\
& + \sum_{i=1}^{\nu} \varphi_{2i}(x)\, \dot{q}_i(t) + e_2 \sum_{i=1}^{\nu} \{(c_1 s_2 c_3 - c_3 s_1)\varphi_{4i}(x) + s_1 c_2 s_3 \varphi_{5i}(x) \\
& + (s_1 s_2 c_3 - s_3 c_1)\varphi_{6i}(x)\}\, \dot{q}_i(t) \\
& + \{e_3 \sum_{i=1}^{\nu} -[(s_1 s_2 s_3 + c_3 c_1)\varphi_{4i}(x) + c_1 c_2 s_3 \varphi_{5i}(x) + (c_1 s_2 c_3 + s_3 s_1)\varphi_{6i}(x)]\, \dot{q}_i(t) \,\rangle \\
& + \mathbf{a}_3 \langle\, v_3 + \omega_1[u_2 + e_2(s_1 s_2 s_3 + c_3 c_1) + e_3(c_1 s_2 s_3 - c_3 s_1)] \\
& - \omega_2[x + u_1 + e_2(s_1 s_2 c_3 - s_3 c_1) + e_3(c_1 s_2 c_3 + s_3 s_1)] + \sum_{i=1}^{\nu} \varphi_{3i}(x)\dot{q}_i(t) \\
& + e_2 \sum_{i=1}^{\nu} [c_1 c_2 \varphi_{4i}(x) - s_1 s_2 \varphi_{5i}(x)]\}\, \dot{q}_i(t) - e_3 \sum_{i=1}^{\nu} \{s_1 c_2 \varphi_{4i}(x) + c_1 s_2 \varphi_{5i}(x)\}\dot{q}_i(t) \,\rangle
\end{aligned}
\tag{3.20}
$$

$$
\begin{aligned}
{}^N\boldsymbol{\omega}^{dB} = {} & \mathbf{b}_1 \left\{ \omega_1 c_2 c_3 + \omega_2 c_2 s_3 - \omega_3 s_2 + \sum_{i=1}^{\nu} [\phi_{4i}(x) - s_2 \phi_{6i}(x)]\dot{q}_i \right\} \\
& + \mathbf{b}_2 \left\{ \left[\omega_1(s_1 s_2 c_3 - s_3 c_1) + \omega_2(s_1 s_2 s_3 + c_3 c_1) + \omega_3 s_1 c_2 + \sum_{i=1}^{\nu} [\phi_{5i}(x) c_1 + \phi_{6i}(x) s_1 c_2]\dot{q}_i \right] \right\} \\
& + \mathbf{b}_3 \left\{ \omega_1(c_1 s_2 c_3 + s_3 s_1) + \omega_2(c_1 s_2 s_3 - c_3 s_1) + \omega_3 c_1 c_2 + \sum_{i=1}^{\nu} [-\phi_{5i}(x) s_1 + \phi_{6i}(x) c_1 c_2]\dot{q}_i] \right\}
\end{aligned}
\tag{3.21}
$$

3.2 Nonlinear Partial Velocities and Partial Angular Velocities for Correct Linearization

$$
\begin{aligned}
{}^N\mathbf{v}_i^C = \mathbf{a}_1 \langle\, & [J(\zeta,t)]^{-1/2}\left\{\varphi_{1i}(x) - \sum_{j=1}^{\nu}\int_0^{\zeta}[J(\sigma,t)]^{-1/2}[\varphi'_{2i}(\sigma)\varphi'_{2j}(\sigma)+\varphi'_{3i}(\sigma)\varphi'_{3j}(\sigma)]d\sigma q_j\right\} \\
& +e_2\{[c_1s_2c_3+s_3s_1]\varphi_{4i}(x)+s_1c_2c_3\varphi_{5i}(x)-[s_1s_2s_3+c_3c_1]\varphi_{6i}(x)\} \\
& +e_3\{[-s_1s_2c_3+s_3c_1]\varphi_{4i}(x)+c_1c_2c_3\varphi_{5i}(x)+[-c_1s_2s_3+c_3s_1]\varphi_{6i}(x)\}\,\rangle \\
& +\mathbf{a}_2\langle\,[\varphi_{2i}(x)+e_2\{[c_1s_2s_3-c_3s_1]\varphi_{4i}(x)+s_1c_2s_3\varphi_{5i}(x)+[s_1s_2c_3-s_3c_1]\varphi_{6i}(x)\} \\
& +e_3\{-[s_1s_2s_3+c_3c_1]\varphi_{4i}(x)+c_1c_2s_3\varphi_{5i}(x)+[c_1s_2c_3+s_3s_1]\varphi_{6i}(x)\,\rangle \\
& +\mathbf{a}_3\langle\,\varphi_{3i}(x)+e_2\{c_1c_2\varphi_{4i}(x)-s_1s_2\varphi_{5i}(x)-e_3\{s_1c_2\varphi_{4i}(x)+c_1s_2\varphi_{5i}(x)\}\,\rangle \\
& i=(1,\ldots,\nu)
\end{aligned}
\tag{3.22}
$$

$$
\begin{aligned}
{}^N\boldsymbol{\omega}_i^{dB} &= \mathrm{b}_1[\phi_{4i}(x)-s_2\phi_{6i}(x)]+b_2[c_1\phi_{5i}(x)+s_1c_2\phi_{6i}(x)]+b_3[-s_1\phi_{5i}(x)+c_1c_2\phi_{6i}(x)] \\
& (i=1,\cdots\nu)
\end{aligned}
\tag{3.23}
$$

To produce linear equations in $q_1,...,q_\nu,\dot{q}_1,...,\dot{q}_\nu$ by the direct linearization method [3,4], to describe small deformations, we can now linearize the nonlinear expressions of partial velocity and partial angular velocity in Eqs. (3.22) and (3.23), respectively. Noting from Eqs. (3.12) and (3.16) that

$$
J(x,t) = 1+\sum_{i=1}^{\nu}\sum_{j=1}^{\nu}\left[\phi'_{2i}(x)\phi'_{2j}(x)+\phi'_{3i}(x)\phi'_{3j}(x)\right]q_iq_j
\tag{3.24}
$$

the linearized version of Eq. (3.24) in modal coordinates is:

$$
\tilde{J}(x,t)=1,\qquad \tilde{J}(\zeta,t)=1
\tag{3.25}
$$

Here and in the following a tilde over a symbol denotes its linearized version. Defining two new symbols,

$$
\begin{aligned}
\beta_{ij} &= \int_0^{\zeta}\phi'_{2i}(\sigma)\phi'_{2j}(\sigma)d\sigma \\
\gamma_{ij} &= \int_0^{\zeta}\phi'_{3i}(\sigma)\phi'_{3j}(\sigma)d\sigma \qquad (i,j=1,\ldots..,\nu)
\end{aligned}
\tag{3.26}
$$

we form linearized the partial velocity expression in Eq. (3.22) as follows.

$$
\begin{aligned}
{}^N\tilde{\mathbf{v}}_i^C = \mathbf{a}_1 & \left\langle \phi_{1i}(x)-\sum_{j=1}^{\nu}(\beta_{ij}+\gamma_{ij})q_j+e_2\left\{\sum_{j=1}^{\nu}[\varphi_{4i}(x)+\varphi_{5j}(x)+\varphi_{5i}(x)\varphi_{4j}(x)]q_j-\phi_{6i}(x)\right\}\right. \\
& \left. +e_3\left\{\sum_{j=1}^{\nu}[\varphi_{4i}(x)\varphi_{6j}(x)+\varphi_{6i}(x)\varphi_{4j}(x)]q_j+\varphi_{5i}(x)\right\}\right\rangle
\end{aligned}
$$

$$+\mathbf{a}_2\left\langle\left\{\phi_{2i}(x)-e_2\sum_{j=1}^{\nu}[\varphi_{4i}(x)\varphi_{4j}(x)+\varphi_{6i}(x)\varphi_{6j}(x)]\right\}q_j\right.$$

$$\left.+e_3\left\{\sum_{j=1}^{\nu}[\varphi_{5i}(x)\varphi_{6j}(x)+\varphi_{6i}(x)\varphi_{5j}(x)]q_j-\varphi_{4i}(x)\right\}\right\rangle$$

$$+\mathbf{a}_3\left\langle\left\{\phi_{3i}(x)+e_2\varphi_{4i}(x)-e_3\{\sum_{j=1}^{\nu}[\varphi_{4i}(x)\varphi_{4j}(x)+\varphi_{5i}(x)\varphi_{5j}(x)]q_j\right\}\right\rangle \quad (i=1,\ldots,\nu)$$

$$(3.27)$$

The linearized partial angular velocity of dB in N follows from Eq. (3.23):

$$
\begin{aligned}
{}^N\tilde{\boldsymbol{\omega}}_i^{dB} &= \mathbf{b}_1\left[\phi_{4i}(x)-\sum_{j=1}^{\nu}\varphi_{5j}(x)\varphi_{5i}(x)q_j\right]\\
&+\mathbf{b}_2\left[\phi_{5i}(x)+\sum_{j=1}^{\nu}\varphi_{4j}(x)\varphi_{6i}(x)q_j\right]\\
&+\mathbf{b}_3\left[\phi_{6i}(x)-\sum_{j=1}^{\nu}\varphi_{4j}(x)\varphi_{5i}(x)q_j\right] \quad (i=1,\ldots\ldots,\nu)
\end{aligned}
\tag{3.28}
$$

3.3 Use of Kane's Method for Direct Derivation of Linearized Dynamical Equations

As per Kane's direct linearization method, the velocity and the angular velocity expressions can now be linearized, because the partial velocity and partial angular velocity expressions have been extracted from their nonlinear counterparts. Linearized velocity of C follows from Eq. (3.20) as:

$$
\begin{aligned}
{}^N\tilde{\mathbf{v}}^C = \mathbf{a}_1&\left\langle v_1+\omega_2e_3-\omega_3e_2+\sum_{k=1}^{\nu}\{\omega_2[\varphi_{3k}(x)+e_2\varphi_{4k}(x)]-\omega_3[\varphi_{2k}(x)-e_3\varphi_{4k}(x)]\}q_k\right.\\
&\left.+\sum_{k=1}^{\nu}\{\phi_{1k}(x)-e_2\varphi_{6k}(x)+e_3\varphi_{5k}(x)\}\dot{q}_k\right\rangle\\
+\mathbf{a}_2&\left\langle v_2+\omega_3x-\omega_1e_3+\sum_{k=1}^{\nu}\{\omega_3[\varphi_{1k}(x)-e_2\varphi_{6k}(x)+e_3\varphi_{5k}(x)]\right.\\
&\left.-\omega_1[\varphi_{3k}(x)+e_2\varphi_{4k}(x)]\}q_k+\sum_{k=1}^{\nu}\{\varphi_{2k}(x)-e_3\varphi_{4k}(x)\}\dot{q}_k\right\rangle
\end{aligned}
\tag{3.29}
$$

$$+\mathbf{a}_3\left\langle v_3-\omega_2 x+\omega_1 e_2+\sum_{k=1}^{\nu}\{\omega_1[\varphi_{2k}(x)-e_3\varphi_{4k}(x)]-\omega_2[\varphi_{1k}(x)\right.$$

$$\left.-e_2\varphi_{6k}(x)+e_3\varphi_{5k}(x)]\}q_k+\sum_{k=1}^{\nu}\{\varphi_{3k}(x)+e_2\varphi_{4k}(x)\}\dot{q}_k\right\rangle$$

The linearized angular velocity of the differential element dB in N follows from Eq. (3.21):

$$\begin{aligned}{}^N\tilde{\boldsymbol{\omega}}^{dB}&=\mathbf{b}_1\left\langle \omega_1+\sum_{k=1}^{\nu}[\omega_2\phi_{6k}(x)-\omega_3\phi_{5k}(x)]q_k+\sum_{k=1}^{\nu}\phi_{4k}(x)\dot{q}_k\right\rangle\\&+\mathbf{b}_2\left\langle \omega_2+\sum_{k=1}^{\nu}[\omega_3\phi_{4k}(x)-\omega_1\phi_{6k}(x)]q_k+\sum_{k=1}^{\nu}\phi_{5k}(x)\dot{q}_k\right\rangle\\&+\mathbf{b}_3\left\langle \omega_3+\sum_{k=1}^{\nu}[\omega_1\phi_{5k}(x)-\omega_2\phi_{4k}(x)]q_k+\sum_{k=1}^{\nu}\phi_{6k}(x)\dot{q}_k\right\rangle\end{aligned}\tag{3.30}$$

Now the acceleration of the generic point C and the angular acceleration of the elemental body dB, both in N, can be expressed by differentiating Eqs. (3.29) and (3.30), and the resulting expressions substituted in the definition of the system generalized inertia force:

$$\tilde{F}_i^*=-\int_0^L \rho^N\mathbf{v}_i^C\cdot{}^N\tilde{\mathbf{a}}^C dx-\int_0^L {}^N\tilde{\boldsymbol{\omega}}_i^{dB}\cdot(\mathbf{I}\cdot{}^N\tilde{\boldsymbol{\alpha}}^{dB}+{}^N\tilde{\boldsymbol{\omega}}^{dB}\times\mathbf{I}.^N\cdot\tilde{\boldsymbol{\omega}}^{dB})\,dx\quad(i=1,\ldots,\nu)\tag{3.31}$$

Here ρdx, $\mathbf{I}dx$ are the mass and central inertia dyadic of dB, ${}^N\tilde{\mathbf{a}}^C, {}^N\tilde{\boldsymbol{\alpha}}^{dB}$ are, respectively, the linearized acceleration of C and angular acceleration of dB in N, and the element central inertia dyadic can be defined as

$$\mathbf{I}dx=\rho(I_{11}\mathbf{b}_1\mathbf{b}_1+I_{22}\mathbf{b}_2\mathbf{b}_2+I_{33}\mathbf{b}_3\mathbf{b}_3)dx\tag{3.32}$$

where I_{11}, I_{22}, I_{33}, expressed in terms of the central principal moments of inertia are:

$$I_{22}=\frac{I_2}{A_0},\quad I_{33}=\frac{I_3}{A_0},\quad I_{11}=\frac{I_2+I_3}{A_0}=I_{22}+I_{33}\tag{3.33}$$

Regarding the generalized active force, since only the elastic restoring forces contribute, those can be derived from a strain energy function U as

$$F_i=-\frac{\partial U}{\partial q_i}\qquad(i=1,\ldots,\nu)\tag{3.34}$$

where U is given by

$$U = \int_0^L \frac{(P_1)^2}{2E_0A_0}dx + \int_0^L \frac{\alpha_2(V_2)^2}{2G_0A_0}dx + \int_0^L \frac{\alpha_3(V_3)^2}{2G_0A_0}dx$$
$$+ \int_0^L \frac{(T_1)^2}{2G\kappa'}dx + \int_0^L \frac{(M_2)^2}{2E_0I_2}dx + \int_0^L \frac{(M_3)^2}{2E_0I_3}dx \tag{3.35}$$

Here the set of forces acting on the cross section of the beam is $P_1\mathbf{a}_1 + V_1\mathbf{a}_2 + V_1\mathbf{a}_3$, and the torque acting on the section is $T_1\mathbf{a}_1 + M_1\mathbf{a}_2 + M_1\mathbf{a}_3$; further, α_2, α_3 are Timoshenko's shear coefficients [5], and k' is an effective torsional constant such that

$$T_1 = G\kappa'\frac{d\theta_1}{dx} \tag{3.36}$$

When effects of shear are included in the analysis, θ_2, θ_3 are related to u_3, u_2 by the following:

$$\frac{\partial u_3(x,t)}{\partial x} = -[\theta_2(x,t) + \psi_2(x,t)]$$
$$\frac{\partial u_2(x,t)}{\partial x} = [\theta_3(x,t) + \psi_3(x,t)] \tag{3.37}$$

where ψ_2, ψ_3 denote angles of shear of a cross section measured at the elastic axis for the directions of $\mathbf{a}_2, \mathbf{a}_3$, respectively. With these relationships, V_2, V_3 appearing in Eq. (3.35) are:

$$V_2 = \frac{A_0G}{\alpha_2}\psi_3 = \frac{A_0G}{\alpha_2}\left[\frac{\partial u_2(x,t)}{\partial x} - \theta_3(x,t)\right]$$
$$V_3 = -\frac{A_0G}{\alpha_3}\psi_2 = \frac{A_0G}{\alpha_2}\left[\frac{\partial u_3(x,t)}{\partial x} + \theta_2(x,t)\right] \tag{3.38}$$

The other force and moment measure numbers appearing in Eq. (3.35) are expressed as:

$$P_1 = E_0A_0\frac{\partial s(x,t)}{\partial x}, \qquad T_1 = G\kappa'\frac{\partial\theta_1(x,t)}{\partial x}$$
$$M_2 = E_0I_2\frac{\partial\theta_2(x,t)}{\partial x}, \qquad M_3 = E_0I_3\frac{\partial\theta_3(x,t)}{\partial x} \tag{3.39}$$

Substitution of these expressions for $P_1, V_2, V_3, T_1, M_2, M_3$, in Eq. (3.35) yields:

$$U = (1/2)\int_0^L \left\{ E_0A_0\left[\frac{\partial s(x,t)}{\partial x}\right]^2 + \frac{GA_0}{\alpha_2}\left[\frac{\partial u_2(x,t)}{\partial x} - \theta_3(x,t)\right]^2 \right.$$
$$\left. + \frac{GA_0}{\alpha_3}\left[\frac{\partial u_3(x,t)}{\partial x} + \theta_2(x,t)\right]^2 \right\} dx$$
$$+ (1/2)\int_0^L \left\{ G\kappa'\left[\frac{\partial\theta_1(x,t)}{\partial x}\right]^2 + E_0I_2\left[\frac{\partial\theta_2(x,t)}{\partial x}\right]^2 + E_0I_3\left[\frac{\partial\theta_3(x,t)}{\partial x}\right]^2 \right\} dx \tag{3.40}$$

Substitution for $s_1, u_2, u_3, \theta_1, \theta_2, \theta_3$, from Eq. (3.16) in Eq. (3.40), and invoking Eq. (3.34) yields:

$$
\begin{aligned}
F_i = -\sum_{j=1}^{\nu} \Bigg\{ \int_0^L & \{E_0 A_0 \phi'_{1i}(x)\phi'_{1j}(x) + \frac{GA_0}{\alpha_3}[\phi_{5i}(x)\phi_{5j}(x) + \phi_{5i}(x)\phi'_{3j}(x) \\
& + \phi_{5j}(x)\phi'_{3i}(x) + \phi'_{3i}(x)\phi'_{3j}(x)] + \frac{GA_0}{\alpha_2}[\phi_{6i}(x)\phi_{6j}(x) - \phi_{6i}(x)\phi'_{2j}(x) \\
& + \phi_{6j}(x)\phi'_{2i}(x) + \phi'_{2i}(x)\phi'_{2j}(x)] + G\kappa'\phi'_{4i}(x)\phi'_{4j}(x) \\
& + E_0[I_2\phi'_{5i}(x)\phi'_{5j}(x) + I_3\phi'_{6i}(x)\phi'_{6j}(x)] \Bigg\} q_j\, dx \qquad (i = 1, \dots, \nu)
\end{aligned}
\tag{3.41}
$$

The term within the brace above can be compactly represented for later reference as:

$$
\tilde{F}_i = -\sum_{j=1}^{\nu} H_{ij} q_j \qquad (i = 1, \dots, \nu) \tag{3.42}
$$

Now if one writes Kane's equations,

$$
F_i + F_i^* = 0, \qquad (i = 1, \dots, \nu) \tag{3.43}
$$

using Eqs. (3.41) and (3.31), and recasts them in a matrix form, one gets the equations of motion:

$$
M\,\ddot{q} + G\,\dot{q} + K\,q = F \tag{3.44}
$$

where the elements of the matrices M, G, K, F are quite lengthy expressions, that are best defined by defining certain intermediate terms, called *modal integrals*.

$$
\breve{W}_{ki} = \int_0^L \rho\phi_{ki}(x)dx \qquad (i = 1, \dots, \nu; \quad k = 1, \dots, 6) \tag{3.45}
$$

$$
W_{klij} = \int_0^L \rho\phi_{ki}(x)\phi_{lj}(x)dx \qquad (i, j = 1, \dots, \nu; \quad k, l = 1, \dots, 6) \tag{3.46}
$$

$$
\breve{X}_{ki} = \int_0^L \rho\, x\, \phi_{ki}(x)dx \qquad (i = 1, \dots, \nu; \quad k = 1, \dots, 6) \tag{3.47}
$$

$$
X_{klij} = \int_0^L \rho\, x\, \phi_{ki}(x)\phi_{lj}(x)dx \qquad (i, j = 1, \dots, \nu; \quad k, l = 1, \dots, 6) \tag{3.48}
$$

$$
\breve{Y}_{ki} = \int_0^L \rho I_{22}\phi_{ki}(x)dx \qquad (i = 1, \dots, \nu; \quad k = 4, 5, 6) \tag{3.49}
$$

$$Y_{klij} = \int_0^L \rho I_{22}\phi_{ki}(x)\phi_{lj}(x)dx \qquad (i,j = 1,\ldots,\nu; \quad k,l = 4,5,6) \tag{3.50}$$

$$\breve{Z}_{ki} = \int_0^L \rho I_{33}\phi_{ki}(x)dx \qquad (i = 1,\ldots,\nu; \quad k = 4,5,6)) \tag{3.51}$$

$$Z_{klij} = \int_0^L \rho I_{33}\phi_{ki}(x)\phi_{lj}(x)dx \qquad (i,j = 1,\ldots,\nu; \quad k,l = 4,5,6) \tag{3.52}$$

$$\mu_{ij} = \int_0^L \rho(\beta_{ij} + \gamma_{ij})dx \qquad (i,j = 1,\ldots,\nu) \tag{3.53}$$

$$\eta_{ij} = \int_0^L \rho x(\beta_{ij} + \gamma_{ij})dx \qquad (i,j = 1,\ldots,\nu) \tag{3.54}$$

$$\begin{aligned} H_{ij} \equiv \int_0^L \Bigg\{ & E_0A_0\phi'_{1i}(x)\phi'_{1j}(x) + \frac{GA_0}{\alpha_2}[\phi_{5i}(x)\phi_{5j}(x) + \phi_{5i}(x)\phi'_{3j}(x) + \phi'_{3i}(x)\phi_{5j}(x) \\ & +\phi'_{3i}(x)\phi'_{3j}(x)] \\ & + \frac{GA_0}{\alpha_3}[\phi_{6i}(x)\phi_{6j}(x) - \phi'_{2j}(x)\phi_{6i}(x) - \phi'_{2i}(x)\phi_{6j}(x) + \phi'_{2i}(x)\phi'_{2j}(x)] + G\kappa'\phi'_{4i}(x)\phi'_{4j}(x) \\ & + E_0I_2\phi'_{5i}(x)\phi'_{5j}(x) + E_0I_3\phi'_{6i}(x)\phi'_{6j}(x) \Bigg\} dx \qquad (i,j = 1,\ldots,\nu) \end{aligned} \tag{3.55}$$

Note that Eq. (3.55) can obtained from a finite element stiffness matrix. Also, regarding the evaluation of the integrals in Eqs. (3.53), (3.54) using Eq. (3.26) by numerical quadrature techniques, integration by parts should be performed in order to bring them into a simpler form.

Now we are ready to express the elements of the M, G, K, F matrices in Eq. (3.44). In what follows certain terms are underlined to facilitate a later discussion on discrepancies between the present formulation and those underlying many public domain software. Briefly, only the underlined terms are present in a conventional theory embedded in many public domain software, but not the massive set of the rest of the terms.

$$\begin{aligned} M_{ij} = \{ & [\underline{\underline{W_{11ij} + W_{22ij} + W_{33ij}}} + Y_{44ij} + Z_{55ij} + Z_{44ij} + Z_{66ij}] \\ & +e_2[W_{44ij} + W_{66ij} - W_{16ij} + W_{34ij} + W_{43ij} - W_{61ij}] \\ & +e_3[W_{15ij} - W_{24ij} - W_{42ij} + W_{51ij} + W_{44ij} + W_{55ij}] - e_2e_3[W_{56ij} + W_{65ij}]\} \\ & (i,j = 1,\ldots,\nu) \end{aligned} \tag{3.56}$$

$$\begin{aligned} G_{ij} = 2\{ & \omega_1\underline{\underline{(W_{32ij} - W_{23ij})}} \\ & +\omega_2[\underline{\underline{W_{13ij} - W_{31ij}}} + Z_{46ij} - Z_{64ij}] + \omega_3[\underline{\underline{W_{21ij} - W_{12ij}}} + Z_{54ij} - Z_{45ij}] \end{aligned}$$

$$
\begin{aligned}
&+e_2[\omega_1(W_{42ij} - W_{24ij}) + \omega_2(W_{14ij} - W_{41ij} + W_{36ij} - W_{63ij}) + \omega_3(W_{62ij} - W_{26ij}) \\
&+e_2\omega_2(W_{46ij} - W_{64ij})] + e_3[\omega_1(W_{43ij} - W_{34ij}) + \omega_2(W_{53ij} - W_{35ij}) \\
&+\omega_3(W_{14ij} - W_{41ij} + W_{25ij} - W_{52ij}) + e_3\omega_3(W_{54ij} - W_{45ij})] \\
&-e_2e_3[\omega_2(W_{45ij} - W_{54ij}) + \omega_3(W_{64ij} - W_{46ij})]\} \qquad (i,j = 1, \ldots, \nu) \qquad (3.57)
\end{aligned}
$$

$$
\begin{aligned}
F_i = -\Bigg\{ &\underline{\underline{(\dot{v}_1 + \omega_2 v_3 - \omega_3 v_2)\breve{W}_{1i} + (\dot{v}_2 + \omega_3 v_1 - \omega_1 v_3)\breve{W}_{2i} + (\dot{v}_3 + \omega_1 v_2 - \omega_2 v_1)\breve{W}_{3i}}} \\
&+\omega_2\omega_3(\breve{Z}_{4i} - \breve{Y}_{4i}) + \dot{\omega}_1(\breve{Y}_{4i} + \breve{Z}_{4i}) + (\dot{\omega}_2 + \omega_3\omega_1)\breve{Y}_{5i} + (\dot{\omega}_3 - \omega_1\omega_2)\breve{Z}_{6i} \\
&\underline{\underline{-(\omega_2^2 + \omega_3^2)\breve{X}_{1i} + (\dot{\omega}_3 + \omega_1\omega_2)\breve{X}_{2i} - (\dot{\omega}_2 - \omega_3\omega_1)\breve{X}_{3i}}} \\
&+e_2[-(\dot{\omega}_3 - \omega_1\omega_2)\breve{W}_{1i} - (\omega_1^2 + \omega_3^2)\breve{W}_{2i} \\
&+(\dot{\omega}_1 + \omega_2\omega_3)\breve{W}_{3i} + (\dot{v}_3 + \omega_1 v_2 - \omega_2 v_1 + \omega_2\omega_3 e_2 + \dot{\omega}_1 e_2)\breve{W}_{4i} - (\dot{\omega}_2 - \omega_3\omega_1)\breve{X}_{4i} \\
&-(\dot{v}_3 + \omega_1 v_2 - \omega_2 v_1 + \omega_2\omega_3 e_2 + \dot{\omega}_1 e_2)\breve{W}_{4i} - (\dot{\omega}_2 - \omega_1\omega_3)\breve{X}_{4i} \\
&-(\dot{v}_1 + \omega_2 v_3 - \omega_3 v_2 + \omega_2\omega_3 e_2 + \dot{\omega}_1 e_2)\breve{W}_{6i} + (\omega_2^2 + \omega_3^2)\breve{X}_{6i}] \\
&e_3[(\dot{\omega}_2 + \omega_3\omega_1)\breve{W}_{1i} - (\dot{\omega}_1 - \omega_2\omega_3)\breve{W}_{2i} - (\omega_1^2 + \omega_2^2)\breve{W}_{3i} \\
&-(\dot{v}_2 + \omega_3 v_1 - \omega_1 v_3 + \omega_2\omega_3 e_3 + \dot{\omega}_1 e_3)\breve{W}_{4i} \\
&-(\dot{\omega}_3 + \omega_1\omega_2)\breve{X}_{4i} + (\dot{v}_1 + \omega_2 v_3 - \omega_3 v_2 + \omega_1\omega_3 e_3 + \dot{\omega}_2 e_3)\breve{W}_{5i} \\
&-(\omega_2^2 + \omega_3^2)\breve{X}_{5i} + e_2e_3[(\omega_3^2 - \omega_2^2)\breve{W}_{4i} - (\dot{\omega}_3 - \omega_1\omega_2)\breve{W}_{5i} \\
&-(\dot{\omega}_2 + \omega_3\omega_1)\breve{W}_{6i}] \Bigg\} \qquad (i = 1, \ldots, \nu) \qquad (3.58)
\end{aligned}
$$

Next we show the expression for K_{ij}, which is quite lengthy:

$$
\begin{aligned}
K_{ij} = \Bigg\{ &\underline{\underline{-(\omega_2^2 + \omega_3^2)W_{11ij} + \omega_1\omega_2(W_{12ij} + W_{21ij})}} - \underline{\underline{(\omega_1^2 + \omega_3^2)W_{22ij} + \omega_1\omega_3(W_{13ij} + W_{31ij})}} \\
&\underline{\underline{-(\omega_1^2 + \omega_2^2)W_{33ij} + \omega_2\omega_3(W_{23ij} + W_{32ij})}} - (\omega_2^2 - \omega_3^2)(Z_{44ij} - Y_{44ij}) + \omega_1\omega_2(Z_{45ij} + Y_{54ij} - Y_{45ij} - Y_{54ij}) \\
&-\omega_1\omega_3(Z_{46ij} + Y_{64ij} - Y_{46ij} - Y_{64ij}) - (\omega_3^2 - \omega_1^2)Y_{55ij} + \omega_2\omega_3(Y_{56ij} + Y_{65ij}) - (\omega_2^2 - \omega_1^2)Y_{66ij} \\
&-\dot{\omega}_1(\underline{\underline{W_{23ij} - W_{32ij}}} + Y_{56ij} + Y_{65ij}) + \dot{\omega}_2(\underline{\underline{W_{13ij} - W_{31ij}}} + Y_{46ij} + Y_{64ij} + Z_{46ij} - Z_{64ij}) \\
&-\dot{\omega}_3(\underline{\underline{W_{12ij} - W_{21ij}}} + Y_{45ij} - Y_{54ij} + Z_{45ij} + Z_{54ij}) - (\dot{v}_1 + \omega_2 v_3 - \omega_3 v_2)\mu_{ij} + (\omega_2^2 + \omega_3^2)\eta_{ij} \\
&+e_2[\omega_1\omega_3(W_{14ij} + W_{41ij}) + (\omega_2^2 + \omega_3^2)(W_{16ij} + W_{61ij}) + \omega_2\omega_3(W_{24ij} + W_{42ij}) - \omega_1\omega_2(W_{26ij} + W_{62ij}) \\
&-(\omega_1^2 + \omega_2^2)(W_{34ij} + W_{43ij}) - \omega_1\omega_3(W_{36ij} + W_{63ij}) - (\dot{v}_2 + \omega_3 v_1 - \omega_1 v_3 - e_2\omega_3^2 + e_2\omega_2^2)W_{44ij} \\
&+(\dot{v}_1 + \omega_2 v_3 - \omega_3 v_2 + \omega_1\omega_2 e_2)(W_{45ij} + W_{54ij}) - e_2\omega_1\omega_3(W_{46ij} + W_{64ij})
\end{aligned}
$$

$$
\begin{aligned}
&-(\dot{v}_2+\omega_3 v_1-\omega_1 v_3-e_2\omega_1^2+e_2\omega_2^2)W_{66ij}+\dot{\omega}_2(W_{14ij}-W_{41ij})-\dot{\omega}_1(W_{24ij}-W_{42ij})-\dot{\omega}_3(W_{26ij}-W_{62ij})\\
&+\dot{\omega}_2(W_{36ij}-W_{63ij})-\dot{\omega}_3 e_2(W_{45ij}+W_{54ij})+\dot{\omega}_2 e_2(W_{46ij}-W_{64ij})+(\dot{\omega}_3-\omega_1\omega_2)\mu_{ij}-(\dot{\omega}_3+\omega_1\omega_2)X_{44ij}\\
&-(\omega_2^2+\omega_3^2)(X_{45ij}+X_{54ij})-(\dot{\omega}_3+\omega_1\omega_2)X_{66ij}]+e_3[-\omega_1\omega_2(W_{14ij}+W_{41ij})-(\omega_2^2+\omega_3^2)(W_{15ij}+W_{51ij})\\
&+(\omega_1^2+\omega_3^2)(W_{24ij}+W_{42ij})+\omega_1\omega_2(W_{25ij}+W_{52ij})-\omega_2\omega_3(W_{34ij}+W_{43ij})+\omega_1\omega_3(W_{35ij}+W_{53ij})\\
&-(\dot{v}_3+\omega_1 v_2-\omega_2 v_1-e_3\omega_2^2+e_3\omega_3^2)W_{44ij}-e_3\omega_1\omega_2(W_{45ij}+W_{54ij})\\
&-(\dot{v}_3+\omega_1 v_2-\omega_2 v_1-e_3\omega_1^2+e_3\omega_3^2)W_{55ij}+(\dot{v}_2+\omega_3 v_1-\omega_1 v_3+\omega_1\omega_3 e_3(W_{56ij}+W_{65ij})\\
&+\dot{\omega}_3(W_{14ij}-W_{41ij}+W_{25ij}-W_{52ij})-\dot{\omega}_1(W_{34ij}-W_{43ij})-\dot{\omega}_2(W_{35ij}-W_{53ij})\\
&-e_3\dot{\omega}_3(W_{45ij}-W_{54ij})+e_3\dot{\omega}_2(W_{46ij}+W_{64ij})-e_3\dot{\omega}_1(W_{56ij}+W_{65ij})-(\dot{\omega}_2+\omega_1\omega_3)\mu_{ij}\\
&+(\dot{\omega}_2-\omega_1\omega_3)(X_{44ij}+X_{55ij})-(\omega_2^2+\omega_3^2)(X_{46ij}+X_{64ij})+(\dot{\omega}_3+\omega_1\omega_2)(X_{56ij}+X_{65ij})\\
&+e_2 e_3[-4\omega_2\omega_3 W_{44ij}+2\omega_1\omega_3(W_{45ij}+W_{54ij})+2\omega_1\omega_2(W_{46ij}+W_{64ij})-\omega_2\omega_3 W_{55ij}\\
&+(\omega_2^2-\omega_1^2)(W_{56ij}+W_{65ij})-\omega_2\omega_3 W_{66ij}-\dot{\omega}_1(W_{55ij}-W_{66ij})+2\dot{\omega}_2 W_{54ij}-2\dot{\omega}_3 W_{64ij}\\
&+\underline{\underline{H}}_{ij}\Bigg\}\quad (i,j=1,\ldots,\nu)
\end{aligned}
\tag{3.59}
$$

3.4 Simulation Results for a Space-Based Robotic Manipulator

One representative problem involving a beam subjected to general base motion is that of simulating the behavior of a space-based robotic manipulator, such as the one shown in Figure 3.3, which consists of three links L_1, L_2, L_3, connected by revolute joints. The outboard link L_3 consists of a base A and two distinct segments B_1, B_2, of which B_1 is 2.667 meters long and has a symmetric box cross section, while B_2 is a 5.333-meter-long channel. Both segments are made of a material for which $E = 6.895 \times 10^{10}$ Nm^2, $G = 2.652 \times 10^{10}$ Nm^2, $\rho/A = 2766.67$kg/m^3. The section properties for B_1 are $A = 3.84 \times 10^{-4}$m^2, and the other parameters are: $I_2 = I_3 = 1.50 \times 10^{-7}$m^4, $\alpha_2 = 2.09$, $\kappa = 2.2 \times 10^{-7}$m^4, $\Gamma = 0$, $e_2 = e_3 = 0$. The corresponding properties for B_1 are $A = 7.3 \times 10^{-5}$m^2, $I_2 = 4.8746 \times 10^{-9}$m^4, $I_3 = 8.2181 \times 10^{-9}$m^4, $\alpha_2 = 3.174$, $\alpha_3 = 1.52$, $\kappa = 2.433 \times 10^{-11}$m^4, $\Gamma = 5.0156 \times 10^{-13}$m^6, $e_2 = 0$, $e_3 = 0.01875$m.

To demonstrate the simulation algorithm, we examine the behavior of L_3 during deployment of the manipulator from a stowed configuration to a fully operational configuration, evolving as follows. The deployment process is presumed to last for 15 seconds, ($T = 15$), during which time values of the angles Ψ_1, Ψ_2, Ψ_3, shown in Figure 3.3, change from 180 deg to 90 deg, 180 deg to 45 deg, and 180 deg to 90 deg, respectively. If the two inboard links (beams) L_1, L_2, each of length l, are treated as rigid, this maneuver results in a prescribed motion of the base A of the outboard link, a motion characterized by the following six temporal functions:

$$v_1 = l(1+c_2)s_3\psi_1\,,\quad v_2 = l\psi_2\,,\quad v_3 = -l(1+c_2)c_3\psi_1 \tag{3.60}$$

$$\omega_1 = \psi_1 s_2 c_3 - \psi_2 s_3\,;\quad \omega_2 = \psi_1 c_2 + \psi_3\,;\quad \omega_3 = \psi_1 s_2 s_3 + \psi_2 c_3 \tag{3.61}$$

$$\text{where } c_i = \cos\psi_i\,;\quad s_i = \sin\psi_i \quad (i = 1,2,3) \tag{3.62}$$

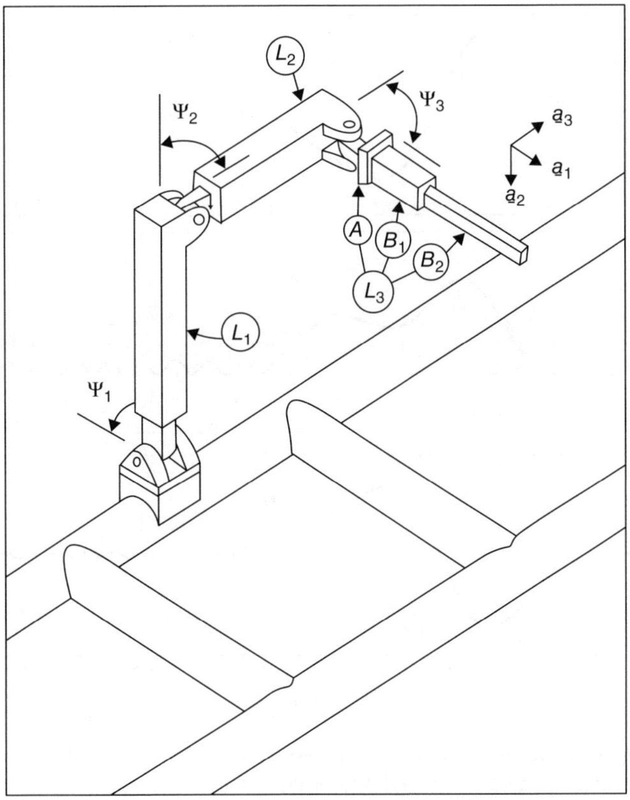

Figure 3.3 Space-Based Robotic Manipulator. Kane *et al.* [1]. Reproduced with permission of the American Institute of Aeronautics and Astronautics, Inc.

We take $l = 8$ m. and let

$$\begin{aligned}\psi_1(t) &= \pi - \frac{\pi}{2T}\left(t - \frac{T}{2\pi}\sin\frac{2\pi t}{T}\right) \text{ rad} && 0 < t < T \\ &= \frac{\pi}{2} \text{ rad} && t > T\end{aligned} \tag{3.63}$$

$$\begin{aligned}\psi_2(t) &= \pi - \frac{3\pi}{4T}\left(t - \frac{T}{2\pi}\sin\frac{2\pi t}{T}\right) \text{ rad} && 0 < t < T \\ &= \frac{\pi}{4} \text{ rad} && t > T\end{aligned} \tag{3.64}$$

$$\begin{aligned}\psi_3(t) &= \pi - \frac{\pi}{T}\left(t - \frac{T}{2\pi}\sin\frac{2\pi t}{T}\right) \text{ rad} && 0 < t < T \\ &= 0 \text{ rad} && t > T\end{aligned} \tag{3.65}$$

Figure 3.4 shows, for the tip of the beam B_2, the transverse displacements and twist, u_2, u_3, θ_1, respectively, during a time interval of 30 seconds, which is twice the deployment time. The first 10 vibration modes of each beam were used to obtain the solution.

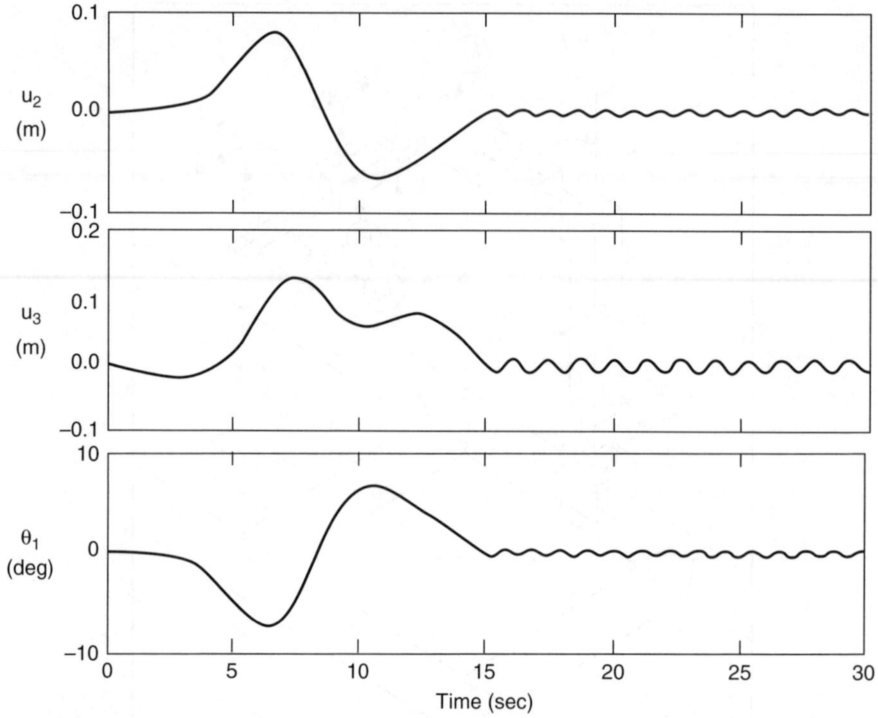

Figure 3.4 Manipulator Deployment Simulation Results. Kane *et al.* [1]. Reproduced with permission of the American Institute of Aeronautics and Astronautics, Inc.

3.5 Erroneous Results Obtained Using Vibration Modes in Conventional Analysis

In passing, we note that conventional solutions used in representative public domain software such as in Refs. [6,7] differ from the preceding analysis in one fundamental issue in that they assume that the elastic deformations in transverse and axial directions are independent spatial functions. That is, the elastic deformation, and the associated velocity of a generic point C in N (see Figure 3.5), are expressed in the conventional approach as:

$$\mathbf{u} = \sum_{j=1}^{\nu} \{\phi_{1j}(x_1, x_2, x_3)\mathbf{b}_1 + \phi_{2j}(x_1, x_2, x_3)\mathbf{b}_2 + \phi_{3j}(x_1, x_2, x_3)\mathbf{b}_3\} q_j(t) \tag{3.66}$$

$${}^N\mathbf{v}^C = {}^N\mathbf{v}^O + {}^N\boldsymbol{\omega}^A x(\mathbf{r}+\mathbf{u}) + \sum_{j=1}^{\nu} \{\phi_{1j}(x_1, x_2, x_3)\mathbf{b}_1 + \phi_{2j}(x_1, x_2, x_3)\mathbf{b}_2 + \phi_{3j}(x_1, x_2, x_3)\mathbf{b}_3\} \dot{q}_j(t) \tag{3.67}$$

This arrangement of using independent spatial functions in the three orthogonal directions is in direct contrast to Eq. (3.11) where the elastic axis stretch is tied to the transverse deformations in two directions. This conventional approach misses the crucial effect of dynamic stiffness,

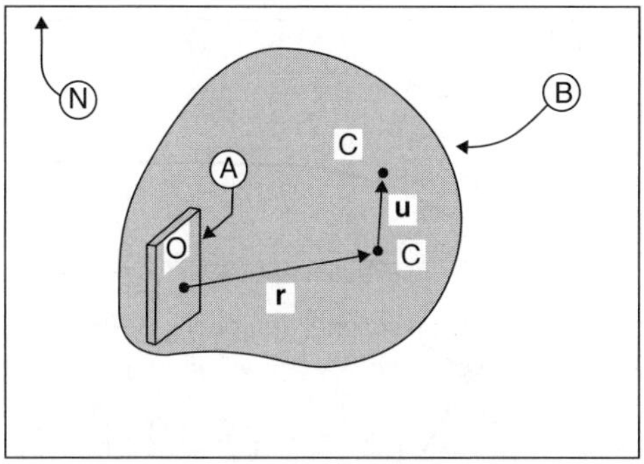

Figure 3.5 Generic Body in a Multibody System. Kane *et al.* [1]. Reproduced with permission of the American Institute of Aeronautics and Astronautics, Inc.

associated with the terms involving μ_{ij}, η_{ij} in the present theory, because of the assumption of independent orthogonal deformations. This is a consequence of premature linearization. This is brought out in Figure 3.6 showing disastrous deflections in transverse directions and axial rotation with time, for a transverse axis spin-up maneuver. Also, when beams or plates are considered, rotations of the elements must be taken into account, which is not done in conventional theories; the difference in the present and conventional theories for beam section axial rotations is also shown in Figure 3.6, for example in axial rotations. Other features of the present theory are in the nature of refinements. These include shear and warping using the concepts of shear area ratio and an effective torsional factor calculated with the aid of a warping factor, and coupled bending and torsion made possible by the introduction of an eccentricity vector from the shear center to the centroid of a cross section. These effects too are not considered in conventional formulations. Finally, one may note that the present chapter also does not consider flexural-torsional coupled vibrations of a beam spinning about the beam axis, such as is discussed in Ref. [8], but rigid body motions of the support in such cases would still require to be treated by the method presented here. Spinning beam vibrations for composite materials have been discussed by dynamic stiffness methods in [9].

Problem Set 3

3.1 Derive the equations of a motion of a rotating cantilever beam, of length L, considering only in-plane bending, neglecting rotatory inertia and shear deformation, and using the vibration modes of a non-rotating cantilever beam [8]:

$$\begin{aligned}\phi_i(x) &= (\sin\beta_i L - \sinh\beta_i L)(\sin\beta_i x - \sinh\beta_i x)\\ &\quad + (\cos\beta_i L + \cosh\beta_i L)(\cos\beta_i x - \cosh\beta_i x);\\ \beta_i &= (m\omega_i^2/EI)^{(1/4)} \quad i = 1, \ldots, n; \quad \beta_1 L = 1.875;\\ \beta_2 L &= 4.694; \quad \beta_3 L = 7.855; \ldots.. \end{aligned} \tag{3.68}$$

where m, E, I, ω are the mass per unit length, modulus of elasticity, section inertia, and vibration frequency. Work out the modal integrals for your choice of beam parameters, and do a simulation using only two modes, that is, $\nu = 2$.

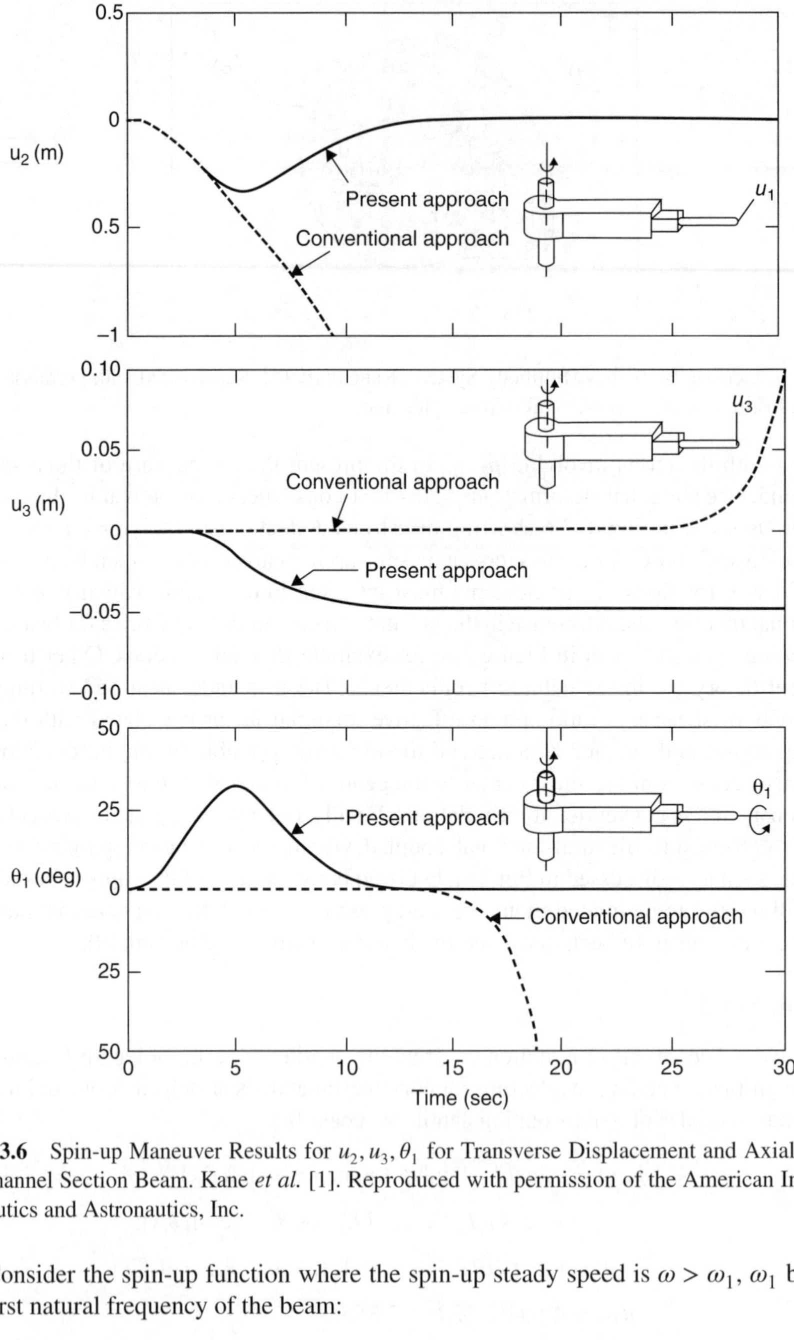

Figure 3.6 Spin-up Maneuver Results for u_2, u_3, θ_1 for Transverse Displacement and Axial Rotation for a Channel Section Beam. Kane *et al.* [1]. Reproduced with permission of the American Institute of Aeronautics and Astronautics, Inc.

3.2 Consider the spin-up function where the spin-up steady speed is $\omega > \omega_1$, ω_1 being the first natural frequency of the beam:

$$\begin{aligned} \omega &= \frac{2}{5}\left[t - \frac{7.5}{\pi}\sin\left(\frac{\pi t}{7.5}\right)\right] && 0 \le t \le 15 \\ &= 6 && 15 \le t \le 30 \end{aligned} \tag{3.69}$$

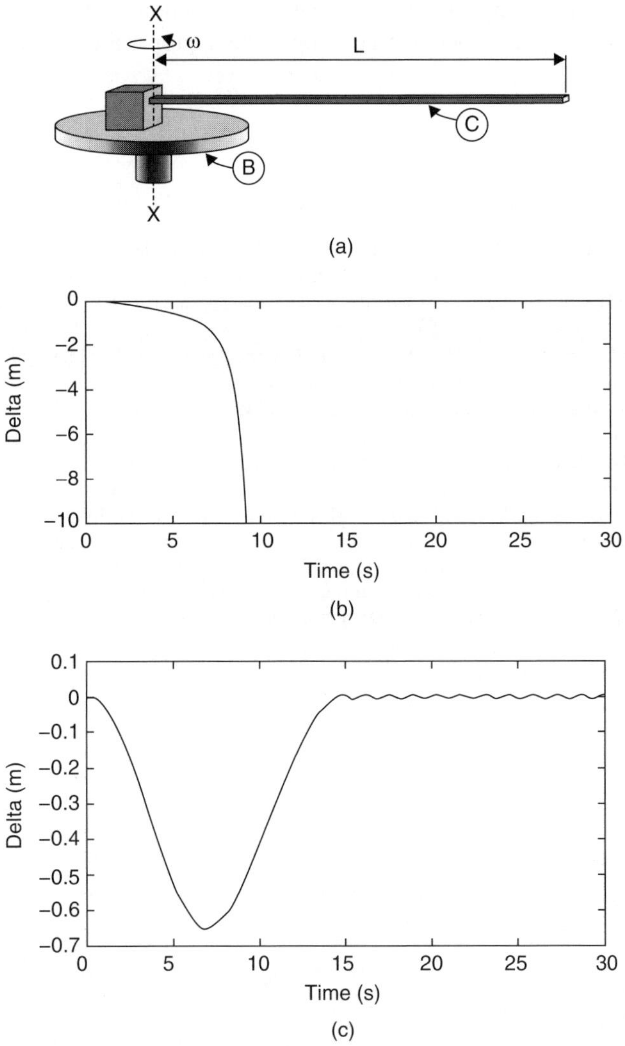

Figure 3.7 (a) Spinning cantilever beam; (b) Conventional analysis using only transverse vibration modes and no fore-shortening, with spin frequency exceeding first mode bending frequency; (c) Spin-up in Present Theory (structural + geometric stiffness, spin freq > first mode freq). Banerjee [10]. Reproduced with permission of the American Institute of Aeronautics and Astronautics, Inc.

Derive the equations of motion using the vibration modes (take one mode) and show that one gets the first curve in Figure 3.7, whereas by doing proper linearization one gets the correct result given by the lower curve (see Ref. [10]).

3.3 Consider a cantilever beam of length L of uniform mass m per length and flexural rigidity EI. The beam can be treated as a Bernoulli-Euler beam, which neglects rotary inertia and shear deformation. Refer to Figure 3.1 and assume that the base of the beam undergoes planar motion with no rotation and only ${}^N\mathbf{v}^O = \mathbf{v}_1\mathbf{a}_1 + \mathbf{v}_1\mathbf{a}_1$ and base acceleration $\dot{v}_1\mathbf{a}_1 + \dot{v}_2\mathbf{a}_2$. Derive the equations of motion of the beam using nonlinear kinematics, and examine the effects of proper and improper linearization using vibration modes.

References

[1] Kane, T.R., Ryan, R.R., and Banerjee, A.K. (1987) Dynamics of a cantilever beam attached to a moving base. *Journal of Guidance, Control, and Dynamics*, **10**(2), 139–151.

[2] Modi, V.J. (1974) Attitude dynamics of satellites with flexible appendages. *Journal of Spacecraft and Rockets*, **11**, 743–751.

[3] Kane, T.R. and Levinson, D.A. (1985) *Dynamics: Theory and Application*, McGraw-Hill.

[4] Kane, T.R., Likins, P.W., and Levinson, D.A. (1983), *Spacecraft Dynamics*, McGraw-Hill.

[5] Timoshenko, S., Young, D.H., and Weaver, W. Jr. (1974) *Vibration Problems in Engineering*, 4th edn, John Wiley & Sons, Inc.

[6] Bodley, C.S., Devers, A.D., Park, A.C., and Frisch, H.P. (1978) A Digital Computer Program for the Dynamic Interaction Simulation of Controls and Structures (DISCOS), vol. I. NASA Technical Paper 1219.

[7] Singh, R.P., van der Voort, R.J., and Likins, P.W. (1984) Dynamics of flexible bodies in tree topology – a computer oriented approach. *Journal of Guidance, Control, and Dynamics*, **8**(5), 584–590.

[8] Meirovitch, L. (1967) *Analytical Methods in Vibrations*, Macmillan Company.

[9] Banerjee, J.R. and Su, H. (2006) Dynamic stiffness formulation and free vibration analysis of a spinning composite beam. *Computers & Structures*, **84**(19–20), 1208–1214.

[10] Banerjee, A.K. (2003) Contributions of multibody dynamics to space flight: a brief review. *Journal of Guidance, Control, and Dynamics*, **26**(3), 385–394.

4

Dynamics of a Plate in Large Overall Motion

This chapter continues on the topic of elastic bodies undergoing small elastic deformation with respect to a reference frame that is itself going through large rotation and translation. More specifically we analyze thin plates executing small elastic vibrations in a situation of large overall motion. We use Kane's method of direct linearization with respect to modal coordinates used with vibration modes, defining the small deformation, as discussed generally in Chapter 1, and in the preceding chapter for the case of beams undergoing large overall motion. The analysis and results presented here are taken from the paper [1].

4.1 Motivating Results of a Simulation

Figure 4.1 shows a rectangular plate whose displacements are measured with respect to a reference frame R in which orthogonal unit vectors $\mathbf{r}_1, \mathbf{r}_2, \mathbf{r}_3$ are fixed. Point O designates a particle of the plate fixed in R, and the plate is assumed to be executing small motions in the frame R, while R is undergoing large and rapid rigid body motions of rotation and translation in a Newtonian reference frame N; that is, both the orientation of the triad $\mathbf{r}_1, \mathbf{r}_2, \mathbf{r}_3$ in N and the position of O in N are unrestricted and can change rapidly. Our objective is to produce an algorithm that can be used to simulate motions of the plate when the motion of R in N is given. Referring to Figure 4.1, consider R as a rigid body, and suppose that a 72-in. x 48-in. x 0.1-in. rectangular plate is simply supported by R along all four edges of the plate, the 72-in. long edge coinciding with X_1, and the 48-in.-long one coinciding with X_2; note that X_1, X_2 align respectively with $\mathbf{r}_1, \mathbf{r}_2$. Let R undergo a spin-up motion about the edge such that X_2 remains fixed in N, and $\omega_2(t)$, the $\mathbf{r}_2$ measure number of the angular velocity of R in N, measured in radius per second, be prescribed as

$$\begin{aligned}\omega_2(t) &= \frac{2\pi n^*}{60t^*}\left(t - \frac{t^*}{2\pi}\sin\frac{2\pi t}{t^*}\right), \quad t \le t^*;\\ &= \frac{2\pi n^*}{60} \quad t > t^*\end{aligned} \tag{4.1}$$

Flexible Multibody Dynamics: Efficient Formulations and Applications, First Edition. Arun K. Banerjee.

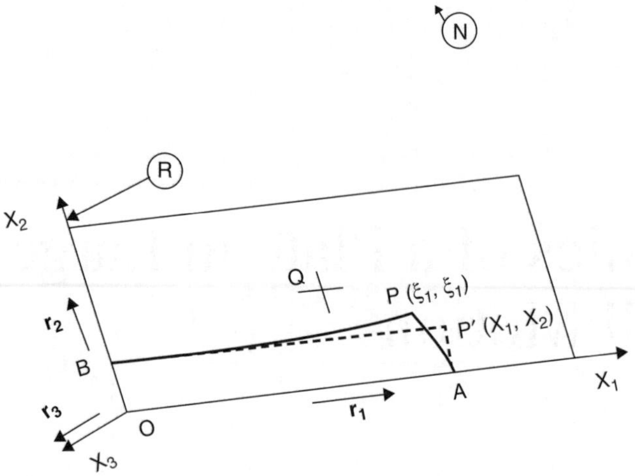

Figure 4.1 Plate Showing Small Elastic Deformation of Line Elements in a Reference Frame R That Undergoes Large Rigid Body Motion in a Newtonian Frame N. Banerjee and Kane [1]. Reproduced with permission of the American Society of Mechanical Engineers.

where n^* is the ultimate spin speed, in revolutions per minute, and t^* is the spin-up time, in seconds. Under these circumstances, Q, the center point of the plate, must experience a deflection, say u_3^Q, parallel to the X_3 axis, and one can predict this deflection in qualitative terms on physical grounds. Initially, u_3^Q is equal to zero. As R begins to rotate, Q must begin to move in the direction of $\mathbf{r}_3$ so that u_3^Q takes on positive values because at first the plate cannot keep up with the motion of the frame R. But soon restoring forces arising from the elasticity of the plate make themselves felt, and eventually becoming sufficiently large not only to arrest the growth of u_3^Q but to cause u_3^Q to decrease. Indeed, sooner or later Q should not only catch up with R – that is u_3^Q should become equal to zero – but Q should overshoot its original position in R, in which event u_3^Q takes on negative values; and after the motion of R has settled down – that is, once ω_2 has reached its maximum value, from which time onward R is rotating with constant angular speed – it is to be expected that a residual "ringing" will take place, that is, that u_3^Q will have an oscillatory character. Thus, a plot of u_3^Q versus time should have the general appearance of the curve in Figure 4.2, which was generated by the theory set forth in the sequel, with $n^* = 12$ revolutions per minute. Figure 4.3 deals with the effect of spin on steady-state response frequency; that is, the ratio of ω_d, the dynamic response frequency when spin-up has been completed, to ω_1, the first natural frequency of the plate, is plotted versus the ratio of ω_s, the spin speed, to ω_1. As can be seen, the theory predicts *dynamic stiffening*, that is, an increase in ω_d with increasing ω_s, and this for all spin speeds.

In what follows we set forth the theory in detail, with conclusions, and an Appendix showing the relationship between certain modal integrals arising in this theory and geometric stiffness coefficients associated with inertia forces of the plate due to large motion of its reference frame.

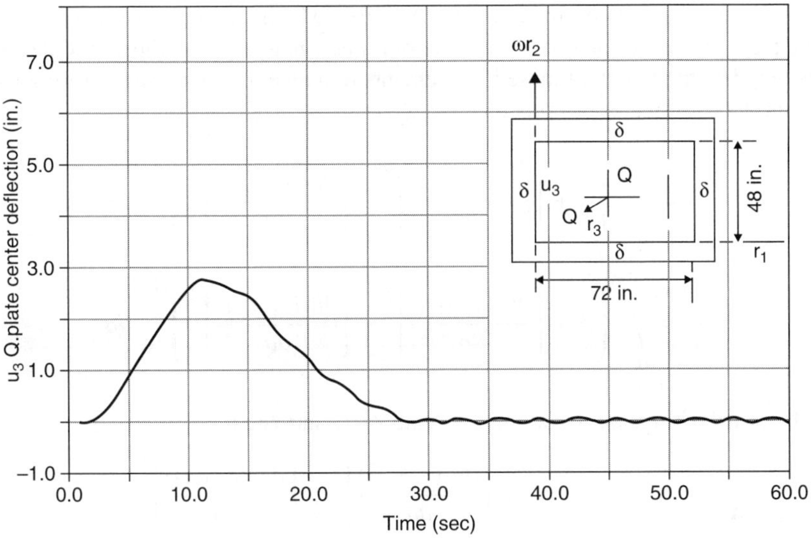

Figure 4.2 Transient Deflection in Inches of the Center Point of a Simply Supported Plate (supports indicated in the inset by *s*) Spun Up from Rest about an Edge along r_2 to a Maximum Spin Rate of 12 rpm. Banerjee and Kane [1]. Reproduced with permission of the American Society of Mechanical Engineers.

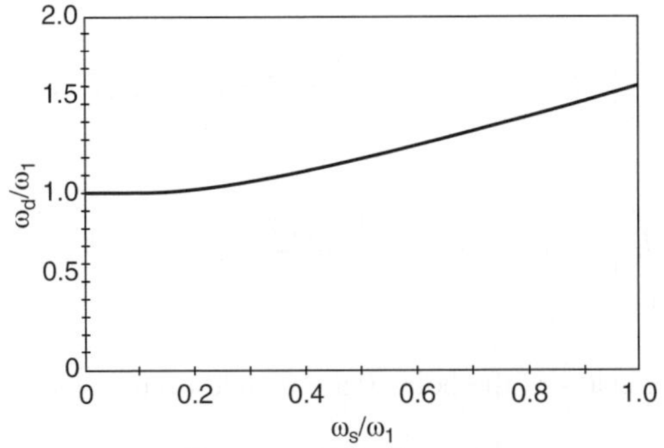

Figure 4.3 Ratio of Dynamic Frequency due to Centrifugal Stiffening to First Natural Frequency, ω_d/ω_1, versus Ratio of Spin Frequency to First Natural Frequency, ω_s/ω_1 of Plate. Banerjee and Kane [1]. Reproduced with permission of the American Society of Mechanical Engineers.

4.2 Application of Kane's Methodology for Proper Linearization

Referring to Figure 4.1, let x_1, x_2 be the coordinates along orthogonal axes parallel to $\mathbf{r}_1, \mathbf{r}_2$ of P', a generic material point P of the plate when the plate is undeformed. Then the displacement of P can be expressed in terms of measure numbers u_1, u_2, u_3 as:

$$\mathbf{u} = u_1(x_1, x_2, t)\mathbf{r}_1 + u_2(x_1, x_2, t)\mathbf{r}_2 + u_3(x_1, x_2, t)\mathbf{r}_3 \tag{4.2}$$

Now consider the line elements in Figure 4.1, AP', BP', which are parallel to x_1, x_2, respectively, when the plate is undeformed. If we restrict our analysis to a plate whose middle surface is inextensible, then the length of these line elements remain constant during deformation of the plate, when the coordinates of P become (ξ_1, ξ_2, t):

$$x_1 = \int_0^{\xi_1} \left\{ 1 + \left[\frac{\partial u_2(\sigma, x_2, t)}{\partial \sigma} \right]^2 + \left[\frac{\partial u_3(\sigma, x_2, t)}{\partial \sigma} \right]^2 \right\}^{1/2} d\sigma \tag{4.3}$$

$$x_2 = \int_0^{\xi_2} \left\{ 1 + \left[\frac{\partial u_1(x_1, \sigma, t)}{\partial \sigma} \right]^2 + \left[\frac{\partial u_3(x_1, \sigma, t)}{\partial \sigma} \right]^2 \right\}^{1/2} d\sigma \tag{4.4}$$

To save labor in writing, we introduce the symbols J_1 and J_2 as:

$$J_1(\sigma, x_2, t) = 1 + \left[\frac{\partial u_2(\sigma, x_2, t)}{\partial \sigma} \right]^2 + \left[\frac{\partial u_3(\sigma, x_2, t)}{\partial \sigma} \right]^2 \tag{4.5}$$

$$J_2(x_1, \sigma, t) = 1 + \left[\frac{\partial u_1(x_1, \sigma, t)}{\partial \sigma} \right]^2 + \left[\frac{\partial u_3(x_1, \sigma, t)}{\partial \sigma} \right]^2 \tag{4.6}$$

Differentiating under the integral sign in Eqs. (4.3) and (4.4) with respect to t, we get for no stretching of the middle surface, that is, x_1, x_2 being constant:

$$0 = \frac{1}{2} \int_0^{\xi_1} [J_1(\sigma, x_2, t)]^{-1/2} \frac{\partial J_1(\sigma, x_2, t)}{\partial t} d\sigma + \dot{\xi}_1 [J_1(\xi_1, x_2, t)]^{1/2} \tag{4.7}$$

$$0 = \frac{1}{2} \int_0^{\xi_2} [J_2(x_1, \sigma, t)]^{-1/2} \frac{\partial J_2(x_1, \sigma, t)}{\partial t} d\sigma + \dot{\xi}_2 [J_2(x_1, \xi_2, t)]^{1/2} \tag{4.8}$$

The instantaneous coordinates of the point P are related to the coordinates of P' by:

$$\xi_1 = x_1 + u_1(x_1, x_2, t) \tag{4.9}$$

$$\xi_2 = x_2 + u_2(x_1, x_2, t) \tag{4.10}$$

Differentiating Eqs. (4.9) and (4.10) with respect to t, and using the results in Eqs. (4.7) and (4.8):

$$\dot{u}_1 = -\frac{1}{2} [J_1(\xi_1, x_2, t)]^{-1/2} \int_0^{\xi_1} [J_1(\sigma, x_2, t)]^{-1/2} \frac{\partial J_1(\sigma, x_2, t)}{\partial t} d\sigma \tag{4.11}$$

$$\dot{u}_2 = -\frac{1}{2} [J_2(x_1, \xi_2, t)]^{-1/2} \int_0^{\xi_2} [J_2(x_1, \sigma, t]^{-1/2} \frac{\partial J_2(x_1, \sigma, t}{\partial t} d\sigma \tag{4.12}$$

Considering points on the line element BP' we express u_2 and u_3 in series expansions:

$$u_2(\sigma, x_2, t) = \sum_{j=1}^{n} \phi_{2j}(\sigma, x_2)\, q_j(t) \tag{4.13}$$

$$u_3(\sigma, x_2, t) = \sum_{j=1}^{n} \phi_{3j}(\sigma, x_2)\, q_j(t) \tag{4.14}$$

Similarly, for the line element AP' we express u_1 and u_3 as:

$$u_1(x_{1,}\sigma, t) = \sum_{j=1}^{n} \phi_{1j}(x_{1,}\sigma)\, q_j(t) \tag{4.15}$$

$$u_3(x_{1,}\sigma, t) = \sum_{j=1}^{n} \phi_{3j}(x_{1,}\sigma)\, q_j(t) \tag{4.16}$$

The spatial functions ϕ_{1j}, ϕ_{2j}, ϕ_{3j} ($j = 1, \ldots, n$) introduced in Eqs. (4.13) and (4.16) are as yet unrestricted; ultimately we will take them to be the vibration mode shapes for a plate with appropriate boundary conditions. The temporal functions $q_j(t)$ ($j = 1, \ldots, n$) play the role of generalized coordinates in the sense of Lagrange, and n is simply any integer indicating how many functions or modes are kept in the series. Differentiating Eqs. (4.5) and (4.6) with respect to t, and using Eqs. (4.13)–(4.16) we have the following:

$$\frac{\partial J_1(\sigma, x_2, t)}{\partial t} = 2\sum_{i=1}^{n}\sum_{j=1}^{n}\left[\phi'_{2i}(\sigma, x_2)\phi'_{2j}(\sigma, x_2) + \phi'_{3i}(\sigma, x_2)\phi'_{2j}(\sigma, x_2)\right]\dot{q}_i(t)\, q_j(t) \tag{4.17}$$

$$\frac{\partial J_2(x_1, \sigma, t)}{\partial t} = 2\sum_{i=1}^{n}\sum_{j=1}^{n}\left[\phi'_{1i}(x_1, \sigma)\,\phi'_{1j}(x_1, \sigma,) + \phi'_{3i}(x_1, \sigma)\phi'_{3j}(x_1, \sigma)\right]\dot{q}_i(t)\, q_j(t) \tag{4.18}$$

where primes denote partial differentiation with respect to the dummy space variable σ. Now, substitution from Eqs. (4.17) and (4.18) into Eqs. (4.11) and (4.12) gives:

$$\begin{aligned}\dot{u}_1 - [J_1(\xi_1, x_2, t)]^{-(1/2)} \int_0^{\xi_1} \Big\{ [J_1(\sigma, x_2, t)]^{-(1/2)} \sum_{i=1}^{n}\sum_{j=1}^{n} \Big[\phi'_{2i}(\sigma, x_2)\phi'_{2j}(\sigma, x_2) \\ + \phi'_{3i}(\sigma, x_2)\phi'_{3j}(\sigma, x_2)\Big]\dot{q}_i q_j \Big\} d\sigma\end{aligned} \tag{4.19}$$

$$\begin{aligned}\dot{u}_2 - [J_2(x_1, \xi_2, t)]^{-(1/2)} \int_0^{\xi_2} \Big\{ [J_2(x_1, \sigma, t)]^{-(1/2)} \sum_{i=1}^{\nu}\sum_{j=1}^{\nu} \Big[\phi'_{1i}(x_1, \sigma)\phi'_{1j}(x_1, \sigma) \\ + \phi'_{3i}(x_1, \sigma)\phi'_{3j}(x_1, \sigma)\Big]\dot{q}_i q_j \Big\} d\sigma\end{aligned} \tag{4.20}$$

In addition, differentiation of Eq. (4.14) with respect to t yields:

$$\dot{u}_3 = \sum_{i=1}^{\nu} \phi_{3i}(x_1, x_2)\dot{q}_i \tag{4.21}$$

The velocity of point P in the Newtonian reference frame N can be formally written as

$$
\begin{aligned}
{}^N\mathbf{v}^P = {}^N\mathbf{v}^O + {}^N\boldsymbol{\omega}^R \times \left\{ \left[x_1 + \sum_{j=1}^{n} \phi_{1j}(x_1,x_2)q_j(t) \right] \mathbf{r}_1 + \left[x_2 + \sum_{j=1}^{n} \phi_{2j}(x_1,x_2)q_j(t) \right] \mathbf{r}_2 \right. \\
\left. + \sum_{j=1}^{n} \phi_{3j}(x_1,x_2)q_j(t)\mathbf{r}_3 \right\} + \dot{u}_1\mathbf{r}_1 + \dot{u}_2\mathbf{r}_2 + \dot{u}_3\mathbf{r}_3
\end{aligned}
\tag{4.22}
$$

where ${}^N\mathbf{v}^O, {}^N\boldsymbol{\omega}^R$, respectively, are the velocity of the point O in N and the angular velocity of the frame R in N, which can be expressed in terms of specified time functions $v_i(t)$ and $\omega(t)$ $(i = 1, 2, 3)$ as follows:

$$
{}^N\mathbf{v}^O = v_1(t)\mathbf{r}_1 + v_2(t)\mathbf{r}_2 + v_3(t)\mathbf{r}_3 \tag{4.23}
$$

$$
{}^N\boldsymbol{\omega}^R = \omega_1(t)\mathbf{r}_1 + \omega_2(t)\mathbf{r}_2 + \omega_3(t)\mathbf{r}_3 \tag{4.24}
$$

Using $\dot{q}_1, \dots, \dot{q}_n$ as generalized speeds, one can now form the ith nonlinear partial velocity of P in N, by inspection of Eqs. (4.19)–(4.22), as:

$$
\begin{aligned}
{}^N\mathbf{v}_i^P = -\mathbf{r}_1 \left\langle [J_1(\xi_1,x_2,t)]^{-(1/2)} \int_0^{\xi_1} \left\{ [J_1(\sigma,x_2,t)]^{-(1/2)} \sum_{j=1}^{n} \left[\phi'_{2i}(\sigma,x_2)\phi'_{2j}(\sigma,x_2) \right. \right. \right. \\
\left. \left. \left. + \phi'_{3i}(\sigma,x_2)\phi'_{3j}(\sigma,x_2) \right] q_j \right\} d\sigma \right\rangle \\
- \mathbf{r}_2 \left\langle [J_2(x_1,\xi_2,t)]^{-(1/2)} \int_0^{\xi_2} \left\{ [J_2(x_1,\sigma,t)]^{-(1/2)} \sum_{j=1}^{n} \left[\phi'_{1i}(x_1,\sigma)\phi'_{1j}(x_1,\sigma) \right. \right. \right. \\
\left. \left. \left. + \phi'_{3i}(x_1,\sigma)\, \phi'_{3j}(x_1,\sigma) \right] q_j \right\} d\sigma \right\rangle \\
+ \mathbf{r}_3 \phi_{3i}(x_1,x_2) \quad (i = 1, \dots, n)
\end{aligned}
\tag{4.25}
$$

So far, no linearizations have been performed; but now that nonlinear expressions for partial velocities are in hand, Kane's method of direct linearization can be used, they can be linearized, along with velocities and accelerations everywhere in $q_j, \dot{q}_j$ $(j = 1, \dots, n)$, to produce the correct equations for small motions of the plate relative to R. In the sequel, we perform such linearizations and place a tilde over a quantity linearized in $q_j, \dot{q}_j$ $(j = 1, \dots, n)$. Thus use of Eqs. (4.13) and (4.14) in Eq. (4.5) leads to:

$$
\tilde{J}_1(\sigma,x_2,t) = 1; \quad \tilde{J}_1(\zeta_1,x_2,t) = 1 \tag{4.26}
$$

Similarly, from Eqs. (4.15) and (4.16) in Eq. (4.6), we have:

$$
\tilde{J}_2(x_1,\sigma,t) = 1; \quad \tilde{J}_2(x_1,\zeta_2,t) = 1 \tag{4.27}
$$

Again, to save labor in writing, we introduce the notations:

$$\alpha_{ij}(x_1,x_2) = \int_0^{x_1} [\phi'_{2i}(\sigma,x_2)\phi'_{2j}(\sigma,x_2) + \phi'_{3i}(\sigma,x_2)\phi'_{3j}(\sigma,x_2)]d\sigma \quad (i,j = 1,\dots,n) \tag{4.28}$$

$$\beta_{ij}(x_1,x_2) = \int_0^{x_2} [\phi'_{1i}(x_1,\sigma)\phi'_{1j}(x_1,\sigma) + \phi'_{3i}(x_1,\sigma)\phi'_{3j}(x_1,\sigma)]d\sigma \quad (i,j = 1,\dots,n) \tag{4.29}$$

Correctly linearized partial velocity of P in N can now be written from Eq. (4.25) as:

$$^N\tilde{\mathbf{v}}_i^P = -\mathbf{r}_1 \sum_{j=1}^n \alpha_{ij}(x_1,x_2)q_j(t) - \mathbf{r}_2 \sum_{j=1}^n \beta_{ij}(x_1,x_2)q_j(t) + \mathbf{r}_3\phi_{3i}(x_1,x_2) \quad (i = 1,\dots,n) \tag{4.30}$$

The linearized velocity of P in N follows from Eqs. (4.22)–(4.24) as:

$$\begin{aligned} ^N\tilde{\mathbf{v}}^P = \mathbf{r}_1 &\left\{ v_1 + \omega_2 \sum_{j=1}^n \phi_{3j}(x_1,x_2)q_j(t) - \omega_2 \left[x_2 + \sum_{j=1}^n \phi_{2j}(x_1,x_2)q_j(t) \right] \right\} \\ + \mathbf{r}_2 &\left\{ v_2 + \omega_3 \left[x_1 + \sum_{j=1}^n \phi_{1j}(x_1,x_2)q_j(t) \right] - \omega_1 \sum_{j=1}^n \phi_{3j}(x_1,x_2)q_j(t) \right\} \\ + \mathbf{r}_3 &\left\{ v_3 + \omega_1 \sum_{j=1}^n [x_2 + \phi_{2j}(x_1,x_2)q_j(t)] - \omega_2 \left[x_1 + \sum_{j=1}^n \phi_{1j}(x_1,x_2)q_j(t) \right] \right. \\ &\left. + \sum_{j=1}^n \phi_{3j}(x_1,x_2)\dot{q}_j(t) \right\} \end{aligned} \tag{4.31}$$

The acceleration of P in N, linearized in $q_1, \dots q_n, \dot{q}_1, \dots \dot{q}_n$, is obtained by differentiating $^N\tilde{\mathbf{v}}^P$ in N and discarding nonlinear terms,

$$^N\tilde{\mathbf{a}}^P = \frac{^Nd}{dt}(^N\tilde{\mathbf{v}}^P) = \frac{^Rd}{dt}(^N\tilde{\mathbf{v}}^P) + {^N\boldsymbol{\omega}^R} \times {^N\tilde{\mathbf{v}}^P} \tag{4.32}$$

and the generalized inertia force corresponding to the ith generalized speed is given by

$$\tilde{F}_i^* = -\int_S {^N\tilde{\mathbf{v}}_i^P} \cdot {^N\tilde{\mathbf{a}}^P} \rho dx_1 dx_2 \quad (i = 1,\dots,n) \tag{4.33}$$

where S is the areal domain of the plate, and ρ is the mass density per unit area. In expanding Eq. (4.33) it is convenient to define the following modal integrals:

$$A_i = \int_S \phi_{3i}\rho dx_1 dx_2 \qquad (i = 1,\dots,n) \tag{4.34}$$

$$B_{ki} = \int_S x_k\phi_{3i}\rho dx_1 dx_2 \qquad (k = 1,2; \quad i = 1,\dots,n) \tag{4.35}$$

$$C_{ij} = \int_S \alpha_{ij}\rho dx_1 dx_2 \qquad (k = 1,2; \quad i,j = 1,\dots,n) \tag{4.36}$$

$$C_{kij} = \int_S x_k \alpha_{ij} \rho dx_1 dx_2 \qquad (k = 1, 2; \quad i, j = 1, \ldots, n) \tag{4.37}$$

$$D_{ij} = \int_S \beta_{ij} \rho dx_1 dx_2 \qquad (i, j = 1, \ldots, n) \tag{4.38}$$

$$D_{kij} = \int_S x_k \beta_{ij} \rho dx_1 dx_2 \qquad (k = 1, 2; \quad i, j = 1, \ldots, n) \tag{4.39}$$

$$E_{kij} = \int_S \phi_{3i} \phi_{kj} \rho dx_1 dx_2 \quad (k = 1, 2, 3; \quad i, j = 1, \ldots, n) \tag{4.40}$$

Using these in Eq. (4.33) and neglecting all nonlinear terms in the generalized coordinates and their derivatives yield:

$$\begin{aligned}
\tilde{F}_i^* = &- \sum_{j=1}^{n} E_{3ij} \ddot{q}_j - \sum_{j=1}^{n} (\omega_1 E_{2ij} - \omega_2 E_{1ij}) \dot{q}_j - \sum_{j=1}^{n} \Big[- (\dot{v}_1 + \omega_2 v_3 - \omega_3 v_2) C_{ij} - (\omega_1 \omega_2 - \dot{\omega}_3) C_{2ij} \\
&+ (\omega_2^2 + \omega_3^2) C_{1ij} - (\dot{v}_2 + \omega_3 v_1 - \omega_1 v_3) D_{ij} - (\dot{\omega}_3 + \omega_1 \omega_2) D_{1ij} - \left(\omega_1^2 + \omega_3^2\right) D_{2ij} \\
&+ (\dot{\omega}_1 + \omega_2 \omega_3) E_{2ij} + (\omega_1 \omega_3 - \dot{\omega}_2) E_{1ij} - \left(\omega_1^2 + \omega_2^2\right) E_{3ij} \Big] q_j - (\dot{v}_3 + \omega_1 v_2 - \omega_2 v_1) A_i \\
&- (\dot{\omega}_1 + \omega_2 \omega_3) B_{2i} - (\omega_1 \omega_3 - \dot{\omega}_2) B_{1i}, \quad (i = 1, \ldots, n)
\end{aligned} \tag{4.41}$$

Each coefficient of $q_j(t)$ $(j = 1, \ldots, n)$ in Eq. (4.41) can be called a *dynamic stiffness coefficient* and terms involving $C_{ij}, C_{1ij}, C_{2ij}, D_{ij}, D_{1ij}, D_{2ij}$ can be related to geometric stiffness associated with translational and rotational inertia loads (Ref. [6], p. 258). This topic is discussed further in the Appendix to this chapter. The generalized active forces (Ref. [2], p. 247) due to material elasticity of the plate are given by:

$$\tilde{F}_i = - \sum_{j=1}^{n} \lambda_{ij} q_j \quad (i = 1, \ldots n) \tag{4.42}$$

Here λ_{ij} is a generic element of the modal stiffness matrix, $\phi^T K \phi$ in the standard notation of structural dynamics, which may or may not be diagonal, depending on the type of component modes (Ref. [4], pp. 467–492) used to represent the elastic deformation of the plate. Finally, the linearized equations of the motion of the plate that is executing small elastic motions in a reference frame undergoing large rigid body motion are obtained by substituting from Eqs. (4.41) and (4.42) into Kane's dynamical equations [2]:

$$\tilde{F}_i + \tilde{F}_i^* = 0, \quad (i = 1, \ldots, n) \tag{4.43}$$

4.3 Simulation Algorithm

A step-by-step procedure for simulating large overall motions of the plate follows.

a. Specify base motion temporal functions, $v_1, v_2, v_3, \omega_1, \omega_2, \omega_3$, in Eqs. (4.23) and (4.24).
b. Specify plate geometry and material constants such as ρ, the mass per unit of area, E, the modulus of elasticity, and ν, Poisson's ratio.

c. Specify the number n of modes to be used.
d. Determine the vibration mode shapes and frequencies of the plate, for example, with boundaries supported and fixed in a Newtonian frame. This is normally done with a finite element code, and yields the mode shapes $\phi_{ij}(x_1, x_2), (i = 1, 2, 3; j = 1, \ldots, n)$ (see Eqs. 4.13–4.15), and the elements of the modal stiffness matrix λ_{ij} $(i, j = 1, \ldots, n)$ (see Eq. 4.43).
e. Select m node points; let x_1^k, x_2^k $(k = 1, \ldots, m)$ denote their coordinates; specify initial displacements $d(x_1^k, x_2^k)$ and initial velocities $v(x_1^k, x_2^k)$ $(k = 1, \ldots, m)$ by assigning numerical values to $d_i(x_1^k, x_2^k)$ and $v_i(x_1^k, x_2^k)$ $(i = 1,2,3;\ k = 1, \ldots, m)$ with

$$\mathbf{d} = d_1\left(x_1^k, x_2^k\right)\mathbf{r}_1 + d_2\left(x_1^k, x_2^k\right)\mathbf{r}_2 + d_3\left(x_1^k, x_2^k\right)\mathbf{r}_3 \tag{4.44}$$

$$\mathbf{v} = v_1\left(x_1^k, x_2^k\right)\mathbf{r}_1 + v_2\left(x_1^k, x_2^k\right)\mathbf{r}_2 + v_3\left(x_1^k, x_2^k\right)\mathbf{r}_3 \tag{4.45}$$

To find the associated initial values of $q_i, \dot{q}_i$ $(i = 1, \ldots, n)$, denoted by $q_i(0), \dot{q}_i(0), i = 1, \ldots, n$ and introduce matrices d, v, ϕ, $q(0)$, $\dot{q}(0)$, as follows:

$$d = \begin{bmatrix} d_1\left(x_1^1, x_2^1\right) \\ d_2\left(x_1^1, x_2^1\right) \\ d_3\left(x_1^1, x_2^1\right) \\ \vdots \\ \vdots \\ \vdots \\ d_1\left(x_1^M, x_2^M\right) \\ d_2\left(x_1^M, x_2^M\right) \\ d_3\left(x_1^M, x_2^M\right) \end{bmatrix}, \quad v = \begin{bmatrix} v_1\left(x_1^1, x_2^1\right) \\ v_2\left(x_1^1, x_2^1\right) \\ v_3\left(x_1^1, x_2^1\right) \\ \vdots \\ \vdots \\ \vdots \\ v_1\left(x_1^M, x_2^M\right) \\ v_2\left(x_1^M, x_2^M\right) \\ v_3\left(x_1^M, x_2^M\right) \end{bmatrix} \tag{4.46}$$

$$\phi = \begin{bmatrix} \phi_{11}\left(x_1^1, x_2^1\right) & \cdots & \phi_{1n}\left(x_1^1, x_2^1\right) \\ \phi_{21}\left(x_1^1, x_2^1\right) & \cdots & \phi_{2n}\left(x_1^1, x_2^1\right) \\ \phi_{31}\left(x_1^1, x_2^1\right) & \cdots & \phi_{3n}\left(x_1^1, x_2^1\right) \\ \cdot & \cdots & \cdot \\ \cdot & \cdots & \cdot \\ \cdot & \cdots & \cdot \\ \phi_{11}\left(x_1^M, x_2^M\right) & \cdots & \phi_{1n}\left(x_1^M, x_2^M\right) \\ \phi_{21}\left(x_1^M, x_2^M\right) & \cdots & \phi_{2n}\left(x_1^M, x_2^M\right) \\ \phi_{31}\left(x_1^M, x_2^M\right) & \cdots & \phi_{3n}\left(x_1^M, x_2^M\right) \end{bmatrix} \tag{4.47}$$

$$q(0) = \begin{Bmatrix} q_1(0) \\ \cdot \\ \cdot \\ \cdot \\ q_n(0) \end{Bmatrix}, \quad \dot{q}(0) = \begin{Bmatrix} \dot{q}_1(0) \\ \cdot \\ \cdot \\ \cdot \\ \dot{q}_n(0) \end{Bmatrix} \tag{4.48}$$

Then $q(0)$ and $\dot{q}(0)$ must satisfy (see Eq. (4.2) and Eqs. (4.13)–(4.15)), the equations

$$\phi q(0) = d, \phi\dot{q}(0) = v \tag{4.49}$$

Hence, solving Eq. (4.49) for least square error by the Moore-Penrose pseudo-inverse, evaluate $q(0)$ and $\dot{q}(0)$

$$q(0) = [\phi^T\phi]^{-1}\phi^T d, \qquad \dot{q}(0) = [\phi^T\phi]^{-1}\phi^T v \tag{4.50}$$

f. Evaluate the modal integrals defined in Eqs. (4.34)–(4.40), employing integration by parts wherever possible (see Appendix), using finite element interpolations (Ref. [5], pp. 234–235).
g. Substitute the modal integrals from step (f) into Eq. (4.40), and the modal stiffness matrix elements from Step (d) into Eq. (4.41).
h. Form Eq. (4.43), expressing them in the matrix form,

$$M\ddot{q} + G\dot{q} + Kq = F \tag{4.51}$$

where M, G, K are (nxn) mass, gyroscopic damping, and overall stiffness matrices, and F is ($nx1$) vector, and integrate Eq. (4.51) numerically. Evaluate the displacement **d** at (x_1, x_2) at time t as:

$$\mathbf{d} = \sum_{j=1}^{n} [\phi_{1j}(x_1, x_2)\mathbf{r}_1 + \phi_{2j}(x_1, x_2)\mathbf{r}_2 + \phi_{3j}(x_1, x_2)\mathbf{r}_3]\, q_j(t) \tag{4.52}$$

We note in passing that all of the numerical results reported in this paper are based on the exact mode shapes and frequencies for natural vibrations of a simply supported plate, that is,

$$\phi_{1j}(x_1, x_2) = \phi_{2j}(x_1, x_2) = 0 \qquad (j = 1, \dots, n) \tag{4.53}$$

$$\phi_{3j}(x_1, x_2) = f_j \sin(i\pi x_1/a) \sin(i\pi x_2/b) \qquad (j = 1, \dots, n) \tag{4.54}$$

where f_j is a normalizing factor tabulated in Blevins (Ref. [6], p. 262) for values of (a/b) for a rectangular plate, where a is the length and b is the width.

4.4 Conclusion

It has been shown that, by retaining nonlinearities up to the partial velocity expression and then linearizing it and the velocity and all else, one can arrive at a theory that captures the phenomenon of dynamic stiffening. The theory assumes that the middle surface of a plate does not stretch. If one started with a linearized velocity, this constraint is missed, as is customarily done in all public domain software, and one obtains dynamic softening, instead, for a rotating plate. The underlying cause is premature linearization. The present theory does suffer from one limitation: namely, it is essentially based on the strip method version of plate theory (see Urugal [7], p. 80). Improvements in plate vibration theory have recently been developed. Incorporation of a plate theory that accounts for section rotation, such as in Ref. [8], and also shear, membrane, and large deformation effects could yield even greater dynamic stiffening, provided the fundamental constraint of the mid-surface deformation or consistent modification of it is taken into consideration.

Appendix 4.A Specialized Modal Integrals

The modal integrals in Eqs. (4.34), (4.35), and (4.40) are standard features in flexible body dynamics – see for example, Refs. [9, 10] – which do not account for dynamic stiffening. Additional integrals arise in the present theory and are given by Eqs. (4.36)–(4.39). Their evaluation is facilitated by employing integration by parts, and the results are listed as follows for a rectangular plate of sides, of length a along X_1-axis and b along X_2-axis:

$$C_{ij} = \rho \int_0^b \left\{ \int_0^a (a - x_1) \left[\frac{\partial \phi_{2i}}{\partial x_1} \frac{\partial \phi_{2j}}{\partial x_1} + \frac{\partial \phi_{3i}}{\partial x_1} \frac{\partial \phi_{3j}}{\partial x_1} \right] dx_1 \right\} dx_2 \tag{4.55}$$

$$C_{1ij} = 0.5\rho \int_0^b \left\{ \int_0^a \left(a^2 - x_1^2\right) \left[\frac{\partial \phi_{2i}}{\partial x_1} \frac{\partial \phi_{2j}}{\partial x_1} + \frac{\partial \phi_{3i}}{\partial x_1} \frac{\partial \phi_{3j}}{\partial x_1} \right] dx_1 \right\} dx_2 \tag{4.56}$$

$$C_{2ij} = \rho \int_0^b x_2 \left\{ \int_0^a (a - x_1) \left[\frac{\partial \phi_{2i}}{\partial x_1} \frac{\partial \phi_{2j}}{\partial x_1} + \frac{\partial \phi_{3i}}{\partial x_1} \frac{\partial \phi_{3j}}{\partial x_1} \right] dx_1 \right\} dx_2 \tag{4.57}$$

$$D_{ij} = \rho \int_0^a \left\{ \int_0^b (b - x_2) \left[\frac{\partial \phi_{1i}}{\partial x_2} \frac{\partial \phi_{1j}}{\partial x_2} + \frac{\partial \phi_{3i}}{\partial x_2} \frac{\partial \phi_{3j}}{\partial x_2} \right] dx_2 \right\} dx_1 \tag{4.58}$$

$$D_{1ij} = 0.5\rho \int_0^a \left\{ \int_0^b \left(b^2 - x_2^2\right) \left[\frac{\partial \phi_{1i}}{\partial x_2} \frac{\partial \phi_{1j}}{\partial x_2} + \frac{\partial \phi_{3i}}{\partial x_2} \frac{\partial \phi_{3j}}{\partial x_2} \right] dx_2 \right\} dx_1 \tag{4.59}$$

$$D_{2ij} = \rho \int_0^a x_1 \left\{ \int_0^b (b - x_2) \left[\frac{\partial \phi_{1i}}{\partial x_2} \frac{\partial \phi_{1j}}{\partial x_2} + \frac{\partial \phi_{3i}}{\partial x_2} \frac{\partial \phi_{3j}}{\partial x_2} \right] dx_2 \right\} dx_1 \tag{4.60}$$

The terms involving $C_{ij}, C_{1ij}, C_{2ij}, D_{ij}, D_{1ij}, D_{2ij}$ in Eq. (4.41) can be related to geometric stiffness due to inertia loading. This is seen as follows. Consider the quantity Q_i defined as:

$$Q_i = \sum_{j=1}^{n} (\dot{v}_1 + \omega_2 v_3 - \omega_3 v_2)\, C_{ij} q_j \tag{4.61}$$

This is one of the terms in Eq. (4.41). Using Eq. (4.55) in Eq. (4.61),

$$Q_i = \rho(\dot{v}_1 + \omega_2 v_3 - \omega_3 v_2) \int_0^b \left\{ \int_0^a (a - x_1) \sum_{j=1}^{n} \left[\frac{\partial \phi_{2i}}{\partial x_1} \frac{\partial \phi_{2j}}{\partial x_1} + \frac{\partial \phi_{3i}}{\partial x_1} \frac{\partial \phi_{3j}}{\partial x_1} \right] q_j dx_1 \right\} dx_2 \tag{4.62}$$

or equivalently,

$$Q_i = \frac{\partial}{\partial q_i}\left\langle \frac{1}{2}\rho(\dot{v}_1 + \omega_2 v_3 - \omega_3 v_2)\int_0^b \int_0^a (a - x_1) \sum_{i=1}^{n}\sum_{j=1}^{n}\left[\frac{\partial \phi_{2i}}{\partial x_1}\frac{\partial \phi_{2j}}{\partial x_1} + \frac{\partial \phi_{3i}}{\partial x_1}\frac{\partial \phi_{3j}}{\partial x_1}\right] q_i q_j dx_1 dx_2 \right\rangle \tag{4.63}$$

Now, in view of Eqs. (4.13) and (4.14), with σ_1 defined as

$$\sigma_1(x_1, x_2) = -\rho(\dot{v}_1 + \omega_2 v_3 - \omega_3 v_2)(a - x_1) \tag{4.64}$$

one can replace Eq. (4.63) with

$$Q_i = -\frac{\partial}{\partial q_i}\left\{\int_0^b \int_0^a \frac{\sigma_1}{2}\left[\left(\frac{\partial u_2}{\partial x_1}\right)^2 + \left(\frac{\partial u_3}{\partial x_1}\right)^2\right] dx_1 dx_2\right\} \tag{4.65}$$

and the term within the brace in this equation can be recognized as the potential energy due to geometric stiffness from a stressed state due to σ_1 (Ref. [4], pp. 338–339), provided the contribution of $(\frac{\partial u_1}{\partial x_1})^2$ to this geometric stiffness is neglected, which is the case for a plate that cannot stretch. Moreover, the form of Eq. (4.65) shows that Q_i can be thought of as a generalized force derivable from a potential function associated with strain energy. Finally, σ_1 as defined in Eq. (4.64) also happens to be the solution of the differential equation, with the boundary condition:

$$\frac{\partial \sigma_1}{\partial x_1} = \rho(\dot{v}_1 + \omega_2 v_3 - \omega_3 v_2) \tag{4.66}$$

$$\sigma_1(a, x_2) = 0 \tag{4.67}$$

Equation (4.66) has the character of a stress equation of motion, for the quantity $\dot{v}_1 + \omega_2 v_3 - \omega_3 v_2$ is a portion of the $\mathbf{r}_1$ measure number of the inertial acceleration of point P' in Figure 4.1. Thus σ_1 defined in Eq. (4.64) plays the role of an acceleration-induced stress or stress due to inertia load; and, proceeding similarly, one finds that $C_{1ij}, C_{2ij}, D_{ij}, D_{1ij}, D_{2ij}$ are also associated with geometric stiffness due to inertia loading.

Problem Set 4

4.1 Fill in the steps of forming the generalized inertia force, uncovering the formation of the modal integrals given here. Code and simulate the motion of a rectangular plate with your choice of prescribed frame motion.

4.2 Consider a panel of a wind turbine spinning up as in Eq. (4.1) with $t^* = 10$, for your choice of blade size. Assume that the plate is mounted on rigid frames moving with the shaft, and use modes for a simply supported plate. Take two modes [3]:

$$
\begin{aligned}
u_3(x,y,t) &= q_{mn}(t)\sin\frac{m\pi x}{a}\sin\frac{m\pi y}{b} \qquad m,n=1,2,..\\
\lambda_{mn} &= \pi^4\left[\left(\frac{m}{a}\right)^2+\left(\frac{n}{b}\right)^2\right]^2\frac{Eh^3}{12\rho\left(1-\nu^2\right)}
\end{aligned}
\tag{4.68}
$$

Here E is the modulus of elasticity; ρ, ν are, respectively, the mass density and Poisson's ratio for the plate; and a, b are its lengths along x- and y-axis; h is plate thickness. Assume wind pressure as follows:

$$
\begin{aligned}
p &= p_0\cos\omega t \quad 0\le t\le \pi/2;\quad 3\pi/2\le t\le 2\pi\\
p &= 0 \quad \pi/2\le t\le 3\pi/2
\end{aligned}
\tag{4.69}
$$

Simulate the response of the wind turbine panel.

4.3 Derive the equations for a "free-free" rectangular plate, with the plane of the plate along the x- and y-axes, being reoriented by a bang-bang torque of magnitude T about the y-axis. Keep only two modes of the plate, so that the generalized speeds are $u_1 = \dot\theta, u_2 = \dot q_1$, $u_3 = \dot q_2$ where θ, q_1, q_2 are the rigid body rotation angle and the two modal coordinates for the vibration modes of the plate. The transverse vibration modes for a square plate of side length a, with all edges free, can be taken from the reference [3]:

$$
\begin{aligned}
u_3(x,y,t) = \sum_{i=1}^{2} &\{[\alpha_i\cos(\beta_i x/a)+\cosh(\beta_i x/a)]+\sin(\beta_i x/a)+\sinh(\beta_i x/a)\}\\
&\{[\alpha_i\cos(\beta_i y/a)+\cosh(\beta_i y/a)]+\sin(\beta_i y/a)+\sinh(\beta_i y/a)\}q_i(t)\\
\alpha_i &= [\sin(\beta_i a)-\sinh(\beta_i a)]/[\cos(\beta_i a)-\cosh(\beta_i a)] \quad i=1,2\\
\beta_1 &= 4.730;\quad \beta_2 = 7.853
\end{aligned}
\tag{4.70}
$$

4.4 Show that if one starts with a linear velocity expression with assumed modes, and uses Kane's (or Lagrange's) method,

$$
\begin{aligned}
{}^N\mathbf{v}^P = {}^N\mathbf{v}^O + {}^N\boldsymbol{\omega}^R \times &\left\{\left[x_1+\sum_{j=1}^{n}\phi_{1j}(x_1,x_2)q_j(t)\right]\mathbf{r}_1+\left[x_2+\sum_{j=1}^{n}\phi_{2j}(x_1,x_2)q_j(t)\right]\mathbf{r}_2\right.\\
&\left.+\sum_{j=1}^{n}\phi_{3j}(x_1,x_2)q_j(t)\mathbf{r}_3\right\}+\sum_{j=1}^{n}\phi_{1j}(x_1,x_2)\dot q_j\mathbf{r}_1+\sum_{j=1}^{n}\phi_{2j}(x_1,x_2)\dot q_j\mathbf{r}_2+\sum_{j=1}^{n}\phi_{3j}(x_1,x_2)\dot q_j\mathbf{r}_3
\end{aligned}
$$

one would obtain an incorrect solution due to premature linearization, which manifests itself as dynamic softening for the spin-up Eq. (4.1). This feature is unfortunately embedded in many public domain software such as Refs. [9, 10].

References

[1] Banerjee, A.K. and Kane, T.R. (1989) Dynamics of a plate in large overall motion. *Journal of Applied Mechanics*, **56**, 887–892.
[2] Kane, T.R. and Levinson, D.A. (1985) *Dynamics: Theory and Applications*, McGraw-Hill.
[3] Banerjee, A.K. and Dickens, J.M. (1990) Dynamics of an arbitrary flexible body in large rotation and translation. *Journal of Guidance, Control, and Dynamics*, **13**(2), 221–227.
[4] Craig, R.R. Jr. (1981), *Structural Dynamics*, John Wiley & Sons, Inc.
[5] Zienkiewicz, O.C. (1977) *The Finite Element Method*, McGraw-Hill.
[6] Blevins, R.D. (1979) *Formulas for Natural Frequency and Mode Shape*, van Nostrand Reinhold Publishing.
[7] Urugal, A.C. (1981) *Stresses in Plates and Shells*, McGraw-Hill.
[8] Boscolo, M. and Banerjee, J.R. (2012) Dynamic stiffness formulation for composite Mindlin plates for exact modal analysis of structures. Part I: Theory. *Computers & Structures*, **96–97**, 61–73.
[9] Bodley, C.S., Devers, A.D., Park, A.C., and Frisch, H.P. (1978) A Digital Computer Program for the Dynamic Interaction Simulation of Controls and Structures (DISCOS). vol. I, NASA Technical Paper 1219.
[10] Singh, R.P., van der Voort, R.J., and Likins, P.W. (1985) Dynamics of flexible bodies in tree topology – a computer oriented approach. *AIAA Journal of Guidance, Control, and Dynamics*, **8**(5), 584–590.

5

Dynamics of an Arbitrary Flexible Body in Large Overall Motion

It has been stated in the preceding chapters on large overall motions of beams and plates that the customary way of describing small elastic deformation in terms of vibration modes with respect to a frame that is itself undergoing large rotation and translation produces incorrect results. To be more explicit, one obtains dynamic softening for a rotating beam or plate where dynamic stiffening is expected. It is pointed out that this error can be traced to premature linearization, in Kane's method, in the form of extracting partial velocities and partial angular velocities with respect to the (vibration) modal coordinates out of linear expressions for velocity and angular velocity. To correct the error, one has to obtain the nonlinear partial velocity and partial angular velocity expressions from the nonlinear expressions for velocity and angular velocity and then linearize these and all other kinematical quantities. It was shown that this process, strictly in that sequence, recovers dynamic stiffening for rotating beams and plates. Unfortunately, for arbitrarily general elastic bodies, it may be impossible to write nonlinear velocity and angular velocity expressions for three-dimensional motion using vibration modes for small elastic deformations. Accordingly, for arbitrary elastic bodies in large overall motion, premature linearization is unavoidable, and one can only compensate for this error a posteriori by ad hoc addition of geometric stiffness due to inertia loads. We showed the effectiveness of this process with some simple examples in Chapter 1. This process will be generalized for arbitrary elastic bodies in this chapter, which is based on the papers by Banerjee and Dickens [1] and Banerjee and Lemak [2]. Before we do so, however, we note that exact, nonlinear theories for large overall motion of continua, including large elastic deformation, were later developed by Simo and Vu-Quoc [3], but their solution procedure, being based on the nonlinear finite method, is very time-consuming. Representation of dynamic stiffening through direct use of geometric stiffness has been made for general structures by Modi and Ibrahim [4], and for beams by Amirouche and Ider [5]. However, the implementation of geometric stiffness in these formulations, based on the instantaneous nodal displacements, is also computationally intensive, even with the use of modes. Our formulation avoids this burden with pre-computed geometric stiffness in a manner to be shown here.

Flexible Multibody Dynamics: Efficient Formulations and Applications, First Edition. Arun K. Banerjee.

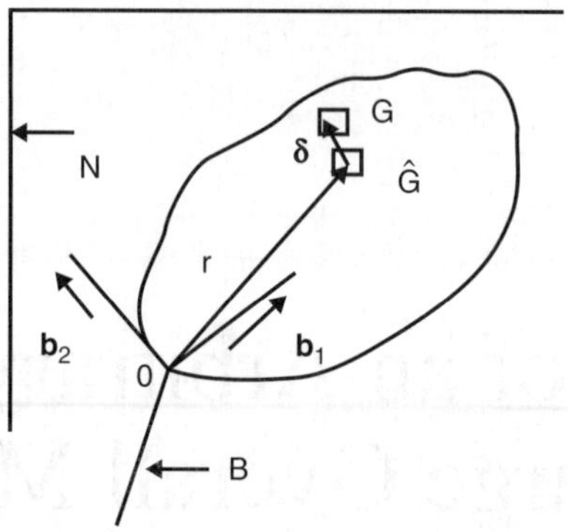

Figure 5.1 Single Flexible Body with "Small" Elastic Displacements in a Flying Reference Frame *B* Undergoing "Large" Rotation and Translation in a Newtonian Frame *N*. Banerjee and Dickens [1].

5.1 Dynamical Equations with the Use of Vibration Modes

Here, to make the chapter self-contained, we will review some of the basic material on kinematics, generalized inertia force, generalized active force [6] for an arbitrary elastic body j going through large overall motion in an inertial frame. Consider the body of Figure 5.1; its elastic motion, assumed "small", is measured with respect to a flying reference frame B with basis vectors $\mathbf{b}_1$, $\mathbf{b}_2$, $\mathbf{b}_3$, attached in the undeformed state of the flexible body at O, and having "large" rigid rotations and translations in a Newtonian frame N. It is assumed that the flexible body is discretized as a system of "nodal" rigid bodies connected by springs as in the finite element modeling methodology. Let G be such a generic nodal body and G^* its mass center. Let the elastic displacement vector of G^* be given by modal superposition,

$$\delta = \sum_{j=1}^{\nu} \phi_j q_j \tag{5.1}$$

where ϕ_j, q_j is the jth vibration mode vector at G^* and corresponding modal coordinates, respectively.

Then, in terms of a vector basis $\mathbf{b}_1$, $\mathbf{b}_2$, $\mathbf{b}_3$, fixed in B, one can characterize the overall motion of G^* by first introducing generalized speeds [6]:

$$u_i = {}^N\mathbf{v}^O \cdot \mathbf{b}_i \qquad (i = 1, 2, 3) \tag{5.2}$$

$$u_{3+i} = {}^N\omega^B \cdot \mathbf{b}_i \qquad (i = 1, 2, 3) \tag{5.3}$$

$$u_{6+i} = \dot{q}_i \qquad (i = 1, ..., \nu) \tag{5.4}$$

Here ${}^N\mathbf{v}^O$ is the velocity of the point O in N, and ${}^N\omega^B$ is the angular velocity of B in N. The velocity of G^* in N is then written as

$$ {}^N\mathbf{v}^{G*} = \sum_{i=1}^{3} u_i \mathbf{b}_i + \sum_{i=1}^{3} u_{3+i} \mathbf{b}_i \times (\mathbf{r} + \delta) + \sum_{j=1}^{\nu} \phi_j u_{6+j} \tag{5.5}$$

where $\mathbf{r}$ is the position from O to $\hat{G}^*$, the location of G^* in the undeformed configuration. The small elastic rotation vector of the nodal rigid body G in B is given by a sum of the jth space-dependent modal rotation vector, multiplied by the generalized coordinates for v modes:

$$\theta = \sum_{j=1}^{v} \mathbf{\Theta}_j q_j \tag{5.6}$$

Here Θ_j is the jth modal vector representing small rotation. One can now write the angular velocity of G in N by invoking the angular velocity addition theorem [6] and doing the time-differentiation of Eq. (5.6) using Eq. (5.4):

$${}^N\boldsymbol{\omega}^G = {}^N\boldsymbol{\omega}^B + {}^B\boldsymbol{\omega}^G = \sum_{i=1}^{3} u_{3+i}\mathbf{b}_i + \sum_{i=1}^{v} \mathbf{\Theta}_i u_{6+i} \tag{5.7}$$

Note that Eqs. (5.5) and (5.7) *are already linear* in the modal generalized speeds, u_{6+i} ($i = 1, \ldots, \nu$), associated with small deformation. As shown in earlier chapters, this is an act of premature linearization, leading to error manifested by loss of stiffness, that can be compensated only *a posteriori* by an *ad hoc* addition of geometric stiffness due to inertia loads. To start with the process, we first note that the ith partial velocity of G^* in N and the ith partial angular velocity of G in N associated with the ith generalized speed can be noted by inspection of Eqs. (5.5) and (5.7) as in Table 5.1.

The acceleration of G^* is developed by differentiating Eq. (5.5) in N:

$$\begin{aligned} {}^N\mathbf{a}^{G*} &= \sum_{j=1}^{3}\left[\dot{u}_j\mathbf{b}_j + \dot{u}_{3+j}\mathbf{b}_j \times (\mathbf{r}+\boldsymbol{\delta}) + \sum_{j=1}^{\nu} \boldsymbol{\phi}_j \dot{u}_{6+j}\right] \\ &+ \sum_{j=1}^{3} u_{3+j}\mathbf{b}_j \times \left\{\sum_{j=1}^{3}[u_j\mathbf{b}_j + u_{3+j}\mathbf{b}_j \times (\mathbf{r}+\boldsymbol{\delta})] + 2\sum_{j=1}^{\nu} \boldsymbol{\phi}_j u_{6+j}\right\} \end{aligned} \tag{5.8}$$

Similarly, the angular acceleration of G follows from Eq. (5.7):

$${}^N\boldsymbol{\alpha}^G = \sum_{j=1}^{3}\left(\dot{u}_{3+j}\mathbf{b}_j + u_{3+j}\mathbf{b}_j \times \sum_{j=1}^{v} \mathbf{\Theta}_j u_{6+j}\right) + \sum_{j=1}^{v} \mathbf{\Theta}_j \dot{u}_{6+j} \tag{5.9}$$

Table 5.1 Vectors of Partial Velocities of G^* and Partial Angular Velocities of G of Figure 5.1.

i	${}^N\mathbf{v}_i^{G*}$	${}^N\boldsymbol{\omega}_i^G$
$1, 2, 3$	$\mathbf{b}_i$	0
$4, 5, 6$	$\mathbf{b}_i \,\mathbf{x} (\mathbf{r} + \boldsymbol{\delta})$	$\mathbf{b}_i$
$7, \ldots, 6 + v$	$\boldsymbol{\phi}_j$	$\mathbf{\Theta}_i$

The generalized inertia force corresponding to the ith generalized speed, with contributions from all differential elements in the entire body, is obtained by integrating over the whole body:

$$F_i^* = -\int_D {}^N\mathbf{v}_i^{G*} \cdot {}^N\mathbf{a}^{G*}\, dm$$
$$- \int_D {}^N\boldsymbol{\omega}_i^G \cdot [\mathbf{dI}^G \cdot {}^N\boldsymbol{\alpha}^G + {}^N\boldsymbol{\omega}^G \times (\mathbf{dI}^G \cdot {}^N\boldsymbol{\omega}^G)] \qquad (i = 1, \ldots, 6 + \nu) \tag{5.10}$$

Here dm is the mass and $\mathbf{dI}^G$ is the central inertia dyadic for the differential element G, respectively. The integrations in Eq. (5.10), performed over the domain D occupied by the flexible body, are straightforward but extremely tedious. This requires doing the integrations in such a way that the integrands do not contain any time-varying quantities. The process of extracting time-invariant groups of terms from the integrand in Eq. (5.10) is performed by invoking certain vector-dyadic identities, and this process yields all the modal integrals reported by Ho and Herber [7], with some additional ones arising because of the inclusion of rotatory inertia of the nodal rigid bodies not considered in Ref. [7]. Details of writing Eq. (5.10) in terms of modal integrals are reported in Ref. [2], with the integrals listed in the Appendix 5.A. However, these details are not central to the issue at hand. It is important, however, to realize that Eq. (5.10) will be of the form

$$-F_i^* = \sum_{j=1}^{6+\nu} M_{ij}\dot{u}_j + C_i \qquad (i = 1, \ldots, 6 + \nu) \tag{5.11}$$

where elements M_{ij} of the so-called "mass matrix" M, with $(i,j = 1, \ldots, 6 + \nu)$, are functions of the modal coordinates, q_j $i = 1, \ldots, \nu$, and C_i, $(i = 1, \ldots, 6 + \nu)$, are functions of the modal coordinates as well as the generalized speeds u_i, $(i = 1, \ldots 6 + \nu)$.

5.2 Compensating for Premature Linearization by Geometric Stiffness due to Inertia Loads

Here we consider generalized active force due to two kinds of internal forces. The first one is associated with standard, structural stiffness. The second one, associated with motion-induced stiffness that compensates for errors due to premature linearization, will be discussed in detail here. To deal with the structural stiffness first, we note that generalized active force due to nominal structural elasticity is commonly written on the basis of mass-normalization of the vibration modes as yielding the elastic vibration frequencies ω_{i-6}, $(i = 7, \ldots., 6 + \nu)$:

$$\begin{aligned} F_i &= 0 & (i = 1, \ldots, 6) \\ &= -\omega_{i-6}^2 q_{i-6} & (i = 7, \ldots 6 + \nu) \end{aligned} \tag{5.12}$$

Motion-induced stiffness is a special case of geometric stiffness that is treated in details by nonlinear finite element theory. Cook [8] has shown that geometric stiffness accounts for the effect of existing forces on bending stiffness, and that it only depends on the element's *geometry*, together with the displacement field, the state of existing stress components, σ_{xx0}, σ_{xy0}, σ_{xz0}, σ_{yy0}, σ_{yz0}, σ_{zz0}, and the (3 × NDOF) matrix of interpolation functions, $[N(x, y, z)]$, where NDOF stands for the number of element degrees of freedom describing the displacement at a point (x, y, z):

$$\{\delta\} = [N(x, y, z)]]\{d\} \tag{5.13}$$

Here $\{\delta\}$ is a column matrix of three components of the elastic displacement at a point within the element and $\{d\}$ the column matrix of NDOF element nodal displacements. Then, the element geometric stiffness matrix is derived from the strain energy induced as the existing stresses do work through the strains representing the nonlinear parts of the Lagrangian strain components. This element geometric stiffness matrix is given in Refs. [8,9] as follows:

$$k_g^e = \int_e \begin{bmatrix} N_{,x}^T & N_{,y}^T & N_{,z}^T \end{bmatrix} \begin{bmatrix} \sigma_{xx0}U_3 & \sigma_{xy0}U_3 & \sigma_{xz0}U_3 \\ \sigma_{xy0}U_3 & \sigma_{yy0}U_3 & \sigma_{yz0}U_3 \\ \sigma_{xz0}U_3 & \sigma_{yz0}U_3 & \sigma_{zz0}U_3 \end{bmatrix} \begin{Bmatrix} N_{,x} \\ N_{,y} \\ N_{,z} \end{Bmatrix} dv \tag{5.14}$$

Here, a variable in the subscripts for the interpolation function $N(x, y, z)$ preceded by a comma indicates differentiation with respect to the variables x,y,z, and U_3 is the (3 × 3) unity matrix. In a finite element structural analysis code such as NASTRAN, computation of geometric stiffness proceeds as follows. First, the distributed loading on the structure is prescribed, and the corresponding linear static equilibrium problem is solved to determine the element stresses; then, element geometric stiffnesses are evaluated following Eq. (5.14), and the standard assembly procedure of the finite element method is used to construct the geometric stiffness matrix K_g of the whole structure. Now the following question arises: If geometric stiffness depends on existing loads, and is to explain the phenomenon of dynamic stiffening with motion, then what is the associated loading? The natural answer is, the loading in the existing state of stress due to the motion itself, namely the stresses due to inertia forces and inertia torques acting throughout the body. This is the reason we call the associated geometric stiffness motion-induced stiffness. We will now proceed to compute the distributed inertia force and inertia torque loading. The inertia force on a nodal rigid body G is given by:

$$\mathbf{f}^* = -dm\,^N\mathbf{a}^{G*} \tag{5.15}$$

Expanding $^N\mathbf{a}^{G*}$ from Eq. (5.8) in the reference configuration in the body $\mathbf{b}_1$, $\mathbf{b}_2$, $\mathbf{b}_3$ basis the components of $\mathbf{f}^*$ can be written in terms of a matrix notation as the load distribution:

$$\begin{Bmatrix} f_1^* \\ f_2^* \\ f_3^* \end{Bmatrix} = -dm\,[U_3 \quad x_1U_3 \quad x_2U_3 \quad x_3U_3] \begin{Bmatrix} A_1 \\ \vdots \\ \vdots \\ \vdots \\ A_{12} \end{Bmatrix} \tag{5.16}$$

Here $A_1, \ldots, A_{12}$, are 12 acceleration terms in the existing or undeformed state, and are as follows:

$$
\begin{aligned}
A_1 &= \dot{u}_1 + z_1 \\
A_2 &= \dot{u}_2 + z_2 \\
A_3 &= \dot{u}_3 + z_3 \\
A_4 &= -\left(u_5^2 + u_6^2\right) \\
A_5 &= \dot{u}_6 + z_4 \\
A_6 &= -\dot{u}_5 + z_5 \\
A_7 &= -\dot{u}_6 + z_4 \\
A_8 &= -\left(u_6^2 + u_4^2\right) \\
A_9 &= \dot{u}_4 + z_6 \\
A_{10} &= \dot{u}_5 + z_5 \\
A_{11} &= -\dot{u}_4 + z_6 \\
A_{12} &= -\left(u_5^2 + u_4^2\right)
\end{aligned} \tag{5.17}
$$

The above terms have been defined using the following intermediate variables:

$$
\begin{aligned}
z_1 &= u_5 u_3 - u_6 u_2 \\
z_2 &= u_6 u_1 - u_4 u_3 \\
z_3 &= u_4 u_2 - u_5 u_1 \\
z_4 &= u_4 u_5 \\
z_5 &= u_6 u_4 \\
z_6 &= u_5 u_6
\end{aligned} \tag{5.18}
$$

Considering now the inertia torque vector on a nodal rigid body G, and neglecting the deformation-related term, for an existing state, in Eq. (5.10), we see that the latter is a relation involving the angular velocity and angular acceleration vectors and the nodal body inertia dyadic:

$$
\begin{aligned}
\mathbf{t}^* &= -[\mathbf{dI}^G . {}^N\alpha^G + {}^N\omega^G \times (\mathbf{dI}^G . {}^N\omega^G)] \\
&= -\{[I_{11}\mathbf{b}_1\mathbf{b}_1 + I_{12}\mathbf{b}_1\mathbf{b}_2 + I_{13}\mathbf{b}_1\mathbf{b}_3 + I_{21}\mathbf{b}_2\mathbf{b}_1 + I_{22}\mathbf{b}_2\mathbf{b}_2 + I_{23}\mathbf{b}_2\mathbf{b}_3 + I_{31}\mathbf{b}_3\mathbf{b}_1 \\
&\quad + I_{32}\mathbf{b}_3\mathbf{b}_2 + I_{33}\mathbf{b}_3\mathbf{b}_3] \cdot (\dot{u}_4 + \dot{u}_5 + \dot{u}_6) + (u_4 + u_5 + u_6) \\
&\quad \times [I_{11}\mathbf{b}_1\mathbf{b}_1 + I_{12}\mathbf{b}_1\mathbf{b}_2 + I_{13}\mathbf{b}_1\mathbf{b}_3 + I_{21}\mathbf{b}_2\mathbf{b}_1 + I_{22}\mathbf{b}_2\mathbf{b}_2 + I_{23}\mathbf{b}_2\mathbf{b}_3 \\
&\quad + I_{31}\mathbf{b}_3\mathbf{b}_1 + I_{32}\mathbf{b}_3\mathbf{b}_2 + I_{33}\mathbf{b}_3\mathbf{b}_3] . (u_4 + u_5 + u_6)\}
\end{aligned} \tag{5.19}
$$

Now the angular velocity and angular acceleration in the undeformed or existing state are given so that the inertia torque in Eq. (5.19) in $\mathbf{b}_1$, $\mathbf{b}_2$, $\mathbf{b}_3$ basis is represented component-wise as

$$\begin{Bmatrix} t_1^* \\ t_2^* \\ t_3^* \end{Bmatrix} = - \begin{bmatrix} I_{11} & I_{12} & I_{13} & I_{13} & I_{33} - I_{22} & -I_{12} & 0 & I_{23} & 0 \\ I_{12} & I_{22} & I_{23} & -I_{23} & I_{12} & I_{11} - I_{33} & 0 & 0 & I_{13} \\ I_{13} & I_{23} & I_{33} & I_{22} - I_{11} & -I_{13} & I_{23} & I_{12} & 0 & 0 \end{bmatrix} \begin{Bmatrix} A_{13} \\ \vdots \\ \vdots \\ A_{21} \end{Bmatrix} \tag{5.20}$$

where the acceleration terms $A_{13}, \ldots, A_{21}$ stand for the following:

$$\begin{aligned} A_{13} &= \dot{u}_4 \\ A_{14} &= \dot{u}_5 \\ A_{15} &= \dot{u}_6 \\ A_{16} &= z_4 \\ A_{17} &= z_6 \\ A_{18} &= z_5 \\ A_{19} &= u_4^2 - u_5^2 \\ A_{20} &= u_5^2 - u_6^2 \\ A_{21} &= u_6^2 - u_4^2 \end{aligned} \tag{5.21}$$

With force and torque loadings defined by Eqs. (5.16) and (5.20), respectively, a nonlinear finite element code can now be used to generate 21 geometric stiffness matrices $K_g^{(i)}$, $i = 1, \ldots 21$, for unit values of each of the acceleration variables A_i, $(i = 1, \ldots 21)$. Motion-induced stiffness due to the actual values of inertia accelerations and angular accelerations corresponding to large translations and rotations can then be obtained by simple multiplication with the current states of accelerations. Generalized active force due to these motion-induced stiffness components can thus be written for ν number of modes as:

$$\begin{aligned} F_i^g &= 0 \qquad (i = 1, \ldots, 6) \\ \begin{Bmatrix} F_7^g \\ \vdots \\ \vdots \\ F_{6+\nu}^g \end{Bmatrix} &= - \sum_{i=1}^{21} S^{(i)} A_i q \end{aligned} \tag{5.22}$$

Here $S^{(i)}$ denote the $\nu \times \nu$ generalized geometric stiffness matrices due to unit values of the ith inertia loading, A_i, $(i = 1, \ldots 21)$, and for each i is given by:

$$S^{(i)} = \Phi^T K_g^{(i)} \Phi \tag{5.23}$$

Here Φ is the matrix of mode shapes and q is the $(\nu \times 1)$ matrix of modal generalized coordinates. With expressions for the generalized inertia forces and generalized actives forces at hand, one can now complete part of Kane's equations of motion in the form:

$$\sum_{j=1}^{6+\nu} M_{ij}\dot{u}_j + C_i = 0 \qquad (i = 1, \ldots, 6) \tag{5.24}$$

$$\sum_{j=1}^{6+\nu} M_{i+6,j}\dot{u}_j + C_{i+6} + \omega_i^2 q_i + \sum_{j=1}^{\nu} \left\{ S_{ij}^{(1)}\dot{u}_1 + S_{ij}^{(2)}\dot{u}_2 + S_{ij}^{(3)}\dot{u}_3 + \left[S_{ij}^{(13)} - S_{ij}^{(9)} - S_{ij}^{(11)}\right]\dot{u}_4 \right.$$

$$\left[S_{ij}^{(14)} + S_{ij}^{(10)} - S_{ij}^{(6)}\right]\dot{u}_5 + \left[S_{ij}^{(16)} + S_{ij}^{(5)} - S_{ij}^{(7)}\right]\dot{u}_6 + S_{ij}^{(1)}z_1 + S_{ij}^{(2)}z_2 + S_{ij}^{(3)}z_3 + \left[S_{ij}^{(5)} + S_{ij}^{(7)}\right]z_4$$

$$\left. + \left[S_{ij}^{(6)} + S_{ij}^{(10)}\right]z_5 + \left[S_{ij}^{(9)} + S_{ij}^{(11)}\right]z_6 + S_{ij}^{(4)}A_4 + S_{ij}^{(8)}A_8 + S_{ij}^{(12)}A_{12} + \sum_{k=16}^{21} S_{ij}^{(k)}A_k \right\} q_j = 0$$

$$(i = 1, \ldots, \nu) \tag{5.25}$$

Note that here we have not considered any generalized active force due to external loads; if these are present, their contribution to the generalized active forces must be appended on the right-hand sides of Eqs. (5.24) and (5.25), in the manner described several times in previous chapters. Equations (5.24) and (5.25) can be written in the compact form as the matrix differential equation,

$$D\dot{U} = R \tag{5.26}$$

where U is the column matrix of generalized speeds $u_1, \ldots.., u_{6+\nu}$. Note from Eqs. (5.24) and (5.25) that the coefficient matrix D is time-varying and unsymmetric. Note also that if rotatory inertia effects of the nodes of a structure can be neglected, which is usually the case, then the second integral in Eq. (5.10) vanishes, and

$$S_{ij}^{(k)} = 0 \qquad (k = 13, \ldots, 21; \quad i,j = 1, \ldots, \nu) \tag{5.27}$$

with the simplification that only 12, not 21, of the motion-induced geometric stiffness matrices in Eq. (5.25) are computed. This completes the formulation of the dynamical equations, with geometric stiffness added to compensate for the error of premature linearization in using modes.

Before leaving this presentation of the dynamical equations, one has to ask again, when is the error of premature linearization acceptable? The answer is given in the next chapter: whenever the rotation frequency of the frame with respect to which the body vibrates is far smaller than the first vibration frequency of the body.

5.2.1 *Rigid Body Kinematical Equations*

To complete the equations of motion, we now consider the kinematical equations for the inertial position of the point O of B, and the attitude of the body B, expressed in terms of Euler parameters (quaternions), $\varepsilon_1, \varepsilon_2, \varepsilon_3, \varepsilon_4$ [10]:

$$\begin{Bmatrix} \dot{x} \\ \dot{y} \\ \dot{z} \end{Bmatrix} = \begin{bmatrix} 1-2(\varepsilon_2^2+\varepsilon_3^2) & 2(\varepsilon_1\varepsilon_2-\varepsilon_3\varepsilon_4) & 2(\varepsilon_3\varepsilon_1+\varepsilon_2\varepsilon_4) \\ 2(\varepsilon_1\varepsilon_2+\varepsilon_3\varepsilon_4) & 1-2(\varepsilon_3^2+\varepsilon_1^2) & 2(\varepsilon_2\varepsilon_3-\varepsilon_1\varepsilon_4) \\ 2(\varepsilon_3\varepsilon_1-\varepsilon_2\varepsilon_4) & 2(\varepsilon_2\varepsilon_3+\varepsilon_1\varepsilon_4) & 1-2(\varepsilon_1^2+\varepsilon_2^2) \end{bmatrix} \begin{Bmatrix} u_1 \\ u_2 \\ u_3 \end{Bmatrix} \tag{5.28}$$

The attitude parameters $\varepsilon_1, \varepsilon_2, \varepsilon_3, \varepsilon_4$ change with time as given by the kinematical equations [10]:

$$\begin{Bmatrix} \dot{\varepsilon}_1 \\ \dot{\varepsilon}_2 \\ \dot{\varepsilon}_3 \\ \dot{\varepsilon}_4 \end{Bmatrix} = \frac{1}{2} \begin{bmatrix} \varepsilon_4 & -\varepsilon_3 & \varepsilon_2 \\ \varepsilon_3 & \varepsilon_4 & -\varepsilon_1 \\ -\varepsilon_2 & \varepsilon_1 & \varepsilon_4 \\ -\varepsilon_1 & -\varepsilon_2 & -\varepsilon_3 \end{bmatrix} \begin{Bmatrix} u_4 \\ u_5 \\ u_6 \end{Bmatrix} \tag{5.29}$$

One can compute the roll-pitch-yaw angles in terms of the Euler parameters. For example, the pitch angle is computed from Ref. [10], for a body 1-2-3 rotation sequence, to the direction cosine matrix in (5.28) as:

$$\theta_2 = \sin^{-1}[2(\varepsilon_3\varepsilon_1 + \varepsilon_2\varepsilon_4)] \tag{5.30}$$

5.3 Summary of the Algorithm

To implement the theory given above, the following step-by-step procedure must be used in sequence:

1. Using a finite element code such as NASTRAN or equivalent, construct the finite element model of the structure with the origin of coordinates constrained to have zero rigid body dof; generate a number ν of column matrices representing a number of vibration modes and frequencies of the structure.
2. On each node of the finite element model set up 12 loads defined in Eq. (5.16) (in terms of the x-, y-, z- coordinates of the node) for unit values of $A_1, \ldots, A_{12}$. If rotatory inertia is deemed significant, then impose 9 torque loads as per Eq. (5.20). Then, for each of these 12 or 21 loads distributed throughout the body, compute the geometric stiffness matrix and convert the latter into generalized force due to motion-induced stiffness, using Eqs. (5.22) and (5.23).
3. All of the above are inputs to a flexible multibody dynamics code. Now undertake the major work of building the large motion dynamics code on the basis of Eqs. (5.24) and (5.25). This involves the work of the construction of the M_{ij} and C_i matrix elements in these two equations. That work, in turn, depends on developing the modal integrals, to be described in detail in Chapter 6, and formed out of the modes generated in step (1).
4. Append kinematical equations for $\dot{q}_i$, $i = 1, \ldots, (6+\nu)$ relating to u_i, $i = 1, \ldots, (6+\nu)$, from Eqs. (5.2)–(5.4), as given by Eqs. (5.28) and (5.29). This involves the relationship between the position coordinates of O, orientation of B, and modal coordinates of the flexible body to their corresponding generalized speeds for this large overall motion problem.

5. Assign initial values to q_i, u_i, $(i = 1, \ldots, 6 + \nu)$. If the initial deformations and their rates are given, the generalized coordinates and generalized speed can be obtained as a best approximation fit by taking a pseudo-inverse (see Ref. [10]).
6. Integrate numerically the dynamical and kinematical equations with these initial conditions.

5.4 Crucial Test and Validation of the Theory in Application

To validate the foregoing theory, which is applicable to any arbitrary flexible body undergoing large overall motion, we compare the motion predicted by this general theory to predictions already in hand from special-purpose theories for beams [11] and plates [12] attached to a moving base. For a crucial test, we use the spin-up function,

$$\begin{aligned} \omega_1 &= \frac{\Omega}{T}\left(t - \frac{T}{2\pi}\sin\frac{2\pi t}{T}\right), \quad & t \leq T \\ &= \Omega, & t > T \end{aligned} \tag{5.31}$$

where ω_1, Ω, T are, respectively, the $\mathbf{b}_1$ component of the spin angular velocity of the body B in N, the final spin-up speed, both in radians per second, and time in seconds to reach the final spin. Figure 5.2 compares the results for the spin-up of a 150-m long WISP antenna attached to the space shuttle, with the antenna tube having a diameter of 0.0635 m and a wall thickness of 0.00254 m, with the first bending mode frequency of $\omega_1 = 0.0546$ rad/sec. In each of the two simulations, with the present general theory and the beam theory, only the first four vibration modes and frequencies are used. Base spin-up is described by Eq. (5.31) with

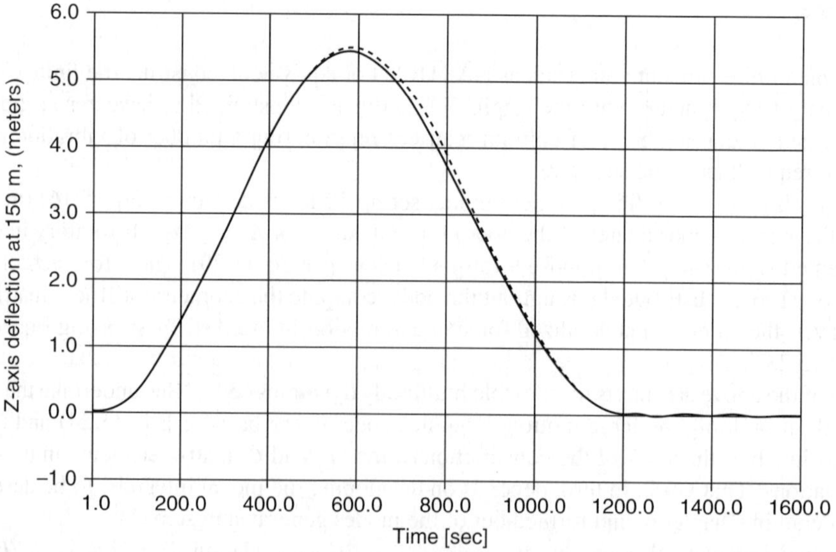

Figure 5.2 Cantilever Beam Tip Deflection at 150 m during Spin-up, Given by the Present, General Theory (solid line) and the Special Beam Theory of Ref. [11]. (Beam first bending frequency 0.546 rad/sec, spin-up speed 0.06 rad/sec, spin-up time 1200 sec.) Banerjee and Dickens [1].

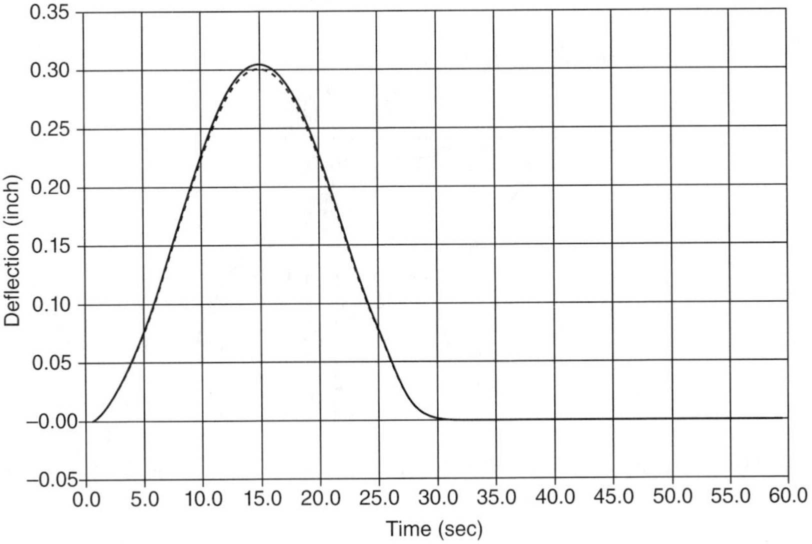

Figure 5.3 Cantilever Plate Corner Deflection in Inch during Spin-up Given by the Present, General Theory (solid line) and the Special Plate Theory of Ref. [12]. (Plate first bending frequency 0.75 rad/sec, spin-up speed 1.25 rad/sec, spin-up time 30 sec.) Banerjee and Dickens [1].

$T = 1200$ sec and $\Omega = 0.06$ rad/sec, meaning that maximum spin frequency is higher than the first mode bending frequency. With the steady spin rate higher than the first bending frequency, premature linearization through the use of vibration modes would produce disastrously wrong results. Compensatory geometric stiffness needed in the present theory is computed by using the code in Ref. [13] for beam elements with existing axial force due to spin and does not include the effect of transverse shear, and this may explain the slight discrepancy between the two curves in Figure 5.2. The overall agreement between the two curves in Figure 5.2 gives an evidence of the adequacy of the present general theory in reproducing dynamic stiffening of a beam given by a beam-specific theory, during and following the spin-up.

Results of comparing the present theory with a special-purpose plate theory [12] for a cantilever plate, again during spin-up about an edge, are shown Figure 5.3. The plate is 72 in. long, 48 in. wide, and 0.1 in. thick, and has a first natural frequency of $\omega_1 = 0.75$ rad/sec. The spin-up is described by Eq. (5.31) with $\Omega = 1.25$ rad/sec, $T = 30$ sec. In this case, the first six vibration modes and frequencies of the plate are used in each simulation. The difference between the results of the general theory and the plate theory is hardly discernible. When one recalls that the special-purpose theories for beam and plate, in Figures 5.2 and 5.3, are obtained by taking the steps for correct linearization, the reproduction of those results by the general theory that compensates for premature linearization due to the use of vibration modes by the ad hoc addition of geometric stiffness due to inertia loading testifies to the correctness of the present theory for any arbitrary flexible body.

Now we *apply this general theory* to the large motion dynamics of two structures, one a rotating paraboloidal shell structure like the Galileo spacecraft, and the other a truss in spin or in translational acceleration. Figure 5.4 shows a dual-spin spacecraft. The paraboloidal,

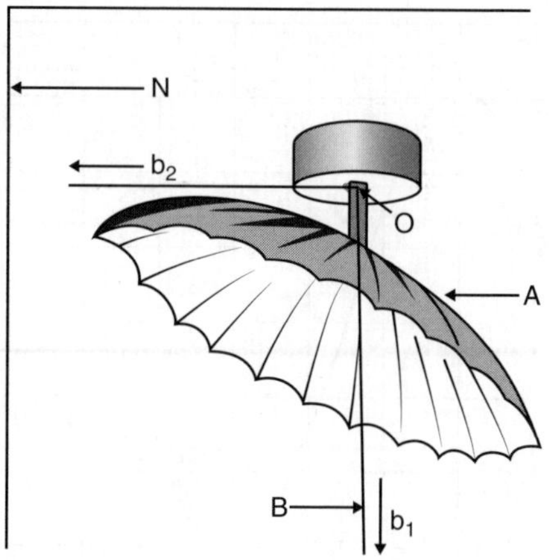

Figure 5.4 Dual-Spin Spacecraft with Flexible, Spinning Offset Paraboloidal Antenna. Banerjee and Dickens [1].

flexible antenna rotates with respect to a rigid base, and we apply the spin-up check with $\Omega = 0.5$ Hz, $T = 30$ sec in Eq. (5.31) and the first natural frequency of 0.433 Hz, lower than the spin-up speed. Figure 5.5 shows what happens when one uses the present theory that incorporates motion-induced stiffness, compared to the predictions of conventional theories [14, 15], using only vibration modes and no geometric stiffness; the latter case is recreated

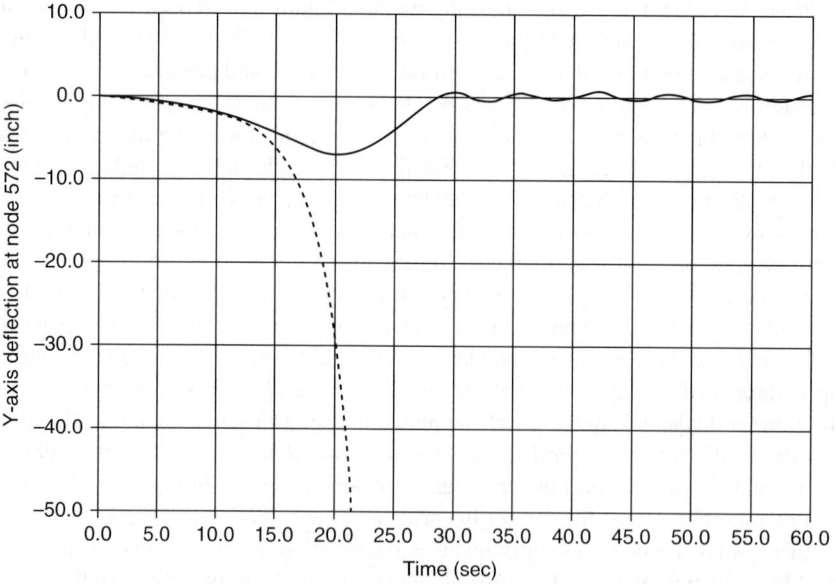

Figure 5.5 Elastic Displacement along Y-Axis at Finite Element Node 572, of a Spinning Offset Paraboloidal Antenna for Spin-up Given by the Present General Theory (solid line) and a Conventional Theory with No Geometric Stiffening (dashed line). (Steady-state spin frequency [0.5 Hz] is greater than the first natural vibration frequency [0.433 Hz]. Spin-up time is 30 sec.) Banerjee and Dickens [1].

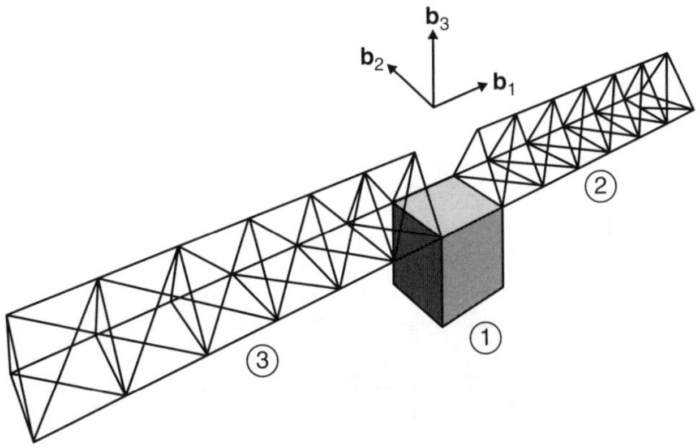

Figure 5.6 A Single "Flexible Body Spacecraft" Made Up of a Rigid Body Fixed to Two Identical Trusses in Spin-up about $\mathbf{b}_3$ or Translational Acceleration along $\mathbf{b}_1$. Banerjee and Lemak [2]. Reproduced with permission of the American Society of Mechanical Engineers.

here by deleting the motion-induced stiffness terms. *A theory with no motion-induced stiffness, predicting unbounded motion at a particular node arbitrarily chosen* (node number 572 in the finite element model), *is clearly erroneous.* For the structure made up of two trusses rigidly attached to a base, in Figure 5.6, we again do a ramp-up to a steady spin or a steady translational speed. Again the vibration modes were obtained by using the finite element code [13], with a first-mode frequency of 3.72 rad/sec; the *translation* speed time-history is again given by Eq. (5.31), with ω now standing for translational speed and Ω the maximum translation speed attained. For this spin-up test, the maximum value of translational speed was carefully chosen at 4 rad/sec to be higher than the first-mode frequency to accentuate the effect of motion-induced stiffness (softness in this situation due to inertia force being compressive). Figure 5.7 shows the tip deflection of a corner of the truss in spin-up, with and without geometric stiffness consideration. Next we do a translation test, using Eq. (5.31) as the speed-up function. The maximum acceleration is then $2\Omega/T$. The critical buckling acceleration for a uniform loading on the truss was calculated by a finite element code [13] to be 53.6 $\mathrm{m/s^2}$. For translation acceleration along $\mathbf{b}_1$ in Figure 5.6, the truss at the right end of the rigid bus is in dynamic compression due to inertia forces, and the truss on the left is in tension. Figure 5.8 shows $\mathbf{b}_3$ direction of the *tip deflection of the truss under compression due to inertia load.* Here we see the evolution of buckling when the maximum acceleration corresponds to the critical acceleration. *The solid line, for including of geometric stiffness (softness in this case), shows buckling as it should,* while the dashed line for no geometric stiffness fails to predict buckling. Figures 5.9 and 5.10 show the results of dynamic softening and stiffening of the trusses when the system is accelerated to 75% of the critical buckling acceleration. Finally, Figure 5.9 *shows tendency to dynamic buckling, that is, loss of stiffness that is expected to occur with translational acceleration*, as in the case of rockets; it is correctly reproduced with the present theory, whereas a theory with no motion-induced stiffness (softness in this case) fails to predict this catastrophe. Figure 5.10 shows the softening of truss-marked body #3 in Fig. 5.6 that is under tension, when no motion-induced stiffness is considered, while the present theory with motion-induced stiffness produces an intuitively correct result.

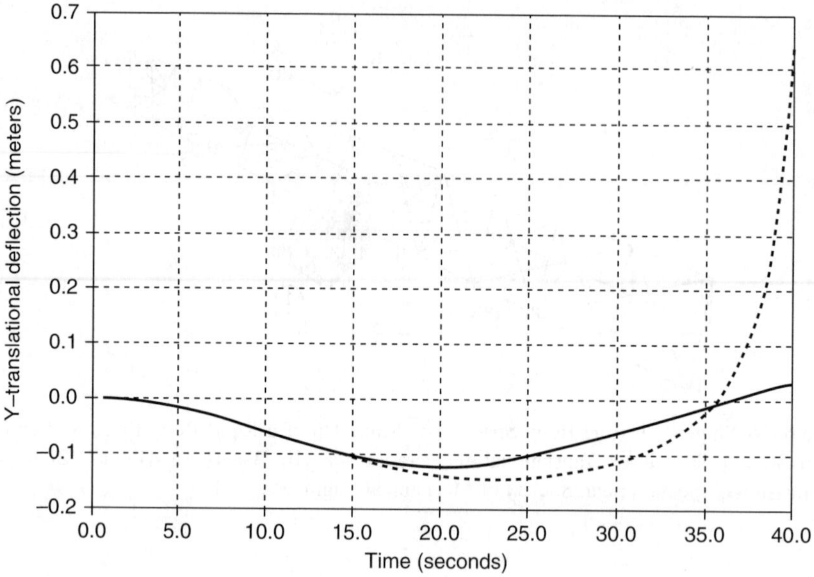

Figure 5.7 Transverse Tip Deflection of a Corner of the Truss in Spin-up Motion (solid line, with geometric stiffness; dashed line, no geometric stiffness). Banerjee and Lemak [2]. Reproduced with permission of the American Society of Mechanical Engineers.

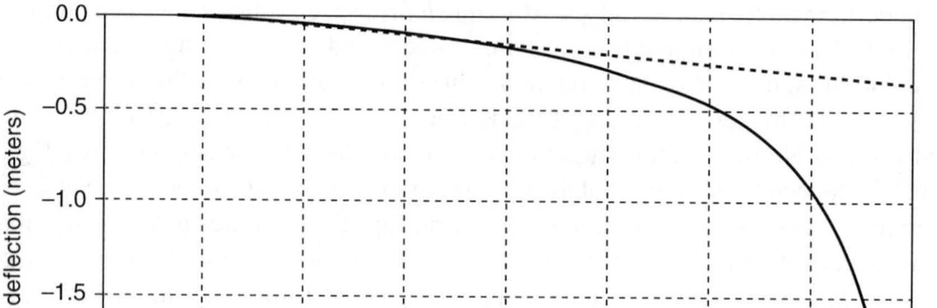

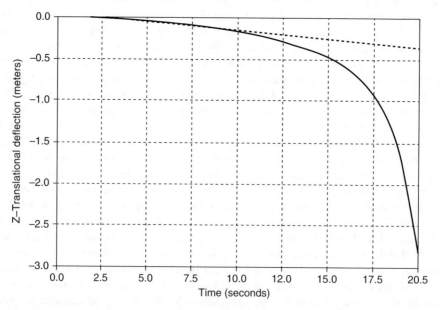

Figure 5.8 Dynamic Buckling of Truss #2 due to Compression under Translational Acceleration (solid line, with geometric stiffness; dashed line, no geometric stiffness). Banerjee and Lemak [2]. Reproduced with permission of the American Society of Mechanical Engineers.

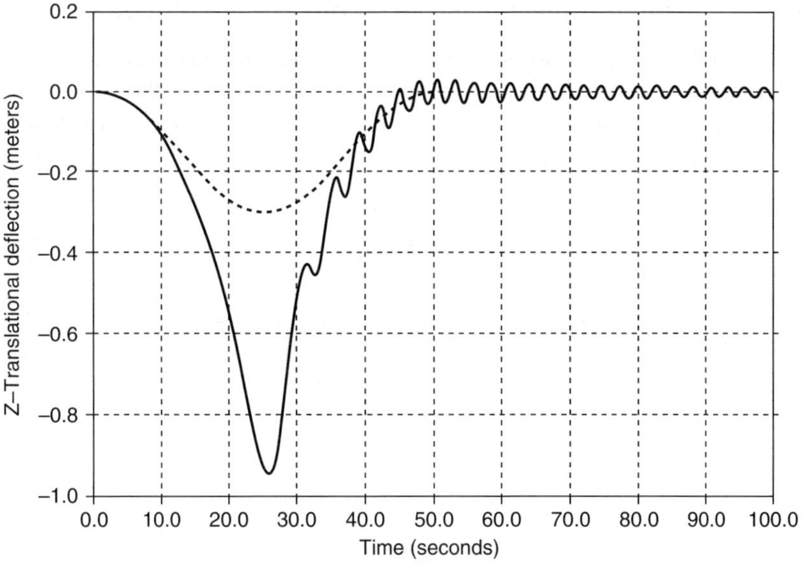

Figure 5.9 Buckling Tendency of the Truss #2, under Compression due to Inertia Load in Subcritical Translational Acceleration along $\mathbf{b}_1$ (solid line with geometric stiffness; dashed line, no geometric stiffness). Banerjee and Lemak [2]. Reproduced with permission of the American Society of Mechanical Engineers.

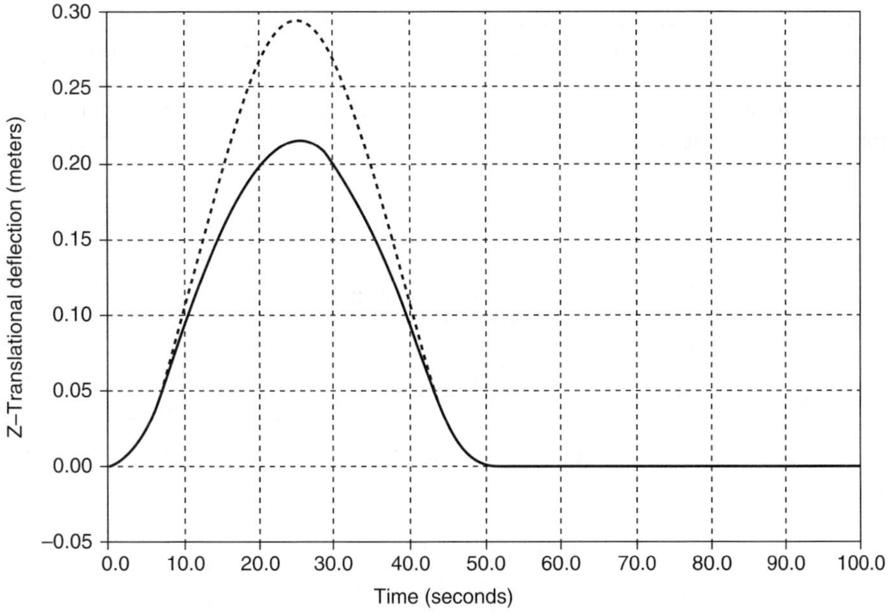

Figure 5.10 Dynamic Stiffening of Truss # 3, in Tension with $\mathbf{b}_1$ Direction Translational Acceleration (solid line with geometric stiffness; dashed line, no geometric stiffness). Banerjee and Lemak [2]. Reproduced with permission of the American Society of Mechanical Engineers.

Appendix 5.A Modal Integrals for an Arbitrary Flexible Body [2]

The following modal integrals arise in general in flexible multibody dynamics equations of motion. The notations follow from the figure shown, where elastic deflection at a point from a point O to a point $\hat{G}$ defined by the position vector $\mathbf{r}^j$, in the undeformed configuration, is denoted by $\boldsymbol{\delta}$ in terms of a linear combination of component modes $\boldsymbol{\phi}_k^j$ $(k = 1, \ldots, \nu, j = 1, \ldots, n)$.

$$\boldsymbol{\delta} = \sum_{k=1}^{\nu} \boldsymbol{\phi}_k^j \eta_k^j \tag{5.32}$$

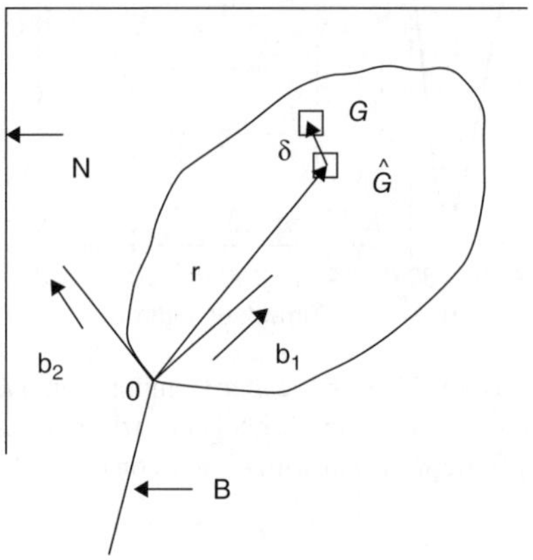

Figure 5.A.1 Single Flexible Body with "Small" Elastic Displacements in a Flying Reference Frame *B* Undergoing "Large" Rotation and Translation in a Newtonian Frame *N*. Banerjee and Dickens [2].

The integrals are evaluated over the whole mass of the jth body, B_j, (the component) of the flexible multibody system.

$$\mathbf{b}_k^j = \int_{B_j} \boldsymbol{\phi}_k^j dm \tag{5.33}$$

$$\mathbf{s}^j = \int_{B_j} \mathbf{r}^j dm + \sum_{k=1}^{M_j} \mathbf{b}_k^j \eta_k^j \tag{5.34}$$

$$\mathbf{c}_k^j = \int_{B_j} \mathbf{r}^j x \boldsymbol{\phi}_k^j dm \tag{5.35}$$

$$\mathbf{d}_{lk}^j = \int_{B_j} \boldsymbol{\phi}_l^j x \boldsymbol{\phi}_k^j dm \tag{5.36}$$

$$\mathbf{g}_k^j = \mathbf{c}_k^j + \sum_{l=1}^{M_j} \mathbf{d}_{lk}^j \eta_l^j \tag{5.37}$$

$$e_{lk}^j = \int_{B_j} \boldsymbol{\phi}_l^j . \boldsymbol{\phi}_k^j dm \tag{5.38}$$

The three integrals, (5.38)–(5.40), produce dyadics that are developed by using the identity

$$\mathbf{a}x(\mathbf{b}x\mathbf{c}) = [(\mathbf{c}.\mathbf{a})\mathbf{U} - \mathbf{ca}].\mathbf{b}$$

Here some dyadics [2] used in the text, showing a tilde, are collected. U is the unity dyadic here. In the text, superscript on a dyadic denotes its transpose. Also, used is a skew-symmetric matrix $\tilde{\psi}_k^j$ formed out of the column vector ψ_k^j for small elastic rotation vector ψ_k^j.

$$\mathbf{N}_l^j = \int_{B_j} \left[\left(\mathbf{r}^j \cdot \boldsymbol{\phi}_l^j \right) \mathbf{U} - .\mathbf{r}^j \boldsymbol{\phi}_l^j \right] dm \tag{5.39}$$

$$\mathbf{I}^j = \int_{B_j} [\mathbf{r}^j \cdot \cdot \mathbf{r}^j U - \mathbf{r}^j \mathbf{r}^j] dm + \sum_{l=1}^{M_j} \left(\tilde{\mathbf{N}}_l^j + \tilde{\mathbf{N}}_l^{j*} \right) \eta_l^j \tag{5.40}$$

$$\tilde{\mathbf{D}}_l^j = \mathbf{N}_l^j + \sum_{l=1}^{M_j} \eta_l^j \int_{B_j} [(\boldsymbol{\phi}_k^j \cdot \boldsymbol{\phi}_l^j) U - \boldsymbol{\phi}_k^j \boldsymbol{\phi}_l^j] dm \tag{5.41}$$

$$\tilde{\mathbf{W}}_{1j} = \int_{B_j} d\tilde{\mathbf{I}}_0, \quad \text{nominal inertia dyadic for rigid body rotation} \tag{5.42}$$

$$\tilde{\mathbf{W}}_{2k}^j = \int_{B_j} d\tilde{\mathbf{I}}_k^j \text{ where } \quad d\tilde{\mathbf{I}}^j = d\tilde{\mathbf{I}}_0^j + \sum_{i}^{M_j} d\tilde{\mathbf{I}}_k^j \eta_k^j \tag{5.43}$$

$$\tilde{\mathbf{W}}_{3k}^j = \int_{B_j} \tilde{\psi}_k^j \times \mathbf{d}\tilde{\mathbf{I}}_0^{j*} \tag{5.44}$$

$$\mathbf{W}_{4k}^j = \int_{B_j} \mathbf{d}\mathbf{I}_0^j \cdot \psi_k^j \tag{5.45}$$

$$\mathbf{W}_{5k}^j = \int_{B_j} \mathrm{d}\mathbf{I}_0^j \times \psi_k^j \tag{5.46}$$

$$\mathbf{W}_{6k}^j = \int_{B_j} \psi_k^j \cdot d\tilde{I}_0^j \tag{5.47}$$

$$\mathbf{W}_{7lk}^j = \int_{B_j} \psi_l^j \cdot d\mathbf{I}_k^j \tag{5.48}$$

$$W_{8lk}^j = \int_{B_j} \tilde{\psi}_l^j \times \tilde{d}\mathbf{I}_0 \cdot \psi_k^j \tag{5.49}$$

$$W_{9lk}^j = \int_{B_j} \left[\psi_l^j \cdot d\mathbf{I}_0^j \cdot \psi_k^j \right] \tag{5.50}$$

$$\tilde{\mathbf{W}}_{10lk}^j = \int_{B_j} \psi_l^j \times \mathrm{d}\tilde{\mathbf{I}}_k^* \tag{5.51}$$

$$\mathbf{W}_{11lk}^j = -\int_{B_j} \left[d\tilde{\mathbf{I}}_0^j \cdot \tilde{\psi}_i^j \right] x \psi_k^j \tag{5.52}$$

$$\mathbf{W}_{12lk}^j = -\int_{B_j} [d\tilde{\mathbf{I}}_0 x \psi_i^j] \cdot \psi_k^j \tag{5.53}$$

Integrals (5.44), (5.48), and (5.50) are obtained by using the vector-dyadic identities, which can be verified by expansion:

$$\mathbf{A} \cdot (\mathbf{b} x \mathbf{c}) = \mathbf{b} \cdot \tilde{\mathbf{C}}\tilde{\mathbf{A}}^*$$
$$\mathbf{c} \cdot \mathbf{d} x \tilde{\mathbf{A}} \cdot \mathbf{b} = -\mathbf{d} \cdot \mathbf{C}\tilde{\mathbf{A}} * \cdot \mathbf{b}$$

Problem Set 5

5.1 Code the general theory for an arbitrary elastic solid in large overall motion with motion-induced stiffness. Verify that the results of the theory reduce to those for special-purpose rotating beam or plate formulations given in preceding chapters.

5.2 Use the code to simulate the spin-up motion of a helicopter rotor blade, with cantilever mode [17]

$$\phi(x) = (\sin\beta_1 L - \sinh\beta_1 L)(\sin\beta_1 x - \sinh\beta_1 x) + (\cos\beta_1 L + \cosh\beta_1 L)(\cos\beta_1 x - \cosh\beta_1 x)$$

5.3 Derive the equations of motion of a free-flying pre-tensioned large solar sail in the sky, of sides a and b, tension S per unit length of edge, γ mass per unit area, with natural frequency and mode shapes as follow [17, 18]:

$$\omega_{ij} = \frac{\lambda_{ij}}{2}\left(\frac{S}{\gamma ab}\right)^{1/2} \qquad i = 1, 2, 3, \ldots;\ j = 1, 2, 3, \ldots; \qquad \lambda_{ij} = \left(i^2\frac{b}{a} + j^2\frac{a}{b}\right)^{1/2}$$

$$\phi_{ij} = \sin\left(\frac{i\pi x}{a}\right)\sin\left(\frac{j\pi y}{b}\right)$$

References

[1] Banerjee, A.K. and Dickens, J.M. (1990) Dynamics of an arbitrary flexible body in large rotation and translation, *Journal of Guidance, Control, and Dynamics*, **13**(2), 221–227.

[2] Banerjee, A.K. and Lemak, M.E. (1991) Multi-flexible body dynamics capturing motion-induced stiffness, *Journal of Applied Mechanics*, **58**(3), 766–775.

[3] Simo, J.C. and Vu-Quoc, L. (1986) On the dynamics of flexible bodies under large overall motions—the plane case: parts I & II, *Journal of Applied Mechanics*, **53**, 849–863.

[4] Modi, V.J. and Ibrahim, A.M. (1988) On the dynamics of flexible orbiting structures, in *Large Space Structures: Dynamics and Control* (eds S.N. Atluri and A.K. Amos), Springer-Verlag, New York, pp. 93–114.

[5] Amirouche, F.M.L. and Ider, S.K. (1989) The influence of geometric nonlinearities in the dynamics of flexible tree-like structures, *Journal of Guidance, Control, and Dynamics*, **12**(6), 830–837.

[6] Kane, T.R. and Levinson, D.A. (1985) *Dynamics: Theory and Application*, McGraw-Hill.

[7] Ho, J.Y.L. and Herber, D.R. (1985) Development of dynamics and control simulation of large flexible space systems, *Journal of Guidance, Control, and Dynamics*, **8**(3), 374–383

[8] Cook, R.D. (1985) *Concepts and Applications of Finite Element Analysis*, McGraw-Hill, pp. 331–341.

[9] Zienkiewicz, O.C. (1977) *The Finite Element Method*, McGraw-Hill, pp. 517–519.

[10] Kane, T.R, Likins, P.W., and Levinson, D.A. (1983) *Spacecraft Dynamics*, McGraw-Hill, pp. 12–13.

[11] Kane, T.R., Ryan, R.R., and Banerjee, A.K. (1987) Dynamics of a cantilever beam attached to a moving base, *Journal of Guidance, Control, and Dynamics*, **10**(2), 139–151.

[12] Banerjee, A.K. and Kane, T.R. (1989) Dynamics of a plate in large overall motion, *Journal of Applied Mechanics*, **56**(6), 887–892.

[13] Wheatstone, W.D. (1983) *EISI-EAL Engineering Analysis Language Reference Manual*, Engineering Transformations Systems, Inc., San Jose, CA, July.

[14] Bodley, C.S., Devers, A.D., Park, A.C., and Frisch, H.P. (1978) A Digital Computer Program for the Dynamic Interaction Simulation of Controls and Structures (DISCOS), vols I & II, NASA TP-1219.

[15] Singh, R.P., van der Voort, R.J., and Likins, P.W. (1985) Dynamics of flexible bodies in tree topology-A computer oriented approach, *Journal of Guidance, Control, and Dynamics*, **8**(5), 584–590.

[16] Banerjee, A.K. and Lemak, M.E. (2008) Dynamics of a Flexible Body with Rapid Mass Loss. Proceedings of the International Congress on Theoretical and Applied Mechanics, August, Adelaide, Australia.

[17] Meirovitch, L. (1967) *Analytical Methods in Vibrations*, McMillan.

[18] Blevins, R.D. (2001) *Formulas for Natural Frequency and Mode Shape*, Krieger Publishing Company, p. 226.

[19] Thomson, W.T. (1963) *Introduction to Space Dynamics*, John Wiley & Sons, Inc., pp. 230–236.

6

Flexible Multibody Dynamics: Dense Matrix Formulation

6.1 Flexible Body System in a Tree Topology

In this chapter we analyze large overall motion of a general, multibody system of hinge-connected rigid and elastic bodies, and in the process extend the theory of capturing motion-induced stiffness to such systems. The material is basically reproduced from Ref. [1]. The interconnection of the bodies in the system is described by a topological tree shown in Figure 6.1, where the bodies are numbered arbitrarily as $1, 2, 3, \ldots, n$ with body numbered 0 being the inertial frame of reference.

Following Huston and Passerello [2], a topological array for the system in Figure 6.1 is defined such that body j has an inboard connecting body $c(j)$ along the path going from body j to body 0. Thus for the system of Figure 6.1, we have the topological tree array:

$$\begin{bmatrix} \mathrm{j} & 1 & 2 & 3 & 4 & 5 & 6 & 7 & 8 & 9 \\ \mathrm{c(j)} & 0 & 1 & 1 & 3 & 4 & 1 & 6 & 6 & 8 \end{bmatrix}$$

6.2 Kinematics of a Joint in a Flexible Multibody Body System

In Figure 6.2, we show two adjacent bodies B_j , $B_{c(j)}$ connected at a hinge allowing both rotation and translation, and having reference frames j and $c(j)$. Frame j is placed at Q_j, a hinge on B_j, and frame $c(j)$ is placed at P_j, the corresponding hinge connection point on $B_{c(j)}$.

Following Kane and Levinson [3], we introduce generalized speeds as motion variables, as many in number as there are total dof in the system. If at Q_j, let there be R_j number of degrees of freedom along rotational hinge axis unit vectors $\mathbf{h}_i^j$, $(i = 1, \ldots, R_j)$, and T_j number of degrees of freedom along translational direction unit vectors $\mathbf{t}_i^j$, $(i = 1, \ldots, T_j)$; note, an underscore is used in the figure to indicate a basis vector, which we denote with a bold sign in the text here. If M_j number of vibration modal coordinates are used to describe the small elastic deformation of body B_j in its body frame basis of $\underline{b}_1^j$, $\underline{b}_2^j$, $\underline{b}_3^j$ in the figure, then the following generalized

Flexible Multibody Dynamics: Efficient Formulations and Applications, First Edition. Arun K. Banerjee.

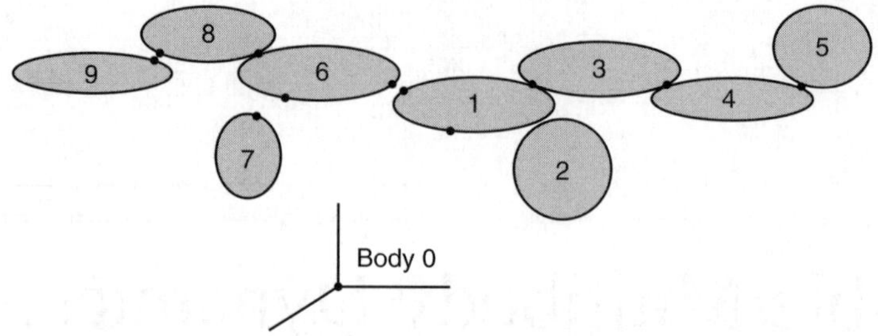

Figure 6.1 A System of Hinge-Connected Rigid and Flexible Bodies in a Tree Topology. Banerjee and Lemak [1]. Reproduced with permission of the American Society of Mechanical Engineers.

speeds are defined for the overall n-body system:

$$\begin{aligned} u_i^j &= {}^{p_j}\boldsymbol{\omega}^j \cdot \mathbf{h}_i^j \quad (j = 1, \ldots, n; i = 1, \ldots, R_j) \\ u_{R_j+i}^j &= {}^{p_j}\mathbf{v}^{Q_j} \cdot \mathbf{t}_i^j \quad (j = 1, \ldots, n; i = 1, \ldots, T_j) \\ u_{R_j+T_j+i}^j &= \dot{\eta}_i^j \quad (j = 1, \ldots, n; i = 1, \ldots, M_j) \end{aligned} \tag{6.1}$$

Here ${}^{p_j}\boldsymbol{\omega}^j$, is the angular velocity vector of frame-j with respect frame-p_j mounted on body $c(j)$ at point P_j, ${}^{p_j}\mathbf{v}^{Q_j}$ is the translational velocity vector of the point Q_j with respect to point P_j, and η_i^j is the ith modal coordinate of body B_j. Note that the total degree of freedom of the n body system, based on Eq. (6.1) for a generic jth body, is ndof, where:

$$\text{ndof} = \sum_{j=1}^{n} (R_j + T_j + M_j) \tag{6.2}$$

6.3 Kinematics and Generalized Inertia Forces for a Flexible Multibody System

To form an expression for generalized inertia force for a multibody system, we will have to sum over the generalized inertia force for individual bodies. Hence, for a generic body B_j, we review briefly the expressions for kinematics and generalized inertia force for an arbitrary

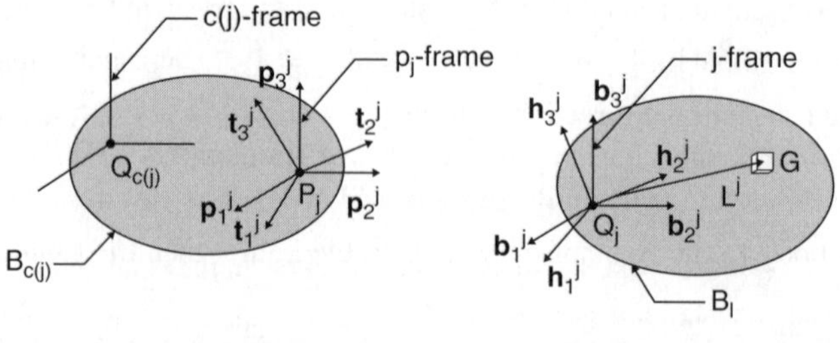

Figure 6.2 Two Hinge-Connected Bodies in a Tree Topology. Banerjee and Lemak [1]. Reproduced with permission of the American Society of Mechanical Engineers.

flexible body that we developed in the preceding chapter. Let the elastic deformation vector of a finite element nodal body at G in Figure 6.2 be expressed in terms of mode shape vector functions $\boldsymbol{\phi}_i^j$ and modal coordinates η_i^j of body B_j as:

$$\mathbf{d}^j = \sum_{i=1}^{M_j} \boldsymbol{\phi}_i^j \eta_i^j \tag{6.3}$$

The velocity of G, in the inertial frame N, located by $\mathbf{r}^j$ from Q_j in undeformed state, is:

$$^N\mathbf{v}^G = {}^N\mathbf{v}^{Q_j} + {}^N\boldsymbol{\omega}^G \times (\mathbf{r}^j + \mathbf{d}^j) + \sum_{i=1}^{M_j} \boldsymbol{\phi}_i^j u^j_{R_j+T_j+i} \tag{6.4}$$

Note here that $^N\mathbf{v}^G, {}^N\boldsymbol{\omega}^G$ are the inertial velocity of G and angular velocity of frame-j. The inertial angular velocity of the nodal rigid body G is expressed in terms of modal rotation vector functions $\boldsymbol{\psi}_i^j$ at G by the angular velocity addition theorem:

$$^N\boldsymbol{\omega}^G = {}^N\boldsymbol{\omega}^j + \sum_{i=1}^{M_j} \boldsymbol{\psi}_k^{\delta} u^j_{R_j+T_j+i} \tag{6.5}$$

At this stage note that Eqs. (6.4) and (6.5) are already linear in the elastic motion modal coordinates, η_i^j, representing premature linearization Partial velocity of point G and partial angular velocity of the nodal body at G follow from Eqs. (6.4) and (6.5):

$$\begin{aligned} {}^N\mathbf{v}_i^G &= {}^N\mathbf{v}_i^{Q_j} + {}^N\boldsymbol{\omega}_i^G \times (\mathbf{r}^j + \mathbf{d}^j) + \boldsymbol{\phi}_i^j \delta_{ik} \ (i = 1, \ldots, \text{ndof}) \\ {}^N\boldsymbol{\omega}_i^G &= {}^N\boldsymbol{\omega}_i^j + \boldsymbol{\psi}_k^j \, \delta_{ik} (i = 1, \ldots, \text{ndof}) \end{aligned} \tag{6.6}$$

where the Kronecker delta δ_{ik} has the value 1 when the ith generalized speed of the system coincides with the kth modal coordinate rate of body B_j and is zero otherwise. The inertial acceleration of point G and the infinitesimal nodal rigid body at G are, respectively, as follows:

$$^N\mathbf{a}^G = {}^N\mathbf{a}^{Q_j} + {}^N\boldsymbol{\alpha}^j \times (\mathbf{r}^j + \mathbf{d}^j) + {}^N\boldsymbol{\omega}^j \times \left[{}^N\boldsymbol{\omega}^j \times (\mathbf{r}^j + \mathbf{d}^j) + 2\sum_{i=1}^{M_j} \boldsymbol{\phi}_i^j u^j_{R_j+T_j+i} \right] + \sum_{i=1}^{M_j} \boldsymbol{\phi}_i^j \dot{u}^j_{R_j+T_j+i} \tag{6.7}$$

$$^N\boldsymbol{\alpha}^G = {}^N\boldsymbol{\alpha}^j + {}^N\boldsymbol{\omega}^j \times \sum_{i=1}^{M_j} \boldsymbol{\psi}_i^j u^j_{R_j+T_j+i} + \sum_{i=1}^{M_j} \boldsymbol{\psi}_k^j \dot{u}^j_{R_j+T_j+i} \tag{6.8}$$

The ith generalized inertia force contribution, to the system, by body B_j is written in two steps:

$$F_i^{j*} = -\int_j {}^N\mathbf{v}_i^G \cdot {}^N\mathbf{a}^G dm - \int_{B_j} {}^N\boldsymbol{\omega}_i^G.[\mathbf{dI}^G \cdot {}^N\boldsymbol{\alpha}^G + {}^N\boldsymbol{\omega}^G \times \mathbf{dI}^G \cdot {}^N\boldsymbol{\omega}^G] \quad (i = 1, \ldots, \text{ndof}) \tag{6.9}$$

First we simply copy from the terms in Eqs. (6.6)–(6.8) in Eq. (6.9); the result is Eq. (6.10). Then we form Eq. (6.11) by compacting some terms in the form of modal integrals. The details are given next. In both equations and later, to save space, we have used the symbols $\dot{\eta}_k^j$, $\ddot{\eta}_k^j$ in place of their notations as generalized speeds as defined in Eq. (6.1), and their derivatives:

$$
\begin{aligned}
F_i^{j*} = &-{}^N\mathbf{v}_i^{Q_j} \cdot \int_{m^j} \left\langle \mathbf{a}^{Q_j} + \alpha^j \times \left(\mathbf{r} + \sum_{k=1}^{M_j} \mathbf{\Phi}_k^j \eta_k^j \right) + \omega^j \times \left[\omega^j \times \left(\mathbf{r} + \sum_{k=1}^{M_j} \mathbf{\Phi}_k^j \eta_k^j \right) + 2 \sum_{k=1}^{M_j} \mathbf{\Phi}_k^j \dot{\eta}_k^j \right] \right. \\
&\left. + \sum_{k=1}^{M_j} \mathbf{\Phi}_k^j \ddot{\eta}_k^j \right\rangle \mathrm{dm} \\
&-\omega_i^j \cdot \int_{m^j} \left(\mathbf{r} + \sum_{k=1}^{M_j} \mathbf{\Phi}_k^j \eta_k^j \right) \times \left\langle \mathbf{a}^{Q_j} + \alpha^j \times \left(\mathbf{r} + \sum_{k=1}^{M_j} \mathbf{\Phi}_k^j \eta_k^j \right) + \omega^j \times \left[\omega^j \times \left(\mathbf{r} + \sum_{k=1}^{M_j} \mathbf{\Phi}_k^j \eta_k^j \right) \right. \right. \\
&\left. \left. + 2 \sum_{k=1}^{M_j} \mathbf{\Phi}_k^j \dot{\eta}_k^j \right] + \sum_{k=1}^{M_j} \mathbf{\Phi}_k^j \ddot{\eta}_k^j \right\rangle \mathrm{dm} \\
&-\delta_{ik} \int_{m^j} \mathbf{\Phi}_k^j \cdot \left\langle \mathbf{a}^{Q_j} + \alpha^j \times \left(\mathbf{r} + \sum_{k=1}^{M_j} \mathbf{\Phi}_k^j \eta_k^j \right) + \omega^j \times \left[\omega^j \times \left(\mathbf{r} + \sum_{k=1}^{M_j} \mathbf{\Phi}_k^j \eta_k^j \right) + 2 \sum_{k=1}^{M_j} \mathbf{\Phi}_k^j \dot{\eta}_k^j \right] \right. \\
&\left. + \sum_{k=1}^{M_j} \mathbf{\Phi}_k^j \ddot{\eta}_k^j \right\rangle \mathrm{dm} \\
&-\omega_i^j \cdot \int_{m^j} \left\langle d\mathbf{I}^G \cdot \left[\alpha^j + \omega^j \times \sum_{j=1}^{M_j} \psi_k^j \dot{\eta}_k^j + \sum_{j=1}^{M_j} \psi_k^j \ddot{\eta}_k^j \right] + \left(\omega^j + \sum_{k=1}^{M_j} \psi_k^j \dot{\eta}_k^j \right) \right. \\
&\left. \times d\mathbf{I}^G \cdot \left(\omega^j + \sum_{k=1}^{M_j} \psi_k^j \dot{\eta}_k^j \right) \cdot \right\rangle \\
&-\delta_{ik} \int_{m^j} \left\langle \psi_k^j . d\mathbf{I}^G \cdot \left[\alpha^j + \omega^j \times \sum_{j=1}^{M_j} \psi_k^j \dot{\eta}_k^j + \sum_{j=1}^{M_j} \psi_k^j \ddot{\eta}_k^j \right] + \left(\omega^j + \sum_{k=1}^{M_j} \psi_k^j \dot{\eta}_k^j \right) \right. \\
&\left. \times\, d\mathbf{I}^G \cdot \left(\omega^j + \sum_{k=1}^{M_j} \psi_k^j \dot{\eta}_k^j \right) \cdot \right\rangle \qquad (i = 1, \ldots, \mathrm{ndof}) \qquad (6.10)
\end{aligned}
$$

Note the integral signs in Eq. (6.10). To simplify Eq. (6.10), first we define the "modal integrals" $\mathbf{s}^j, \mathbf{b}_k^j, \mathbf{c}_k^j, \mathbf{d}_k^j, e_{ik}^j, \mathbf{g}_k^j$ in the form of vectors and scalars, and if rotational nodal inertia is to be considered then the following dyadics and vectors, all given in Appendix 5.A, are used:

$$\mathbf{I}^j, \mathbf{N}_k^{j*}, \mathbf{D}_k^j, \tilde{\mathbf{W}}_1^j, \tilde{\mathbf{W}}_{2k}^j, \tilde{\mathbf{W}}_{3k}^j, \mathbf{W}_{4k}^j, \tilde{\mathbf{W}}_{5k}^j, \mathbf{W}_{6k}^j, \mathbf{W}_{7lk}^j, \mathbf{W}_{8lk}^j, \mathbf{W}_{9lk}^j, \tilde{\mathbf{W}}_{10lk}^j, \mathbf{W}_{11lk}^j, \mathbf{W}_{12lk}^j$$

Evaluation of these modal integrals leads to the following form of Eq. (6.10), where we have put down a left superscript N on kinematical vectors to explicitly indicate that these are to be referenced to the Newtonian frame N:

$$\begin{aligned}
F_i^{j*} = & -{}^N\mathbf{v}_i^{Q_j}.\left\{ m^j\,{}^N\mathbf{a}^{Q_j} - \mathbf{s}^j \times {}^N\alpha^j + \sum_{k=1}^{M_j} \mathbf{b}_k^j \ddot{\eta}_k^j + {}^N\omega^j \times \left[{}^N\omega^j \times \mathbf{s}^j + 2\sum_{k=1}^{M_j} \mathbf{b}_k^j \dot{\eta}_k^j \right] \right\} \\
& -{}^N\omega_i^j.\left\{ \mathbf{s}^j \times {}^N\mathbf{a}^{Q_j} + \mathbf{I}^j.{}^N\alpha^j + \sum_{k=1}^{M_j} \mathbf{c}_k^j \ddot{\eta}_k^j + {}^N\omega^j \times \mathbf{I}^j.{}^N\omega^j + 2\sum_{k=1}^{M_j} \mathbf{N}_k^{j*}\dot{\eta}_k^j.{}^N\omega^j \right\} \\
& -\delta_{ik}\left\{ \mathbf{b}_k^j.{}^N\mathbf{a}^{Q_j} + \mathbf{g}_k^j.{}^N\alpha^j + \sum_{k=1}^{M_j} e_{ik}^j \ddot{\eta}_k^j - {}^N\omega^j.\mathbf{D}_k^j.{}^N\omega^j + 2\sum_{k=1}^{M_j} \mathbf{d}_{ik}^j \dot{\eta}_k^j.{}^N\omega^j \right\} \\
& -{}^N\omega_i^j.\left\{ \begin{array}{l} \left(\tilde{\mathbf{W}}_1^j + \sum\limits_{l=1}^{M_j} \tilde{\mathbf{W}}_{2k}^j \eta_k^j \right).{}^N\alpha^j + {}^N\omega^j.\sum\limits_{l=1}^{M_j} \tilde{\mathbf{W}}_{3k}^j \dot{\eta}_k^j + \sum\limits_{l=1}^{M_j} \mathbf{W}_{4k}^j \ddot{\eta}_k^j \\ +{}^N\omega^j \times \left[\left(\tilde{\mathbf{W}}_1^j + \sum\limits_{l=1}^{M_j} \tilde{\mathbf{W}}_{2k}^j \eta_k^j \right).{}^N\omega^j + \sum\limits_{k=1}^{M_j} \mathbf{W}_{4k}^j \dot{\eta}_k^j \right] - \sum\limits_{k=1}^{M_j} \tilde{\mathbf{W}}_{5k}^j \dot{\eta}_k^j.{}^N\omega^j \end{array} \right\} \\
& +\delta_{ik}\left\{ \begin{array}{l} \left(\mathbf{W}_{6k}^j + \sum\limits_{l=1}^{M_j} \mathbf{W}_{7lk}^j \eta_k^j \right).{}^N\alpha^j + {}^N\omega^j.\sum\limits_{l=1}^{M_j} \mathbf{W}_{8lk}^j \dot{\eta}_l^j + \sum\limits_{l=1}^{M_j} \mathbf{W}_{9lk}^j \ddot{\eta}_l^j \\ -{}^N\omega^j.\left[\tilde{\mathbf{W}}_{3k}^j.{}^N\omega^j + \sum\limits_{l=1}^{M_j} \tilde{\mathbf{W}}_{10lk}^j \eta_k^j.{}^N\omega^j - \sum\limits_{l=1}^{M_j} (\mathbf{W}_{11lk}^j - \mathbf{W}_{12lk}^j)\dot{\eta}_l^j \right] \end{array} \right\} \\
& i = 1, \ldots, \text{ndof},\; j = 1, \ldots, n
\end{aligned} \tag{6.11}$$

Here $\mathbf{I}^j, \mathbf{N}_k^{j*}, \mathbf{D}_k^j$ are dyadics, with $*$ denoting a transpose. If rotatory inertia is ignorable, then terms within the two elongated braces at the end of (6.11) are excluded. The system generalized inertia force corresponding to the ith generalized speed is obtained by summing up over all bodies $j = 1, \ldots, n$.

$$F_i^* = \sum_{j=1}^{N} F_i^{j*} \qquad (i = 1, \ldots, \text{ndof}; \quad j = 1, \ldots, n) \tag{6.12}$$

6.4 Kinematical Recurrence Relations Pertaining to a Body and Its Inboard Body

In this section we develop the inter-relationships between ${}^N\boldsymbol{\omega}^j$, ${}^N\mathbf{v}^{Q_j}$, ${}^N\boldsymbol{\alpha}^j$, ${}^N\mathbf{a}^{Q_j}$ of body j in Eq. (6.11), in terms of those variables for the inboard body $c(j)$, $B_{c(j)}$in Figure 1(b), that is, ${}^N\boldsymbol{\omega}^{c(j)}$, ${}^N\mathbf{v}^{Q_{c(j}}$, ${}^N\boldsymbol{\alpha}^{c(j}$, ${}^N\mathbf{a}^{Q_{c(j}}$. Introducing an orthogonal triad of basis vectors $\mathbf{p}_i^j$ fixed to the frame p_j and assuming R_j number of rotations about possibly nonorthogonal axes, angular velocity of frame-j is given by the angular velocity addition theorem:

$${}^N\boldsymbol{\omega}^j = {}^N\boldsymbol{\omega}^{c(j)} + \sum_{i=1}^{M_{c(j)}} \boldsymbol{\psi}_i^{c(j)}(P_j)\eta_i^{c(j)} + \sum_{i=1}^{3}\sum_{k=1}^{R_j} \mathbf{p}_i^j G_{ik}^j \dot{\theta}_k^j, \tag{6.13}$$

Here the first two terms give the angular velocity of a frame mounted at the terminal point P_j of body $c(j)$, with respect to which the hinge rotations for the frame of body j takes place. We also have $G_{ik}^j = \mathbf{p}_i^j \cdot \mathbf{h}_k^j$, where $\mathbf{h}_k^j$ is the hinge kth rotation axis vector, and $\boldsymbol{\psi}_i^{c(j)}(P_j), \dot{\theta}_k^j$ are, respectively, the ith modal rotation at P_j and the kth relative rotation rate of the hinge at Q_j. The velocity of point Q_j can be written by assuming T_j number of translational degrees of freedom from P_j along possibly nonorthogonal axes as follows:

$$\begin{aligned}
{}^N\mathbf{v}^{Q_j} &= {}^N\mathbf{v}^{Q_{c(j)}} + {}^N\boldsymbol{\omega}^{c(j)} \times \left[\mathbf{r}^{Q_{c(j)}P_j} + \sum_{k=1}^{M_{c(j)}} \boldsymbol{\phi}_k^{c(j)}(P_j)\eta_k^{c(j)}\right] + \sum_{k=1}^{M_{c(j)}} \boldsymbol{\phi}_k^{c(j)}(P_j)\dot{\eta}_k^{c(j)} \\
&+ \left[{}^N\boldsymbol{\omega}^{c(j)} + \sum_{k=1}^{M_{c(j)}} \boldsymbol{\psi}_k^{c(j)}(P_j)\dot{\eta}_k^{c(j)}\right] \times \sum_{i=1}^{3}\sum_{k=1}^{T_j} \mathbf{p}_i^j L_{ik}^j \tau_k^j + \sum_{i=1}^{3}\sum_{k=1}^{T_j} \mathbf{p}_i^j \left[\dot{L}_{ik}^j \tau_k^j + L_{ik}^j \dot{\tau}_k^j\right]
\end{aligned}$$

$$\text{where} \quad L_{ik}^j = \mathbf{p}_i^j \cdot \boldsymbol{\tau}_k^j \tag{6.14}$$

Here $\boldsymbol{\tau}_k^j$denote the kth translation vector at the hinge Q_j, wih τ_k^j being its magnitude. Equations (6.13) and (6.14) provide the mechanism for developing the expressions for inertial (or N-frame) velocities of a hinge point Q_j and the angular velocity of frame j of body B_j for all values of $j = 1, \ldots, n$, in the tree topology of Figure 1(a), starting at body 1. The partial velocities and partial angular velocities needed in Eq. (6.11) are obtained from their defining relations (assuming no prescribed motion) as

$$
\begin{aligned}
{}^N\mathbf{v}^{Q_j} &= \sum_{i=1}^{n} {}^N\mathbf{v}_i^{Q_j} u_i \\
{}^N\boldsymbol{\omega}^j &= \sum_{i=1}^{n} {}^N\boldsymbol{\omega}_i^j u_i
\end{aligned}
\tag{6.15}
$$

and the acceleration and angular accelerations of Q_j and frame-j are expressed from Eq. (6.15) in terms of terms involving derivatives of the generalized speeds and remainder acceleration terms:

$$
\begin{aligned}
{}^N\mathbf{a}^{Q_j} &= \sum_{i=1}^{3} \mathbf{v}_i^{Q_j} \dot{u}_i + \mathbf{h}^j \\
{}^N\alpha^j &= \sum_{i=1}^{3} \boldsymbol{\omega}_i^j \dot{u}_i + \mathbf{f}^j
\end{aligned}
\tag{6.16}
$$

6.5 Generalized Active Forces due to Nominal and Motion-Induced Stiffness

This section represents a multibody application of the work in Chapter 5. Here we will express generalized active forces due to nominal structural stiffness and motion-induced geometric stiffness, and we will do so in matrix form because both these quantities are typically developed by finite element methods, which use matrix tools. Generalized active force due to nominal elasticity is customarily represented on the basis of component modes [4] for the *j*th body in the matrix form:

$$
F_n^j = -[\Phi^j]^T K_n^j \Phi^j \eta^j = -\Lambda_n^j \eta^j \tag{6.17}
$$

Here Φ^j, K_n^j, η^j are, respectively, the mode shape matrix (with superscript T denoting matrix transpose), nominal structural stiffness matrix, and the column matrix of modal coordinates for body j. The matrix Λ_n^j is diagonal if mass-normalized vibration modes are used for body j, and non-diagonal if constraint modes are used in addition to the normal modes, as in the Craig-Bampton method [4]. It should be noted that modes here reflect deformations about the initial state of stress of the body, and motion-induced stiffness is due to a special case of initial state of stress [5] caused by inertia loading. It has its origin in the strain energy term,

$$
P_{NL} = \int_B \varepsilon_{NL}^T \sigma_0 dv \tag{6.18}
$$

where σ_0 is initial the stress state column matrix, and ε_{NL} is the Lagrangian strain tensor:

$$\sigma_0 = (\sigma_{110}, \sigma_{120}, \sigma_{130}, \sigma_{220}, \sigma_{230}, \sigma_{330})^T$$

$$\varepsilon_{NL} = \begin{Bmatrix} 0.5\left(w_{1,1}^2 + w_{2,1}^2 + w_{3,1}^2\right) \\ 0.5\left(w_{1,2}^2 + w_{2,2}^2 + w_{3,2}^2\right) \\ 0.5\left(w_{1,3}^2 + w_{2,3}^2 + w_{3,3}^2\right) \\ w_{1,3}w_{1,2} + w_{2,3}w_{2,2} + w_{3,3}w_{3,2} \\ w_{1,3}w_{1,1} + w_{2,3}w_{2,1} + w_{3,3}w_{3,1} \\ w_{1,2}w_{1,1} + w_{2,2}w_{2,1} + w_{3,2}w_{3,1} \end{Bmatrix} \tag{6.19}$$

Here w_1, w_2, w_3, are 1-, 2-, 3- orthogonal components of the elastic displacement, and differentiation with respect to a component is indicated by a subscripted comma. In the standard procedure of the finite element theory (see Zienkiewicz [6]), one assumes interpolation functions $N(x,y,z)$ between the nodal displacements d for the displacement of a point within an element, and derives elemental stiffness matrices based on the potential energy functions. In this case, gradient of the potential energy given by Eq. (6.18) leads to the geometric or initial stress stiffness matrix k_g^e as before:

$$\{w\} = \left[N(x_1, x_2, x_3)\right]$$

$$\mathbf{k}_g^e = \int_e \left[N_{,x}^T \quad N_{,y}^T \quad N_{,z}^T\right] \begin{bmatrix} \sigma_{xx0}I_3 & \sigma_{xy0}I_3 & \sigma_{xz0}I_3 \\ \sigma_{xy0}I_3 & \sigma_{yy0}I_3 & \sigma_{yz0}I_3 \\ \sigma_{xz0}I_3 & \sigma_{yz0}I_3 & \sigma_{zz0}I_3 \end{bmatrix} \begin{Bmatrix} N_{,x} \\ N_{,y} \\ N_{,z} \end{Bmatrix} \mathbf{dv} \tag{6.20}$$

Here I_3 is a 3 × 3 unity matrix. In a finite element code such as NASTRAN or EISI-EAL, the latter [7] used to generate the results in this chapter, the reference stress stiffness matrix is computed by first evaluating the stresses due to a distributed loading, computing the element geometric stiffness matrices as per Eq. (6.20), and finally assembling the latter into a global geometric stiffness matrix for the structure. For structures in large overall motion, the distributed loading is a system of inertia forces and torques corresponding to the motion. To produce a theory that is fully linear in the modal coordinates, consistent with the small displacement assumption, we consider inertia force and torques that involve only zeroth-order terms in the modal coordinates. In other words, we neglect terms related to elastic displacement in Eqs. (6.6)–(6.8) to form expressions for the inertia force and torque on a nodal rigid body G in body j, that produce the "existing" state [5], incremental deformations with respect to which give rise to geometric stiffness, that is, we write:

$$\begin{aligned} f^{jG*} &= -dm\left[{}^N\mathbf{a}^{Q_j} + {}^N\boldsymbol{\alpha}^j \times \mathbf{r}^j + {}^N\boldsymbol{\omega}^j \times ({}^N\boldsymbol{\omega}^j \times \mathbf{r}^j)\right] \\ t^{jG*} &= -\left[\mathbf{dI}^G \cdot {}^N\boldsymbol{\alpha}^j + {}^N\boldsymbol{\omega}^j \times \mathbf{dI}^G \cdot {}^N\boldsymbol{\omega}^j\right] \end{aligned} \tag{6.21}$$

Equation (6.21) is now written in matrix form in terms of the j-basis vector elements, and direction cosine elements between body 1 and body j:

$$[\mathbf{b}^j]^T = \left[\mathbf{b}_1^j \ \mathbf{b}_2^j \ \mathbf{b}_3^j\right] \qquad (j = 1, \ldots, N)$$

$$C_{1j}(i,k) = \mathbf{b}_i^1 \cdot \mathbf{b}_k^j \tag{6.22}$$

We now define a (nx1) column matrix U formed by stacking the generalized speeds given by Eq. (6.1) for all the bodies in the system. Then Eq. (6.14) is rewritten in matrix form in terms of the partial velocity for point Q_j and the partial angular velocity matrix from frame-j, expressed in *body 1 basis*:

$$
\begin{aligned}
{}^N\boldsymbol{\omega}^j &= [\mathbf{b}^1]^T \omega_u^j U \\
{}^N\mathbf{v}^{Q_j} &= [\mathbf{b}^1]^T v_u^{Q_j} U
\end{aligned}
\tag{6.23}
$$

Equation (6.16) is then rewritten in terms of matrices as:

$$
\begin{aligned}
{}^N\mathbf{a}^{Q_j} &= [\mathbf{b}^1]^T C_{1j}^T \left(v_u^{Q_j} \dot{U} + h^j \right) \\
{}^N\boldsymbol{\alpha}^j &= [\mathbf{b}^1]^T C_{1j}^T \left(\omega_u^j \dot{U} + f^j \right)
\end{aligned}
\tag{6.24}
$$

At this stage we introduce a set of intermediate scalar variables, $z_i^j, \quad i = 1, \dots 15;\ j = 1, \dots, N$ to facilitate expressing angular velocity of the frame j, and the acceleration of a generic nodal body:

$$
\begin{aligned}
{}^N\boldsymbol{\omega}^j &= z_1^j \mathbf{b}_1^j + z_2^j \mathbf{b}_2^j + z_3^j \mathbf{b}_3^j \\
\tilde{\omega}^j &= \begin{bmatrix} 0 & -z_3^j & z_2^j \\ z_3^j & 0 & -z_1^j \\ -z_2^j & z_1^j & 0 \end{bmatrix} \\
\begin{bmatrix} z_4^j & z_5^j & z_6^j \\ z_5^j & z_7^j & z_8^j \\ z_6^j & z_8^j & z_9^j \end{bmatrix} &= C_{1j}^T \tilde{\omega}^j \tilde{\omega}^j C_{1j} \\
z_{10}^j &= z_1^j z_2^j \\
z_{11}^j &= z_2^j z_3^j \\
z_{12}^j &= z_3^j z_1^j \\
z_{13}^j &= \left(z_1^j\right)^2 - \left(z_2^j\right)^2 \\
z_{14}^j &= \left(z_2^j\right)^2 - \left(z_3^j\right)^2 \\
z_{15}^j &= \left(z_3^j\right)^2 - \left(z_1^j\right)^2 \\
\mathbf{r}^j &= x_1^j \mathbf{b}_1^j + x_2^j \mathbf{b}_2^j + x_3^j \mathbf{b}_3^j
\end{aligned}
\tag{6.25}
$$

In terms of the above-defined scalars, the matrix form of the inertia force in Eq. (6.21) can be written as representing the following set of twelve inertial acceleration loadings, as used before:

$$\begin{Bmatrix} f_1^{j*} \\ f_2^{j*} \\ f_3^{j*} \end{Bmatrix} = -dm \begin{bmatrix} I_3 & x_1^j I_3 & x_2^j I_3 & x_3^j I_3 \end{bmatrix} \begin{Bmatrix} A_1^j \\ \vdots \\ \vdots \\ A_{12}^j \end{Bmatrix} \tag{6.26}$$

where I_3 is the 3×3 unity matrix, A_i^j $(i = 1, \ldots 12)$ are expressed below in terms of the previous definitions, and e_i is the ith row of the unity matrix I_3, as follows:

$$\begin{aligned}
&\begin{Bmatrix} A_1^j \\ A_2^j \\ A_3^j \end{Bmatrix} = C_{1j}^T \left\{ v_u^{Q_j} \dot{U} + h^j \right\} \\
&A_4^j = z_4^j \\
&\begin{Bmatrix} A_5^j \\ A_6^j \\ A_7^j \end{Bmatrix} = \begin{Bmatrix} z_5^j + e_3 C_{1j}^T \left(v_u^{Q_j} \dot{U} + f^j \right) \\ z_6^j - e_2 C_{1j}^T \left(v_u^{Q_j} \dot{U} + f^j \right) \\ z_5^j - e_3 C_{1j}^T \left(v_u^{Q_j} \dot{U} + f^j \right) \end{Bmatrix} \\
&A_8^j = z_7^j \\
&\begin{Bmatrix} A_9^j \\ A_{10}^j \\ A_{11}^j \end{Bmatrix} = \begin{Bmatrix} z_8^j + e_1 C_{1j}^T \left(\omega_u^j \dot{U} + f^j \right) \\ z_6^j + e_2 C_{1j}^T \left(\omega_u^j \dot{U} + f^j \right) \\ z_8^j - e_1 C_{1j}^T \left(\omega_u^j \dot{U} + f^j \right) \end{Bmatrix} \\
&A_{12}^j = z_9^j
\end{aligned} \tag{6.27}$$

The inertia torque on a nodal rigid body in body-j, given in Eq. (6.21), can be similarly rewritten in terms of the moment of inertia components of the nodal body in frame-j, as constituting the set of nine torque loadings distributed over the body:

$$\begin{Bmatrix} t_1^* \\ t_2^* \\ t_3^* \end{Bmatrix} = - \begin{bmatrix} I_{11} & I_{12} & I_{13} & I_{13} & I_{33} - I_{22} & -I_{12} & 0 & I_{23} & 0 \\ I_{12} & I_{22} & I_{23} & -I_{23} & I_{12} & I_{11} - I_{33} & 0 & 0 & I_{13} \\ I_{13} & I_{23} & I_{33} & I_{22} - I_{11} & -I_{13} & I_{23} & I_{12} & 0 & 0 \end{bmatrix} \begin{Bmatrix} A_{13}^j \\ \vdots \\ \vdots \\ A_{21}^j \end{Bmatrix} \tag{6.28}$$

Here, for the jth body in the multibody case the scalars A_i^j, $(i = 13, \ldots, 21)$ are:

$$\begin{Bmatrix} A_{13}^j \\ A_{14}^j \\ A_{15}^j \end{Bmatrix} = -C_{1j}^T \left(\omega_u^j \dot{U} + h^j \right)$$
$$A_{15+i}^j = z_{9+i}^j \qquad (i = 1, \ldots, 6) \tag{6.29}$$

With the force and torque loadings at the nodes now defined by Eqs. (6.26) and (6.28), these can be input to a finite element code to generate the corresponding geometric stiffness matrices $K_g^{(i)}, i = 1, \ldots 21$, for unit values of the time-varying quantities, A_i $i = 1, \ldots, 21$. These geometric stiffness matrices are post-multiplied by the global modal matrix Φ and the result pre-multiplied by Φ^{j^T}, and the instantaneous values of the A_i giving generalized active force for motion-induced geometric stiffness. The time-varying nature of A_i are input from the multibody computer program. Dealing with a "generic jth body" for the ith inertia loading at a time, the generalized geometric stiffness and corresponding generalized active force contributions are:

$$S^{j(i)} = \Phi^{j^T} K_g^{j(i)} \Phi^j \qquad i = 1, \cdots 21$$
$$F_g^j = -\sum_{i=1}^{21} A_i^j S^{j(i)} \eta^j \tag{6.30}$$

Combining contributions from the nominal and geometric stiffness for the jth body in the ith generalized generalized speed, using Eqs. (6.17) and (6.30), we have the scalar form of the generalized active force due to stiffness:

$$F_i^j = -\delta_{jk} \sum_{k=1}^{M_j} \left[\Lambda_n^{j(i)} + \sum_{m=1}^{21} A_i^j S^{j(i)} \right] \eta_k^j \qquad (i = 1, \ldots, \text{ndof}) \tag{6.31}$$

If rotatory inertia of the nodal bodies can be ignored, then the second summation in Eq. (6.31) goes only up to 12. System general active forces are obtained by summing over all N bodies j:

$$F_i = \sum_{j=1}^{N} F_i^j \qquad (i = 1, \ldots \text{ndof}) \tag{6.32}$$

Finally to complete the system equations of motion, we invoke Kane's dynamical equations [3]:

$$F_i^* + F_i = 0 \qquad (i = 1, \ldots, \text{ndof}) \tag{6.33}$$

Considering the forms of Eq. (6.15) embedded in Eq. (6.10), and of Eqs. (6.26) and (6.28) in Eq. (6.29), it can be shown that that the dynamical equations take the matrix form:

$$D\dot{U} + E = 0 \tag{6.34}$$

The coefficient matrix D in Eq. (6.34) is unsymmetric because of acceleration dependent stiffness, and is a function of the generalized coordinates, and E is a function of both generalized coordinates and generalized speeds.

6.6 Treatment of Prescribed Motion and Internal Forces

Here we consider two types of problems that can be treated alike. In one, relative motion between two contiguous bodies may be prescribed and the interaction force is to be determined to realize this motion; in another, the nonworking internal forces at a joint in a structure may be of interest. If some of the generalized coordinates are prescribed, which amounts to prescribing corresponding generalized speeds and their derivatives, then Eq. (6.34) can be written in partitioned matrix form as follows, with a new term added to represent the generalized active force F representing interaction forces needed to realize the prescribed motion:

$$\begin{bmatrix} D_{ff} & D_{fp} \\ D_{pf} & D_{pp} \end{bmatrix} \begin{Bmatrix} \dot{U}_f \\ \dot{U}_p \end{Bmatrix} = -\begin{Bmatrix} E_{fp} \\ E_{pf} \end{Bmatrix} - \begin{Bmatrix} 0 \\ F \end{Bmatrix} \tag{6.35}$$

Here subscripts f and p denote free and prescribed variables. The solution is:

$$\begin{aligned} D_{ff}\dot{U}_f &= -D_{ff}\dot{U}_p - E_{fp} \\ F &= -\begin{bmatrix} D_{pf} & D_{pp} \end{bmatrix} \begin{Bmatrix} \dot{U}_f \\ \dot{U}_p \end{Bmatrix} - E_{pf} \end{aligned} \tag{6.36}$$

For our choice of generalized speeds, which are time-derivatives of relative angles and distances, see Eq. (6.1), the elements of the column matrices F are directly the forces and torques that do work over the prescribed motion. In a problem like antenna shape control prescribing the elastic displacement, $F = \Phi^t f$, f being the column matrix of required actuator force distribution, at the finite element nodal degrees of freedom, can be solved in the least-square sense by using a pseudo-inverse.

If the internal force at a hinge is of interest, the hinge is prescribed extra degrees of freedom and corresponding interaction forces and torques are exposed so that these forces and torques do work when motions involving those degrees of freedom takes place. This prescribed motion is subsequently set equal to zero. The interaction forces are then found from Eq. (6.36).

6.7 "Ruthless Linearization" for Very Slowly Moving Articulating Flexible Structures

When the rotation rate or translation rate is an order-of-magnitude less than the first natural frequency of a system, then certain drastic simplifications can be made that make the computer simulations run faster. This is the case of flexible robotic manipulators that go through large angle change between component members, but at a relatively very slow rotation rate compared to the first modal frequency of the component. In these cases, all geometric stiffness terms due to motion-induced inertia loads can be ignored. What is more, terms in Eq. (6.11), involving modal coordinates in the "mass matrix," products of the modal coordinates and their rates, and products of the rigid body rotation / translation rates with the modal coordinates or their rates are neglected. This is the approximation for slow motion recommended by Sincarsin and Hughes [8], and employed for simulation of the Canada Arm space manipulator. Such a process has been called "ruthless linearization" by Padilla and von Flotow [9]. Later, Sharf and Damaren [10] showed that adding terms containing products of rigid body rotation rates and derivatives of modal coordinates to the ruthlessly linearized equations removes the deficiency in many cases, and by keeping modal coordinate dependent terms in the mass matrix produced accurate results for a wide range of maneuvers and joint rotational speeds. Ghosh [11] has implemented online control with the dynamics of a ruthlessly linearized model for a very slowly rotating system with articulating appendages, and simulation results match well with measurements. On a separate issue, geometric stiffness effects due to interbody forces on a component may not be ignorable for a very slowly moving, articulated system.

6.8 Simulation Results

Having established the effects of including or excluding geometric stiffness on the spin-up or translational acceleration response of a composite truss-type structure in the previous chapter, we present here some new results on slewing maneuver of an articulated system of flexible bodies. Figure 6.2 shows a space crane consisting of a rigid body with two hinge-connected articulated trusses. We assume that the rigid body undergoes a simple rotation through an angle q_1 in inertial space about the axis shown, while the trusses rotate through the relative angles q_2 and q_3 as shown. In a slewing maneuver, the commanded values of these angles are q_{1c}, q_{2c}, q_{3c}, respectively, and the following joint toques are applied at the hinges to realize the desired step commands:

$$T_i = -k_i(q_{ic} - q_i) - c_i\dot{q}_i \qquad (i = 1, 2, 3) \tag{6.37}$$

The control gains in Eq. (6.36) are chosen for the worst case, or largest overall outlying inertia to turn at the respective revolute joints, for a bandwidth of 0.1 Hz and a damping factor of 0.707. This controller bandwidth is below the first natural frequency of the combined system in the initial configuration. Each truss is 12 m long and consists of tubular members interconnected as shown at 21 joints; the cross section of the truss is a right isosceles triangle, the members at the right angle being 2 m long. Young's modulus for all members is 10^9 N/m^2, the cross-sectional area of each member is 8.46×10^{-4}, and each member has a mass density 10^4 kg/m^3. The rigid base has a mass of 10^4 kg, and a rotation axis moment of inertia of 1200 kg-m^2. While both trusses are essentially the same, a change is made to the inner truss by assigning a lumped stiffness at the free end to model a possible gimbal fixture that attaches to all the three nodes of the truss tip, so that all three nodes are in one plane. The system is modeled by the finite element code [7], and the modal representation for the inner truss includes the first ten fixed-fixed vibration modes and six constraint modes [4] representing

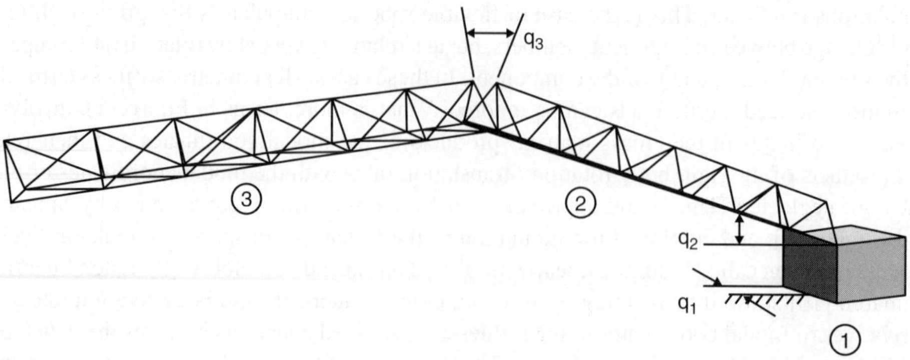

Figure 6.3 A Space Crane of Two Hinged Trusses with 3 Rigid Body dof in a Commanded Slewing Maneuver. Banerjee and Lemak [1]. Reproduced with permission of the American Society of Mechanical Engineers.

unit displacements for the end planes of the truss. With 3 rigid body dof and 10 modes for the outboard truss, the system in Figure 6.3 has 29 dof.

Figure 6.4 shows the rotation due to bending at the free end of the second truss, body number 3, during the slewing maneuver. Figure 6.5 displays the angle of twist at the end of the inner truss at the elbow. Plots such as these can obviously be used to assess

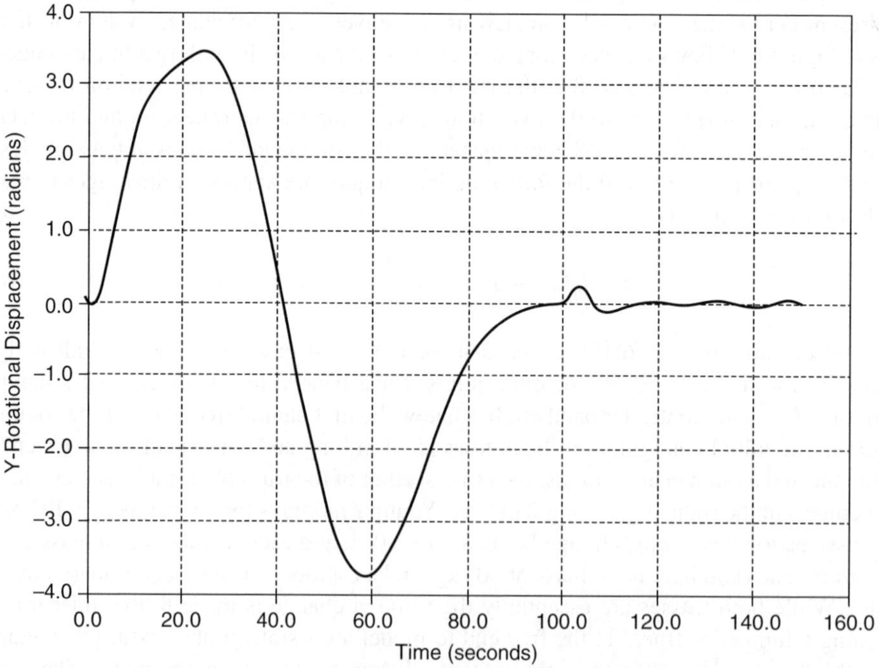

Figure 6.4 Elastic Rotation due to Bending at the Tip of the End Truss During the Slewing. Banerjee and Lemak [1]. Reproduced with permission of the American Society of Mechanical Engineers.

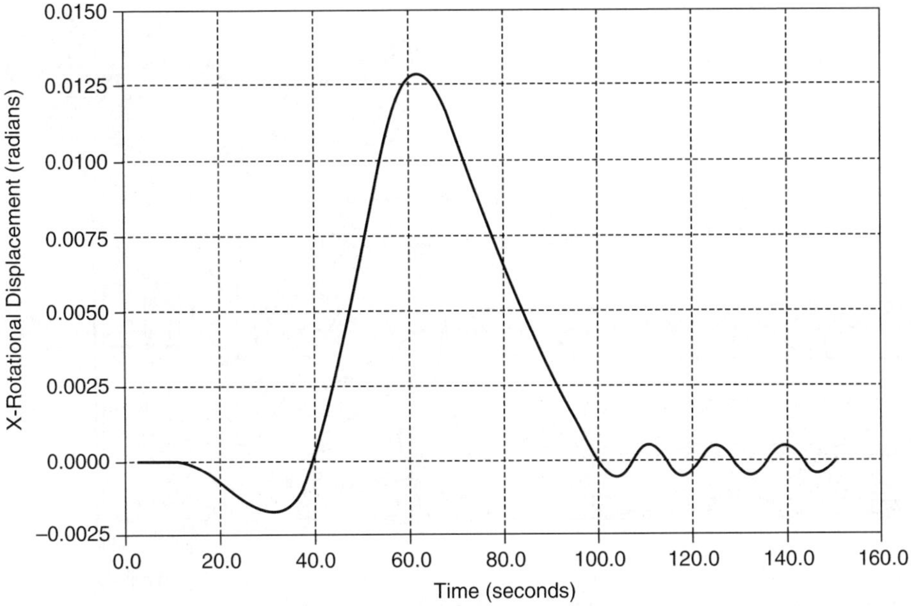

Figure 6.5 Angle of Twist at the End of the Inner Truss During the Slewing Maneuver. Banerjee and Lemak [1]. Reproduced with permission of the American Society of Mechanical Engineers.

control-structure interaction corresponding to various control law designs for meeting a specified control objective. Finally, an animation of the motion of the crane during the slewing maneuver is shown in Figure 6.6, where the structure, initially in the plane of the maneuver, comes out of the plane during the maneuver. Before leaving this chapter, we note that flexible multibody dynamics has been the subject of several public domain software [12, 13] and research by several authors. Later, Meirovitch and Temple [14] analyzed the problem in terms of quasi-coordinates and partial differential equations but did not give any solutions thereof. None of these references consider motion-induced stiffness effects, and thus would predict dynamic softening when stiffening is physically expected, and vice versa.

Problem Set 6

6.1 Develop the multibody equations for two initially colinear and overlying beams connected by massive revolute joints, with the system tumbling at a high altitude under gravity, and an initial small angle, 5 degrees, between the beams increasing to 90 degrees, due to aerodynamic force. Use free-free bending modes for one and free-hinged modes for the other [15],

$$\text{free-free } \phi_i^{(1)}(x) = \cosh\left(\frac{\lambda_i x}{L}\right) + \cos\left(\frac{\lambda_i x}{L}\right) - o_i\left[\sinh\left(\frac{\lambda_i x}{L}\right) + \sin\left(\frac{\lambda_i x}{L}\right)\right]$$

$$o_1 = 0.9825;\ \lambda_1 = 4.7300, o_1 = 1.0008;\ \lambda_2 = 7.8352;\ o_3 = 0.9999;\ \lambda_3 = 10.9956;$$

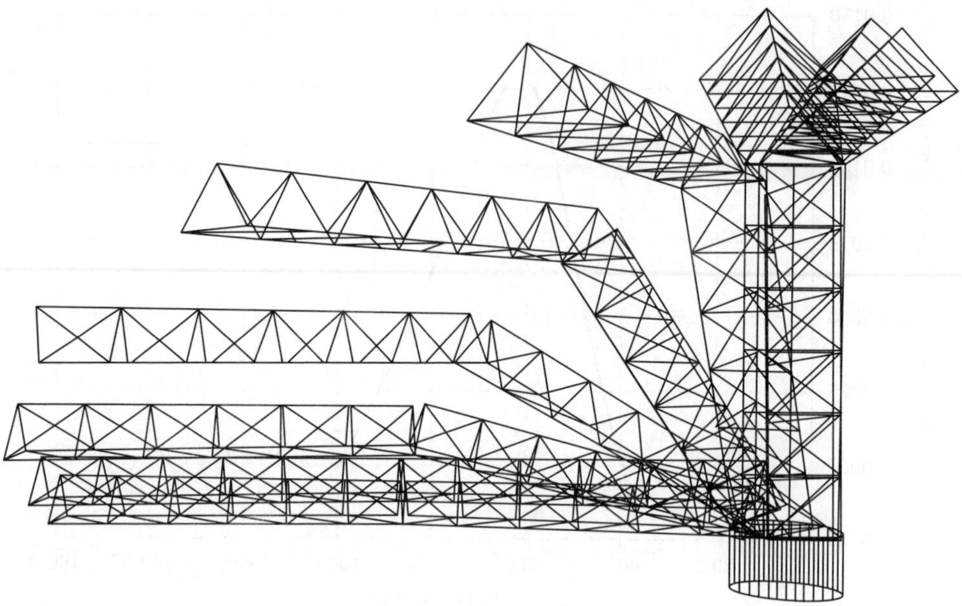

Figure 6.6 Animation of the Slewing Maneuver of the Space Crane. Banerjee and Lemak [1]. Reproduced with permission of the American Society of Mechanical Engineers.

$$\text{free-hinged } \phi_i^{(1)}(x) = \cosh\left(\frac{\lambda_i x}{L}\right) + \cos\left(\frac{\lambda_i x}{L}\right) - o_i\left[\sinh\left(\frac{\lambda_i x}{L}\right) + \sin\left(\frac{\lambda_i x}{L}\right)\right];$$

$$o_1 = 1.0008;\ \lambda_1 = 3.9266,\ o_1 = 1.0000;\ \lambda_2 = 7.0685; o_3 = 1.0000;\ \lambda_3 = 10..2102.$$

6.2 Code the equations and use numerical values of your choice for a simulation for a uniform pressure load on both beams, and plot the free end deflections from 5 degree to the 90 degree for the two beams.

6.3 Figure 6.7 shows the space station Canada Arm manipulator [15]. Do this exercise departing from the details of this figure. Assume that one end of the manipulator is attached to a heavy object on the ground. The first link is short and rotates in "roll" with respect to the attaching base; the second link is also short and rotates in "yaw" with respect to its inboard body. These two links, of same length, can be modeled as rigid. Links 3 and 4 are both flexible beams, of same length, and they both rotate in "pitch." Link 5 is a short one of same length as link 1, modeled as rigid, and rotates in yaw. Finally, link 6 is a long beam of same length as link 3; it rotates in roll, and attached to a rigid body called the "end-effector." Derive the equations of motion of this 6 dof robotic manipulator, ignoring all geometric stiffness terms. Use parameter values of your choice to simulate the motion, keeping one mode per beam.

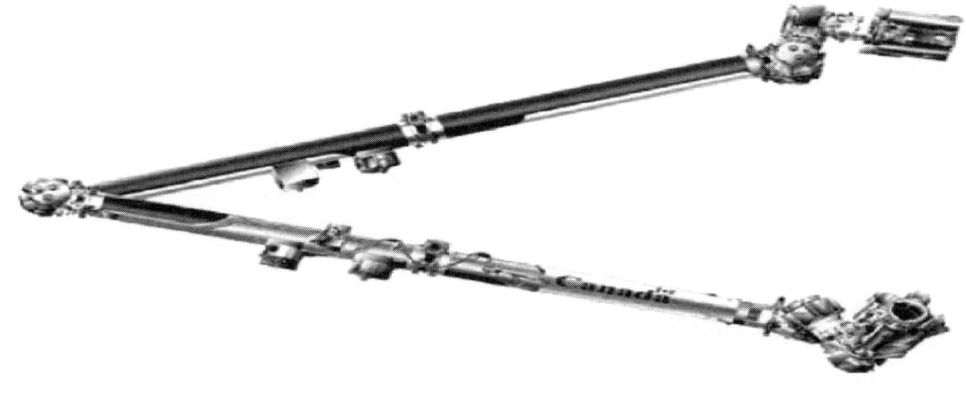

Figure 6.7 A 6 dof Robotic Manipulator.

References

[1] Banerjee, A.K. and Lemak, M.E. (1991) Multi-flexible body dynamics capturing motion-induced stiffness. *Journal of Applied Mechanics*, **58**, 766–775.

[2] Huston, R.L. and Passerello, C.E. (1980) Multibody structural dynamics including translation between bodies. *Computers and Structures*, **12**, 713–720.

[3] Kane, T.R. and Levinson. D.A. (1985) *Dynamics: Theory and Application*, McGraw-Hill.

[4] Craig, R.R. Jr. (1981) *Structural Dynamics, An Introduction to Computer Methods*, John Wiley & Sons, Inc.

[5] Cook, R.D. (1974) *Concepts and Applications of Finite Element Analysis*, John Wiley & Sons, Inc.

[6] Zienkiewicz, O.C. (1977), *The Finte Element Method*, McGraw-Hill.

[7] Whetstone, W.D. (1983) *EISI-EAL, Engineering Analysis Language Reference Manual*, Engineering Information Systems, Inc.

[8] Sincarsin, G.B. and Hughes, P.C. (1989) Dynamics of an elastic multibody chain: Part-A- body motion equations. *Dynamics and Stability of Systems*, **3**, 209–226.

[9] Padilla, C.E. and von Flotow, A.H. (1992) Nonlinear strain-displacement relations and flexible multibody dynamics. *Journal of Guidance, Control, and Dynamics*, **15**(1), 128–136.

[10] Sharf, I. and Damaren, C. (1992) Simulation of Flexible-Link Manipulators: Basis Functions and Nonlinear Terms in the Motion Equations. Proceedings of the 1992 IEEE International Conference on Robotics and Automation, Nice, France, May 1992.

[11] Ghosh, T.K. (2013) Correspondence on Simulation of Space Robotic Manipulators at NASA Johnson Space Center.

[12] Bodley, C.S., Devers, A.D., Park, A.C., and Frisch, H.P. (1978) A Digital Computer Program for the Dynamic Interaction Simulation of Controls and Structures (DISCOS). vols. **I** and II, NASA TP-1219.

[13] Singh, R.P., van der Voort, R.J., and Likins, P.W. (1985) Dynamics of flexible bodies in tree topology – A computer oriented approach. *Journal of Guidance, Control, and Dynamics*, **8**(5), 584–590.

[14] Meirovitch, L. and Stemple, T. (1995) Hybrid equations of motion for flexible multibody systems using quasicoordinates. *Journal of Guidance, Control, and Dynamics*, **18**(4), 678–688.

[15] Blevins, R.D. (2001) *Formulas for Natural Frequency and Mode Shape*, Krieger Publishing, Florida.

References

7

Component Mode Selection and Model Reduction: A Review

We review in this chapter selection of modes and model reduction for a component of a flexible multibody system, to represent the deformations of the component with good fidelity using as few modes as possible. We must note that for a multibody system with hinge degrees of freedom locked or driven, vibration modes of the combined system change with the relative change in configuration between components, and thus we can only speak of the modes for the component. While the higher number of modes always models deformations more accurately, this also makes the system model more expensive in simulation or for use in online control. Model reduction by modal truncation is thus a necessary step for time-efficient simulation. This, of course, entails errors in representing deformation, and we will show a way of compensating for errors due to modal truncation. Our main reference for component mode selection is Craig [1], which is complemented by Wijker [2]. We will first discuss two widely used methods of model representation, namely Craig-Bampton method and the method of static condensation. In model reduction we will discuss Guyan reduction [3] and modal effective mass methods. Modal truncation vectors will be presented as a catch-all tool of compensation for modal truncation. Next we will review synthesis of component modes for a multibody system. Further model reduction, after a system model is synthesized, will be discussed in terms of singular value decomposition, which we will illustrate with an example.

7.1 Craig-Bampton Component Modes for Constrained Flexible Bodies

A physical component is typically discretized into many nodes, each node having six physical dof. Using the notation u for elastic displacement, not generalized speeds as in previous chapters, the n physical degrees of freedom (dof) of a component can always be divided into a set of interior dof, u_i, and a set of juncture or "boundary" dof, u_b, when the component is part

Flexible Multibody Dynamics: Efficient Formulations and Applications, First Edition. Arun K. Banerjee.

of a multibody system. The equations of free vibration of a component are written in terms of its mass and stiffness matrices obtained from a finite element model, in the form:

$$\begin{bmatrix} m_{ii} & m_{ib} \\ m_{ib}^t & m_{bb} \end{bmatrix} \begin{Bmatrix} \ddot{u}_i \\ \ddot{u}_b \end{Bmatrix} + \begin{bmatrix} k_{ii} & k_{ib} \\ k_{ib}^t & k_{bb} \end{bmatrix} \begin{Bmatrix} u_i \\ u_b \end{Bmatrix} = \begin{Bmatrix} 0 \\ 0 \end{Bmatrix}, \text{ or } M\ddot{u} + Ku = 0 \tag{7.1}$$

These n physical dof u are reduced by a Rayleigh-Ritz approximation by m number of modes as,

$$u = \psi q \tag{7.2}$$

where ψ, is a $(n \times m)$ matrix of m "component modes" and q is a $(m \times 1)$ matrix of "modal coordinates." Component modes can be of various kinds, like fixed-fixed or fixed-free vibration modes, rigid body modes, and the so-called constraint modes and attachment modes, described below [1]. First, however, we review the definition of **vibration modes** of the component represented by Eq. (7.1) in terms of an eigenvalue problem for the form set up from Eq. (7.1) in the following operational sequence of steps:

$$u = \phi\eta$$

$$M\phi\ddot{\eta} + K\phi\eta = 0$$

$$\phi^t M\phi\ddot{\eta} + \phi^t K\phi\eta = 0$$

$$\text{Let } \hat{M} = \phi^t M\phi; \quad \hat{K} = \phi^t K\phi \tag{7.3a}$$

$$\hat{M}\ddot{\eta} + \hat{K}\eta = 0$$

$$\text{Let } \eta = \hat{\phi}\sin\omega t$$

$$(\hat{K} - \omega^2\hat{M})\hat{\phi}\sin\omega t = 0$$

The last statement defines the following algebraic eigenvalue problem:

$$(\hat{K} - \omega^2\hat{M})\hat{\phi} = 0 \tag{7.3b}$$

The eigenvalues are the squares of the natural frequencies, $\omega_i^2, i = 1, \ldots, n$, and are obtained from the above by solving the determinant equation,

$$\det(\hat{K} - \omega^2\hat{M}) = 0 \tag{7.3c}$$

for non-zero values of the modes; the modes are then "normalized", or scaled, to the element masses $m_i, i = 1, \cdots, n$, as follows:

$$\hat{\phi}^t M\hat{\phi} = \hat{M} = diag[m_1, m_2, \ldots, m_n]; \quad \hat{\phi}^t K\hat{\phi} = \hat{K} = diag\left[\omega_1^2, \omega_2^2, \ldots, \omega_n^2\right] \tag{7.4}$$

Usually one keeps only a few number, k, of the modes. In most flexible multibody dynamics representations, component modes consist of a mixture of fixed-interface vibration modes together with a set of **constraint modes**. The latter, also called in the industry as **boundary node functions**, are defined as follows [1]:

Let the elastic dof u of the component be partitioned into a set B, usually the boundary dof, in terms of which constraint modes are to be defined, and let I be the *complement* of the set B, usually the interior dof. A constraint mode or boundary node function is defined as the static deformation shape that results by imposing unit displacement on one boundary coordinate of the B set at a time, while holding the remaining coordinates of the B set fixed to zero displacement. From this definition, the set of constraint modes satisfies the matrix equation:

$$\begin{bmatrix} k_{ii} & k_{ib} \\ k_{ib}^t & k_{bb} \end{bmatrix} \begin{bmatrix} \Psi_{ib} \\ I_{bb} \end{bmatrix} = \begin{bmatrix} 0_{ib} \\ R_{bb} \end{bmatrix} \tag{7.5}$$

Here I_{bb} is a bxb unity matrix, b being the number of elements in the set B, and R_{bb} is the set of reaction forces, at the dof of the B set, brought into play by prescribing unit displacements in the dofs of the B set, in the manner described above. Solving from the top row gives:

$$\Psi_{ib} = -k_{ii}^{-1} k_{ib} \tag{7.6}$$

The set of "constraint modes" is then completely defined as:

$$\Psi_c = \begin{bmatrix} \Psi_{ib} \\ I_{bb} \end{bmatrix} = \begin{bmatrix} -k_{ii}^{-1} k_{ib} \\ I_{bb} \end{bmatrix} \tag{7.7}$$

Most flexible multibody dynamics component mode representations use a set of normal modes augmented by a set of constraint modes. When the component is an intermediate body between two outlying bodies, the normal modes are for boundary condition with both ends fixed, and the constraint modes use six unit displacements for each end. For a terminal body, cantilever modes for interior points, with respect to the attaching body, are augmented by constrained modes generated by six constraint modes from the attaching body. Craig-Bampton modes for the component can now be formed by augmenting a truncated set of fixed-interface normal vibration modes with the constraint modes, or boundary node functions, associated with unit boundary displacements. Thus the physical dof of the component, described in this manner, are called **Craig-Bampton modes**, and are represented as:

$$\begin{Bmatrix} u_i \\ u_b \end{Bmatrix} = \begin{bmatrix} \phi_k & -k_{ii}^{-1} k_{ib} \\ 0 & I_{bb} \end{bmatrix} \begin{Bmatrix} \eta_k \\ u_b \end{Bmatrix} \tag{7.8}$$

An alternative to the use of Craig-Bampton modes is **attachment modes**, which are described in Ref. [1]. Two other methods, component modes by Guyan reduction and modal effective mass, are discussed in the next section. In the author's experience with two major aerospace organizations, flexible multibody dynamics codes use Craig-Bampton modes more commonly because of their accounting of component boundary conditions.

7.2 Component Modes by Guyan Reduction

Guyan reduction [3], also known as **static condensation**, is a method of obtaining component modes when inertia effects of certain dof can be ignored. Consider a model with given loading:

$$\begin{bmatrix} m_{aa} & m_{ae} \\ m_{ea} & m_{ee} \end{bmatrix} \begin{Bmatrix} \ddot{u}_a \\ \ddot{u}_e \end{Bmatrix} + \begin{bmatrix} k_{aa} & k_{ae} \\ k_{ea} & k_{ee} \end{bmatrix} \begin{Bmatrix} u_a \\ u_e \end{Bmatrix} = \begin{Bmatrix} F_a \\ F_e \end{Bmatrix} \tag{7.9}$$

Here the subscript a refers to "active" dof, and the subscript e refers to to-be-eliminated dof that may be assumed as inactive, motion-wise, in that they do not respond significantly to the loading. The choice of $u_a, u_e$is then solely based on the assumption that the inertia loads $m_{aa}\ddot{u}_a$are significantly larger than the other inertia loads, and the load F_e in Eq. (7.9) is small relative to F_a. The implied approximation to the above equation is then:

$$\begin{bmatrix} m_{aa} & m_{ae} \\ 0 & 0 \end{bmatrix} \begin{Bmatrix} \ddot{u}_a \\ \ddot{u}_e \end{Bmatrix} + \begin{bmatrix} k_{aa} & k_{ae} \\ k_{ea} & k_{ee} \end{bmatrix} \begin{Bmatrix} u_a \\ u_e \end{Bmatrix} = \begin{Bmatrix} F_a \\ 0 \end{Bmatrix} \tag{7.10}$$

From the bottom row we have:

$$u_e = -k_{ee}^{-1} k_{ea} u_a \tag{7.11}$$

This yields the reduction transformation for the overall dof:

$$\begin{Bmatrix} u_a \\ u_e \end{Bmatrix} = \begin{bmatrix} I \\ -k_{ee}^{-1} k_{ea} \end{bmatrix} \{u_a\} \equiv T u_a \tag{7.12}$$

Substituting Eq. (7.12) in the force-free vibration equation corresponding to Eq. (7.9), written as $M\ddot{u} + Ku = 0$, and pre-multiplying by T^T, where T is defined in Eq. (7.12), yields:

$$T^T M T \ddot{u}_a + T^T K T u_a = 0 \tag{7.13}$$

One can compute eigenvalues and eigenvectors corresponding to Eq. (7.13) with:

$$\begin{aligned} u_a &= \phi_a \sin \omega t \\ [T^T M T - \omega_a^2 T^T K T]\phi_a &= 0 \end{aligned} \tag{7.14}$$

Here $\omega_a, \phi_a$are, respectively, the natural frequency and mode shape for the active dof. From Eqs. (7.12) and (7.14) then, we have the component mode shapes from static condensation,

$$\begin{bmatrix} \phi_a \\ \phi_e \end{bmatrix} = \begin{bmatrix} I \\ -k_{ee}^{-1} k_{ea} \end{bmatrix} [\phi_a] \tag{7.15}$$

or, in more compact notation for all dof with $u = \psi q$,

$$\psi = T\phi_a \tag{7.16}$$

7.3 Modal Effective Mass

Modal effective mass [2] is a measure to assign importance to individual modes from the viewpoint of excitability of structures with base acceleration. A mode with a high modal effective mass means a high transmitted force at the base, due to that mode. This then becomes a basis for reducing the number of modes. Any component structure can be thought of as having two sets of degrees of freedom (dof): a set of boundary dof and the internal elastic dof. For the purpose of computing modal effective mass, the boundary dof are given rigid body motion, one at a time for a maximum of six in number, and internal elastic deformations are obtained by keeping all other juncture dof set to zero. With u_j, u_e denoting juncture and elastic dof,

$$u = \phi_r u_j + \phi_e u_e \tag{7.17}$$

$$\text{For rigid body modes, } \phi_r^T M \phi_r = M_{rr} \tag{7.18}$$

Here M is the $n \times n$ nodal mass matrix for the component, and M_{rr} is the 6×6 rigid body matrix, of mass m, mass moment, a 3×1 matrix s with $\tilde{s}$, a skew-symmetric matrix formed out of its elements, and mass moment of inertia I, a 3×3 matrix, both about the about the boundary dof:

$$M_{rr} = \begin{bmatrix} mU & -\tilde{s} \\ \tilde{s} & I \end{bmatrix} \tag{7.19}$$

On the other hand, if the kth elastic mode is orthogonalized by mass, then:

$$\phi_k^t M \phi_k = m_k \tag{7.20}$$

Now *define* the 1×6 row matrix L_k, rigid-elastic mode coupling vector, for the kth mode:

$$L_k = \phi_k^t M \phi_r \tag{7.21}$$

Modal effective mass for the kth elastic mode is then defined as the following 6×6 matrix:

$$M_{\text{eff},k} = \frac{L_k^t L_k}{m_k} \tag{7.22}$$

The goal of model reduction, keeping p number of modes, with $p < n$, is that:

$$M_{rr} \cong \sum_{k=1}^{p} M_{\text{eff},k} \tag{7.23}$$

This is true exactly if p is the number n of all possible modes, because then from Eqs. (7.20)–(7.22):

$$M_{\text{eff}} = [\phi_r^t M \phi_k]\,[\phi_k^t M \phi_k]^{-1}\,[\phi_k^t M \phi_r] = [\phi_r^t M \phi_k]\,[\phi_k^{-1} M^{-1} \phi_k^{-t} \phi_k^t M \phi_r = \phi_r^t M \phi_r = M_{rr} \tag{7.24}$$

A mode with a high modal effective mass is excited strongly by base motion, giving rise to a high reaction force at the base. This is because if $\ddot{u}_b$ represents a 6×1 column matrix of unit base acceleration in the 6 base dof, then $(-M_{rr}\ddot{u}_b)$ represents the reaction force transmitted to the base providing the base acceleration. Thus any mode that contributes strongly to the sum in Eq. (7.23) contributes strongly to the transmitted force, and has to be kept in a reduced-order model, while a mode with low modal effective mass can be eliminated for a multibody system.

7.4 Component Model Reduction by Frequency Filtering

Another technique of model reduction, reported by Ghosh [4], is based on the highest frequency of a component that is consistent with a desirable time step for stable numerical integration. As a rule of thumb, 10 integration steps are used to track a harmonic of period T. With $T = 2\pi/\omega$, this allows a highest frequency of vibration for a period of step-size, dt, as:

$$\omega_{\text{max}} = (2\pi/dt)/10 \tag{7.25}$$

For a forced vibration model described by the sequence of steps,

$$\begin{aligned}
&[M]\{\ddot{u}\} + [K]\{u\} = [F]\{g(t)\}\\
&\{u\} = [\Phi]\{\eta\}\\
&[\Phi]^t[M][\Phi]\{\ddot{\eta}\} + [\Phi]^t[K][\Phi]\{\eta\} = [\Phi]^t[F]\{g(t)\}\\
&\{\ddot{\eta}\} + [\Omega^2]\{\eta\} = [\Phi]^t[F]\{g(t)\}
\end{aligned} \tag{7.26}$$

the ith component mode equation follows, after keeping n modes and adding viscous damping, as:

$$\ddot{\eta}_i + 2\xi_i\omega_i\dot{\eta}_i + \omega_i^2\eta_i = \phi_i^t[F]\{g(t)\} \qquad i = 1, \ldots., n \tag{7.27}$$

If $\omega_i \leq \omega_{\text{max}}$, then Eq. (7.27) is used. If $\omega_i \geq \omega_{\text{max}}$, then the frequency-filtering technique is turned on by using, instead of Eq. (7.27), the following:

$$\ddot{\eta}_i + 2\xi_i\dot{\eta}_i + \omega_{\text{max}}^2\eta_i = \left(\frac{\omega_{\text{max}}^2}{\omega_i^2}\right)\phi_i^t[F]\{g(t)\} \qquad i = 1, \ldots., n \tag{7.28}$$

This method has been found satisfactory in the Canada Arm Manipulator dynamics simulation.

7.5 Compensation for Errors due to Model Reduction by Modal Truncation Vectors

Use of the model reduction given by the Craig-Bampton modes, or the attachment modes, or static condensation of the component model equations, or the modal effective mass method requires the user to decide how many vibration modes to keep. Modal truncation causes error due to incomplete representation of deformation, and a scheme for some kind of compensation for modes not kept was devised by Dickens [5]. To see how the method works, consider the basic problem of small vibration under space- and time-dependent loading:

$$M\ddot{u} + Ku = F(x, y, z)g(t) \tag{7.29}$$

Here u refers to the internal dof of a component body, with its boundary dof locked; when the body is a terminal body, only its attachment dof are locked. In Eq. (7.29), $F(x,y,z)$ is a matrix representing spatial distribution of the load and $g(t)$ a column matrix of time-functions. In terms of component modes ψ and a modal coordinate ξ, the displacement is represented as:

$$u = \psi\xi \tag{7.30}$$

Recall that the associated eigenvalue problem for r number of eigenvalues and eigenvectors is given by the following, where the second equality is due to the diagonal nature of Ω^2:

$$K\psi = \Omega^2 M\psi = M\psi\Omega^2; \quad \psi = \left[\psi_1 \cdots\cdots \psi_r\right]; \quad \Omega^2 = diag\left[\omega_1^2, \ldots., \omega_r^2\right] \tag{7.31}$$

Now customarily one keeps component frequencies up to twice that of the system frequency cutoff of interest. Model reduction brings in error for not keeping all the modes. Dickens [5] devised a scheme for dynamic compensation for the modes **not** kept. The scheme uses a deflection shape to represent deformation due to the spatial part of the residual load, which is **not** represented by the modes that are kept. The idea is, the spatial distribution of the load should be representable, in the sense of a Fourier series, by a sum of modes times the modal participation factors of those modes. The spatial distribution of the force that *can* be represented by the modes retained is given by the following sequence of computational steps:

Spatial load representable by the deformation Eq. (7.30) is given from statics by:

$$F = K\psi\xi \tag{7.32}$$

From the eigenvalue solver, Eq. (7.31), Eq. (7.32) becomes:

$$F = M\psi\Omega^2\xi. \tag{7.33}$$

Pre-multiplying by ψ^T gives:

$$\psi^T F = \psi^T M\psi\Omega^2\xi \tag{7.34}$$

Assuming that the modes are normalized (scaled) by the mass of the structure, $\psi^T M\psi = mU$, where m is the total mass of the structure and U is a unity matrix of size given by the number of modes kept, one obtains from Eq. (7.34):

$$\psi^T F = m\Omega^2\xi \tag{7.35}$$

Equation (7.35) leads to the modal participation factor in the spatial decomposition of F:

$$\xi = \left[m\Omega^2\right]^{-1} \psi^T F \tag{7.36}$$

Therefore, the spatial load represented by the modes kept follows from Eqs. (7.33) and (7.36), with subscript k indicating the contribution of the modes kept:

$$F_k = M\psi\Omega^2 \left[m\Omega^2\right]^{-1} \psi^T F = M\psi m^{-1}\psi^T F \tag{7.37}$$

Hence the spatial distribution of the load F that **cannot** be represented by the number of modes kept, and the corresponding static deflection is given from Eq. (7.37), in sequence, by:

$$\begin{aligned} F_{\text{res}} &= [I - M\psi m^{-1}\psi^T]F \\ Ku_{\text{res}} &= F_{\text{res}} \end{aligned} \tag{7.38}$$

Dickens [5] calls this displacement vector, u_{res}, the modal truncation vector associated with the spatial distribution load F. In large overall motion of flexible bodies, Eq. (5.16) in Chapter 5 reveals that one flexible body is subjected to a set of 12 distributed inertia loads from just the rigid body or frame rotation/translation – that is, F, u_{res} are each matrices with 12 columns [10]. Now the spatial distribution of the *inertia loads* on a component undergoing large motion of the reference frame, identified in Eq. (5.16), is specified as:

$$\begin{Bmatrix} f_1^* \\ f_2^* \\ f_3^* \end{Bmatrix} = -dm \begin{bmatrix} U_3 & x_1U_3 & x_2U_3 & x_3U_3 \end{bmatrix} \begin{Bmatrix} A_1 \\ \vdots \\ \vdots \\ A_{12} \end{Bmatrix} \tag{7.39}$$

Here U_3 is the 3×3 unity matrix and $A_1, \ldots., A_{12}$ are the time-varying components of the acceleration at a node of mass dm, located at the point with coordinates x_1, x_2, x_3 from the origin of the body coordinates. The effects of all external and working internal forces for the system are equivalently represented, to a zeroth order, by the rigid body inertia forces. Note that here we have neglected rotational inertia effects, which would change the total number of loads from 12 to 21, described in Chapter 5. This requires the generation of 12 modal truncation vectors, and we want these vectors to be orthogonal between themselves and to the normal vibration modes. This is achieved by the following sequence of steps, including a construction of a 12×12 eigenvalue problem for the modal truncation vectors:

$$\begin{aligned} M^* &= u_{\text{res}}^T M u_{\text{res}} \\ K^* &= u_{\text{res}}^T K u_{\text{res}} \\ K^*\Phi &= M^*\Phi\Omega_{\text{MTV}}^2 \\ \Psi_{\text{MTV}} &= u_{\text{res}}\Phi \end{aligned} \tag{7.40}$$

Here $\Psi_{\text{MTV}}, \Omega_{\text{MTV}}$ are called, respectively, the modal truncation vectors and the "frequencies" associated with these modal truncation vectors. Note M and K are the original finite element model mass and stiffness matrices, and the dimensions of u_{RES}, Φ are ndof $\times$ 12 and 12×12, respectively.

Now the modal basis is extended by adding the modal truncation vectors to the normal vibration modes to get a more complete modal basis that compensates for the deleted modes, in two sets of column matrices,

$$[\Psi] = \begin{bmatrix} \psi & \psi_{MTV} \end{bmatrix} \tag{7.41}$$

and this modal basis $[\Psi]$ is used to apply the mode *displacement* solution. Note that these very loads as shown in Eq. (7.39) are also used in computing load-dependent geometric stiffness required to counteract the effect of premature linearization endemic with the use of modes [6]. Thus a typical component model should ideally consist of a chosen number of vibration modes plus the modal truncation vectors corresponding to the spatial loads represented by Eq. (7.39). Of course, in actual practice, depending on the strength of the acceleration terms in $A_1, \ldots\ldots, A_{12}$, one can reduce the total number of the modal truncation vectors.

7.6 Role of Modal Truncation Vectors in Response Analysis

Dickens et al. [7] have shown the superiority in frequency response over a wide range of frequencies, of augmentation of the modal basis by modal truncation vectors as in Eq. (7.41), over the modal acceleration method. This is significant because the mode acceleration method [2] gives a response that converges as $(1/\omega_i^2)$ with ith frequency modes, with higher modes contributing less, as seen below:

$$\begin{aligned} M\ddot{u} + Ku &= F(x, y, z)g(t) \\ u &= K^{-1}[F(x, y, z)g(t) - M\ddot{u}] \\ &= K^{-1}F(x, y, z)g(t) - K^{-1}\sum_{i=1}^{N} M\phi_i\ddot{\eta}_i \\ &= K^{-1}F(x, y, z)g(t) - \sum_{i=1}^{N} \frac{1}{\omega_i^2}\phi_i\ddot{\eta}_i \end{aligned} \tag{7.42}$$

where $\ddot{\eta}_i + \omega_i^2\eta_i = \phi_i^T F(x, y, z)g(t)$.

Note that the definition of the eigenvalues, Eq. (7.3a), was used here in the summation term, and the mode-displacement solution to the forced vibration problem was used in the last step.

Dickens [7] shows response analysis with modal truncation vectors by the following example in Fig. 7.1.

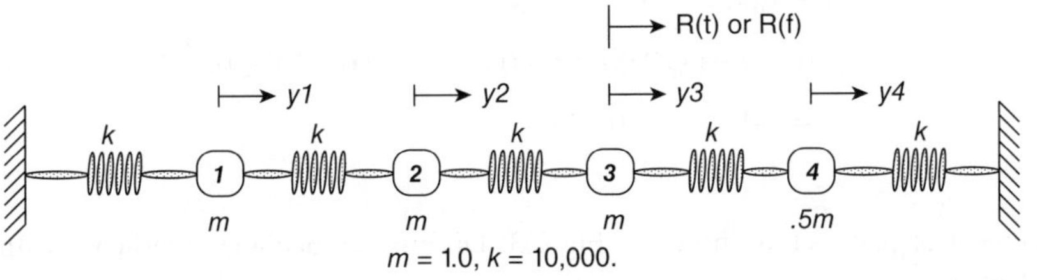

Figure 7.1 Simple Spring-Mass System with Space- and Time-dependent Force on Mass #3.

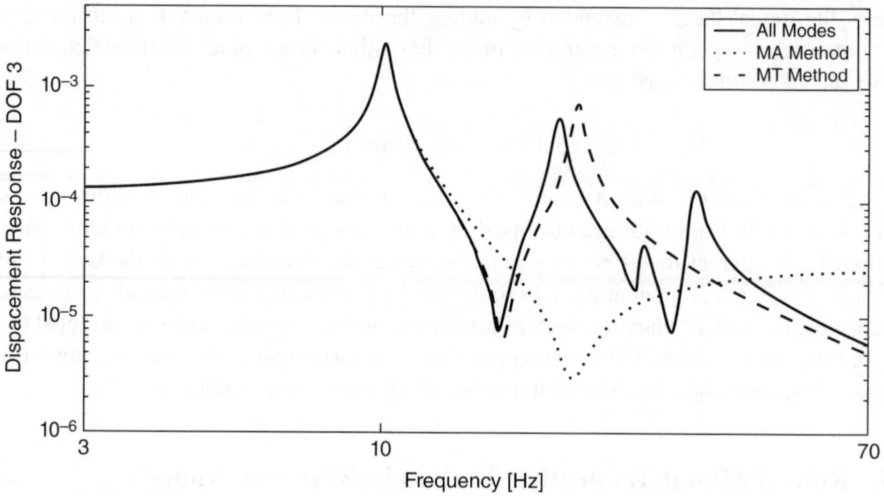

Figure 7.2 Frequency Response of the System in Figure 7.1, with Assumed 2% Damping.

$$M = \begin{bmatrix} 1.0 & 0 & 0 & 0 \\ 0 & 1.0 & 0 & 0 \\ 0 & 0 & 1.0 & 0 \\ 0 & 0 & 0 & 0.5 \end{bmatrix} K = 10000.0 \begin{bmatrix} 2.0 & -1.0 & 0 & 0 \\ -1.0 & 2.0 & -1.0 & 0 \\ 0 & -1.0 & 2.0 & -1.0 \\ 0 & 0 & -1.0 & 2.0 \end{bmatrix}$$

$$\psi_1 = \begin{Bmatrix} 0.39948 \\ 0.63631 \\ 0.61408 \\ 0.34183 \end{Bmatrix} \psi_{\text{MTV}} = \begin{Bmatrix} -0.53176 \\ -0.41943 \\ 0.71885 \\ 0.22164 \end{Bmatrix} \quad R(t) = \begin{Bmatrix} 0.0 \\ 0.0 \\ 1.0 \\ 0.0 \end{Bmatrix} r(t); \text{[}R(t) \text{ on } y3; \text{ see Figure 7.1]}$$

Frequency response of the system, by keeping all modes, modal acceleration (MA) method, and the modal truncation (MT) vector method, is shown in Fig. 7.2.

For a loading applied on mass # 3, $R(t)$, shown before, the time-dependent function is:

$$\begin{aligned} r(t) &= 1.00 \qquad \leq t \leq 0.03 \\ &= 1.0 - (1/0.003)(t - 0.03) \qquad 0.03 \leq t \leq 0.033 \\ &= 0.0 \qquad t > 0.033 \end{aligned}$$

Displacement at mass #1 is shown in Fig 7.3. Differences become significant with higher modal density.

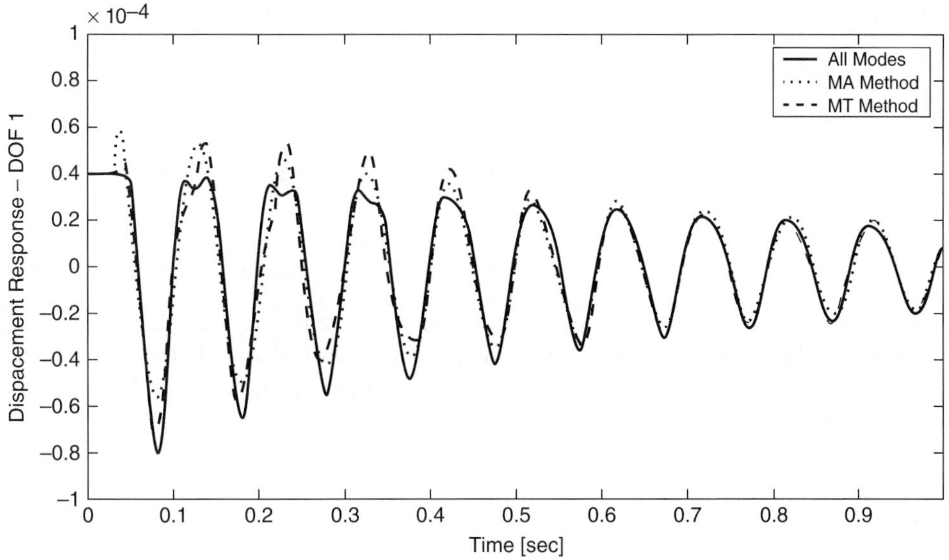

Figure 7.3 Time Response of the System at Mass #1 with Loading at Mass #3.

7.7 Component Mode Synthesis to Form System Modes

This section follows the work of Craig described in Ref. [1]. Let the system be composed of two components that have a common interface, with sets of displacement degrees of freedom, u_1, u_2. Kane's equations with undetermined multipliers, as discussed in Chapter 1, for small vibration (i.e., no large overall motion) can be shown to be given by the following dynamical equations and the common interface displacement continuity constraint:

$$\begin{bmatrix} M_1 & 0 \\ 0 & M_2 \end{bmatrix} \begin{Bmatrix} \ddot{u}_1 \\ \ddot{u}_2 \end{Bmatrix} + \begin{bmatrix} K_1 & 0 \\ 0 & K_2 \end{bmatrix} \begin{Bmatrix} u_1 \\ u_2 \end{Bmatrix} = \begin{bmatrix} C_1^T \\ C_2^T \end{bmatrix} \lambda$$
$$\begin{bmatrix} C_1 & C_2 \end{bmatrix} \begin{Bmatrix} \ddot{u}_1 \\ \ddot{u}_2 \end{Bmatrix} = \{0\} \tag{7.43}$$

Let the physical displacements be represented by the Craig-Bampton component modes:

$$\begin{aligned} u_1 &= \psi_1 q_1 \\ u_2 &= \psi_2 q_2 \end{aligned} \tag{7.44}$$

Let the component modes be normalized such that:

$$\begin{aligned} \mu_1 &= \psi_1^T M_1 \psi_1 \quad & \kappa_1 &= \psi_1^T K_1 \psi_1 \\ \mu_2 &= \psi_2^T M_2 \psi_2 \quad & \kappa_2 &= \psi_2^T K_2 \psi_2 \end{aligned} \tag{7.45}$$

Defining

$$p = \begin{Bmatrix} q_1 \\ q_2 \end{Bmatrix}; \quad \mu = \begin{bmatrix} \mu_1 & 0 \\ 0 & \mu_1 \end{bmatrix}; \quad \kappa = \begin{bmatrix} \kappa_1 & 0 \\ 0 & \kappa_1 \end{bmatrix}; \quad c = [C_1\psi_1 C_2\psi_2] \tag{7.46}$$

enables one to write the dynamical and constraint equations of Eq. (7.43) as follows:

$$\begin{aligned} \mu\ddot{p} + \kappa p &= c^T\sigma \\ c\ddot{p} &= 0 \end{aligned} \tag{7.47}$$

Now partition the set of coordinates in p as a set of coordinates p_i that are independent and the complement of the set p_i as dependent dof, p_d. The constraint equation in Eq. (7.47) can then be written as:

$$\begin{bmatrix} c_{dd} & c_{di} \end{bmatrix} \begin{Bmatrix} p_d \\ p_i \end{Bmatrix} = 0 \tag{7.48}$$

This is always possible because of the constraint equation in Eq. (7.48) expressing the fact that there is a common interface between the structures. Now define a reduction transformation:

$$p = \begin{Bmatrix} p_d \\ p_i \end{Bmatrix} = \begin{bmatrix} -c_{dd}^{-1}c_{di} \\ I_{ii} \end{bmatrix} \{p_i\} \equiv Sp_i \equiv S\eta \tag{7.49}$$

Now the dynamical equation for the *system* in Eq. (7.43) becomes

$$M\ddot{\eta} + K\eta = 0 \tag{7.50}$$

where

$$M = S^T\mu S; \quad K = S^T\kappa S \tag{7.51}$$

The right-hand-side zero in Eq. (7.47) is explained by the fact that:

$$cS \equiv \begin{bmatrix} c_{dd} & c_{di} \end{bmatrix} \begin{Bmatrix} -c_{dd}^{-1}c_{di} \\ I_{ii} \end{Bmatrix} = 0 \tag{7.52}$$

When there are more than two components involved, modal synthesis for the system modes can be done by extending the same idea, by considering how the components are attached to each other. In articulated robot dynamics formulations with possibly nonlinear constraints on the motion, the above approach is generalized as follows, because the formulations end in equations of the form:

$$\begin{aligned} &[M(q)]\,\ddot{q} = C^T(q,t)\lambda + G(q,\dot{q},t) \\ &[C(q,t)]\dot{q} = B(q,t) \quad \Rightarrow \quad [C(q,t)]\ddot{q} + \dot{C}\dot{q} = \dot{B} \quad \Rightarrow \quad [C(q,t)]\ddot{q} = D(q,\dot{q},t) \end{aligned} \tag{7.53}$$

Here the first equation is that of dynamics and the second string expresses the constraints expressed in the velocity form, differentiated to get the acceleration form, of the constraint equations. Eliminating the constraint forces (and torques) λ, one arrives at the final form of the equations of motion (suppressing the arguments shown above):

$$M\ddot{q} = C^T \left[CM^{-1}C^T\right]^{-1} \left\langle D - CM^{-1}G\right\rangle + G \tag{7.54}$$

7.8 Flexible Body Model Reduction by Singular Value Decomposition of Projected System Modes

Further reduction of component modes is possible after doing component mode synthesis. One approach, based on load transformation matrices relating forces and displacements, is given in Ref. [3]. Here we show the effectiveness of a simpler method. Consider the example of three bars, with one fixed to the ground and the other two hinged to it, as shown in Figure 7.1. Let the elastic deformations of the three bars in Figure 7.1 be given in terms of component modes by:

$$\delta_1 = \varphi_1 q_1; \quad \delta_2 = \varphi_2 q_2; \quad \delta_3 = \varphi_3 q_3 \tag{7.55}$$

Component mode synthesis of the system would relate the component modal coordinates, q_i to the multibody system modal coordinates η of the system, with hinges between components locked, in the following matrix form. The locking of hinges is a practical approach, because the frequencies and modes actually change with relative orientation change between components:

$$\begin{Bmatrix} q_1 \\ q_2 \\ q_3 \end{Bmatrix} = \begin{bmatrix} A \\ B \\ C \end{bmatrix} \begin{Bmatrix} \eta \end{Bmatrix} \tag{7.56}$$

Here A, B, C are projections of system eigenvectors on the subspace of component modal coordinates, meaning as many columns for "component mode representation" as the number of system modes. By the singular value decomposition of a matrix A, it is meant that A is broken down as shown in the first line of Eq. (7.57) below. Singular value of a matrix shows how many columns of a matrix are linearly independent. (Singular value decomposition of the matrices A, B, C can be obtained by the *matlab command*, SVD.)

$$\begin{aligned} &\text{svd}(A) = [U_1, \Sigma_1, V_1]; \Rightarrow A = U_1 \Sigma_1 V_1^T \\ &\text{svd}(B) = [U_2, \Sigma_2, V_2] \\ &\text{svd}(C) = [U_3, \Sigma_3, V_3] \end{aligned} \tag{7.57}$$

Now keeping as many columns in U_1, U_2, U_3 as a chosen number of largest singular values in $\Sigma_1, \Sigma_2, \Sigma_3$, and denoting these as $\hat{U}_1, \hat{U}_2, \hat{U}_3$, we make the deformation approximations

$$
\begin{aligned}
\delta_1 &= \varphi_1 q_1 = \varphi_1 \hat{U}_1 q_1 = \hat{\varphi}_1 \eta_1 \\
\delta_2 &= \varphi_2 q_2 \equiv \varphi_1 \hat{U}_2 q_2 = \hat{\varphi}_2 \eta_2 \\
\delta_3 &= \varphi_3 q_3 \equiv \varphi_3 \hat{U}_3 q_3 = \hat{\varphi}_3 \eta_3
\end{aligned} \tag{7.58}
$$

where $\hat{\phi}_1, \hat{\phi}_2, \hat{\phi}_3$ may be called the reduced modal sets for components 1,2,3.

Before closing our discussion of component mode synthesis, it is noteworthy that representation of individual components of a flexible multibody system by component modes may give rise to overall high frequencies, and thus simulations may become time-consuming. Thus for speed of simulation in certain online applications [7] of multibody dynamics, system modes are formed, reduced in number, out of component modes, and kept frozen for a chosen duration, apportioning component deformations during articulation corresponding to these system modes. After the chosen interval, system modes are updated and component modes apportioned therefrom for further articulation.

In Figure 7.4 shown below we consider a system for applying the component model reduction by singular value decomposition (SVD) of the synthesized system model, as presented by Eqs. (7.56) and (7.57). Out of 15 system modes, the SVD model reduction method kept 15 modes in matrix A of body 1, 5 modes in matrix B of body 2, and 9 modes in matrix C of body 3 (top branch) to get a (15+5+9) dof model. Figure 7.5 shows results of comparing a 29 dof model designed to match first 15 frequencies of a 68 dof articulated flexible three-link system.

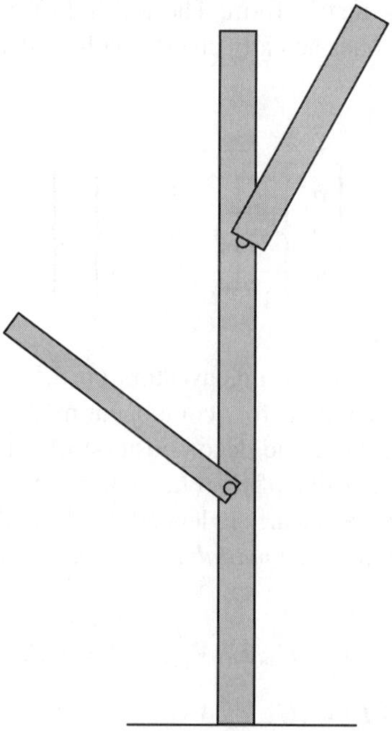

Figure 7.4 Three-Bar System with 68 Modes to Demonstrate Component Mode Selection.

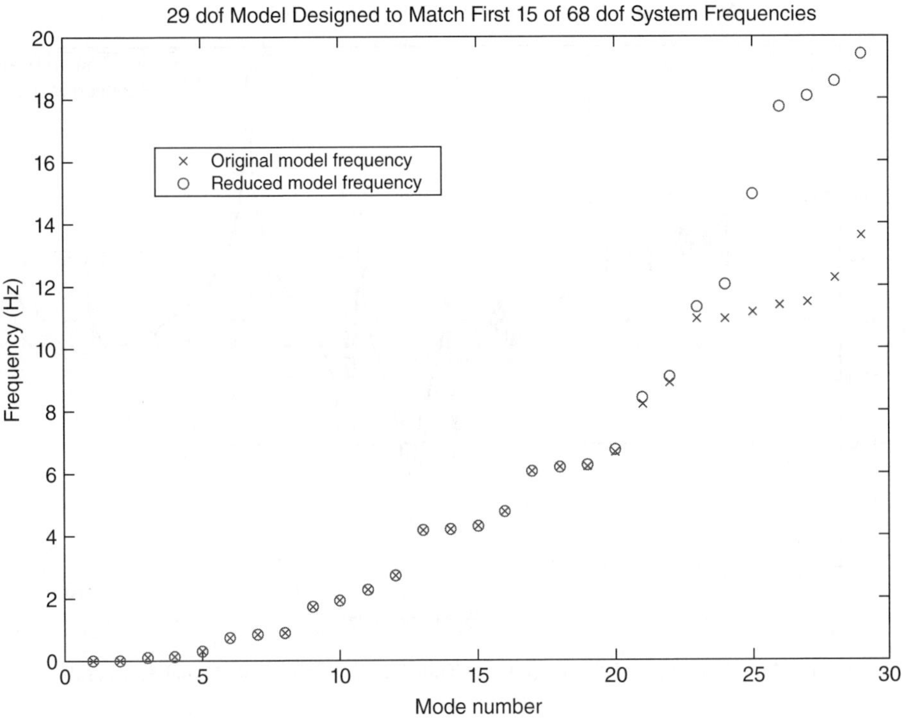

Figure 7.5 29 dof Model Matching First 15 of 68 dof Frequencies of the Three-Bar System.

Figure 7.6 shows the result of applying the same reduction technique to an actual spacecraft application. In it we overlay the Bode plots of magnitude of a transfer function from an input torque to an output angle in an instrument mounted in the spacecraft. The curve slightly raised, originally in blue, is for the full-order system, and the slightly lower curve is for the reduced-order system. This shows the effectiveness of model reduction of the components from the full system, based on the singular values of the projection of the system modes to the component modal degrees of freedom.

7.9 Deriving Damping Coefficient of Components from Desired System Damping

Follow the self-explanatory computational sequence given below, after choosing a desired system level damping:

$$
\begin{aligned}
&M\ddot{X} + C\dot{X} + KX = 0\\
&X = \Phi q\\
&\ddot{q} + \Phi^t C \Phi \dot{q} + \Omega^2 q = 0
\end{aligned}
$$

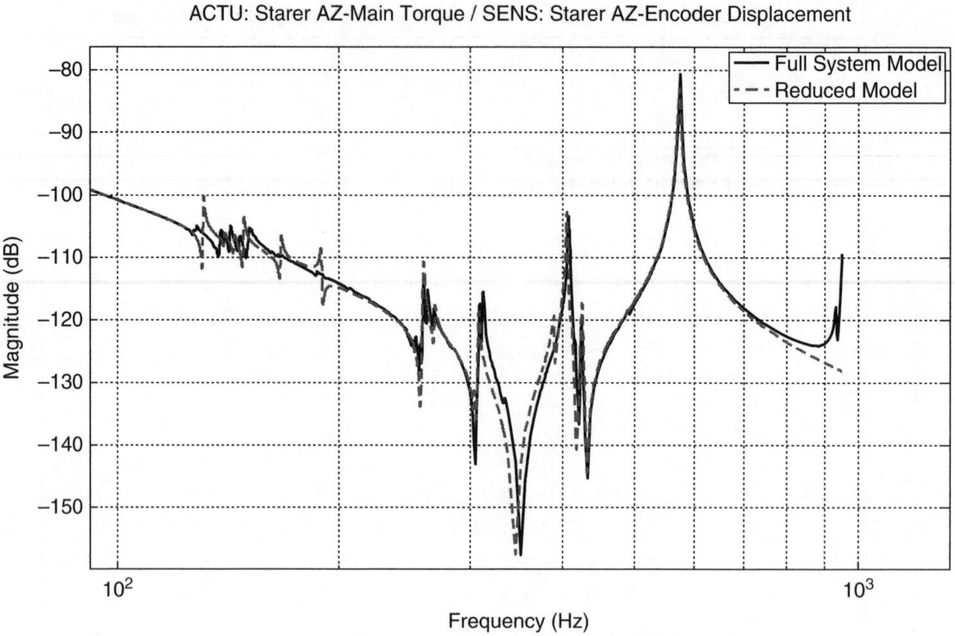

Figure 7.6 Bode Plot of Toque to Angle Transfer Function for an Actual Spacecraft.

$$\Phi^t C\Phi = 2\Xi\Omega$$
$$C = \Phi^{-t} 2\Xi\Omega\Phi^{-1}$$
$$\Phi^{-1} = (\Phi^t\Phi)^{-1}\Phi^t$$

Note that we have used the pseudo-inverse of the system modal matrix in the last line. Component level damping is derived for a jth component by the selection matrix S_j that identifies the body in the system:

$$c_j = S_j C S_j^T$$

Results of re-synthesis of system damping with derived component damping for an actual flexible multibody spacecraft is given in Figure 7.7.

Problem Set 7

7.1 Work out the steps in Ref. [1] of component mode synthesis of two components, one L/3, the other 2L/3 for a fixed-fixed beam of length L (choose your value of L, E, I), with each component described by a few fixed-fixed modes and two constraint modes corresponding to unit displacement and unit rotation. Keep three modes for one component and six for the other.

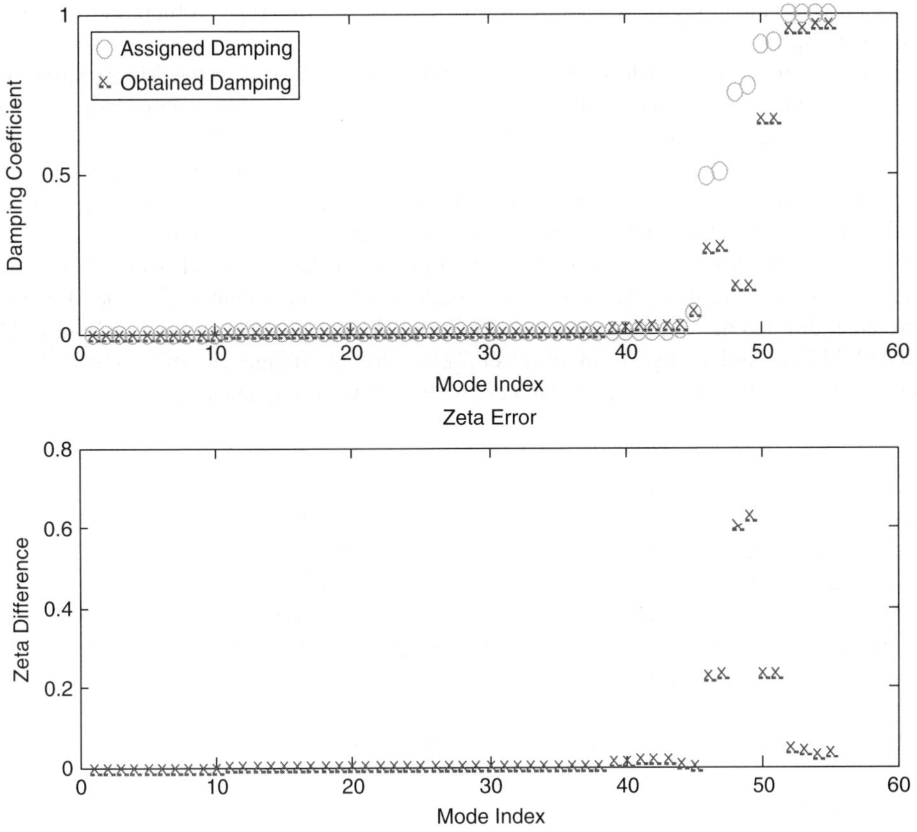

Figure 7.7 Results of System Damping Re-Synthesis with Derived Component Damping.

7.2 Consider a simple cantilever beam with a tip load. Consider a few cantilever beam vibration modes to describe its deformation in a forced vibration analysis. Then construct a modal truncation vector corresponding to the number of vibration modes kept.

7.3 A massless cantilever beam of length L has two discrete masses with mass m at $L/2$ and L. Assume a transverse mode shape of $\phi(x) = (x/L)^3$. Calculate the modal effective mass.

7.4 Repeat the example in Section 7.6, using two vibration modes and a modal truncation vector.

Appendix 7.A Matlab Codes for Structural Dynamics

7.A.1 [Codes Contributed by Dr. John Dickens, Dickens Associates, Fremont, Ca]

A solution to Problem 1 stated above is given later, complete with codes that would be useful to more general problems in structural dynamics. To solve problem 1, in Matlab, "cd" to the directory where the m-files given below are typed and saved, and type problem_set_7_1 at the Matlab prompt.

In the subroutines given below, the function "problem_set_7_1.m" is the driver routine. The function "beam_2d.m" calculates the stiffness and mass for a two-dimensional beam element having transverse and rotation degrees of freedom. "beam_string.m" will create a beam model by repeating and assembling several beam elements together. "eigen_mk.m" calculates the eigenvalues/frequencies given the mass and stiffenss matrices. "Blevins_beam_freqs.m" is a tabulation of the beam frequencies from Blevins [8] of interest in this analysis.

"beam_model_reduction.m" reduces a multi-element beam model using component modes, modal truncation vectors ("mtv_vectors.m"), and boundary node functions ("boundar_dof_fcn.m"). "problem_set_7_1.m" also does the synthesis both without MTVs and with MTVs. And finally, "print_checks.m" compares the frequencies of the reduced beam models, clamped-free and clamped-clamped, with analytical frequencies given by Blevins.

```
%%%%%%%%%%%%%%%%%%%%%%%%%%%%%%%%%
function problem_set_7_1
% Synthesis of 2 beam models.  Synthesis is fixed-Fixed
%   Model_1 of  L/3 and 10 elements,  Left Model
%   Model 2 of 2L/3 and 20 elements,  Right Model
% Script developed by John Dickens using Matlab 6.0.0.88 (R12)
fprintf('\nProblem Set 7.1\n');
clear
% Section = 1.1 X 5.5 rectangular, density .3
rho = .3;
E   = 30E+06;
A   = 1.1*5.5;
m_l = rho*A;             % mass/unit length
I   = 1.1*5.5^3/12;      % 1.1 x 5.5 rectangular section
EI  = E*I;
L     = 100;  % Total Beam length
n_elt = 30;
n1_elt = n_elt/3;
n2_elt = 2*n_elt/3;
L_e    = L/n_elt;
bm_elmt = beam_2d (EI,L_e,m_l, 'Consistent');  % One beam element
% Create the full clamped-clamped beam without Substructures ——-
model   = beam_string(bm_elmt.stif, bm_elmt.mass, n_elt);
lst     = 3:size(model.mass,1)-2;    % Fix both Ends
model.stif = model.stif(lst,lst);
model.mass = model.mass(lst,lst);
sys          = eigen_mk(model.mass,model.stif);
Blev_freq_sys = Blevins_beam_freqs (E,I, A,rho, L);
%————————————————————-
% Construct Substructure 1, Left end Fixed, n1_elt = 10 elements,
Retain
```

```
% 3 modes and Include MT Vectors
chk = 1;
n1_mod = 3;
sub1 = beam_model_reduction (bm_elmt,n1_elt,n1_mod,'Left');
if chk==1                                               % Check
With Blevins
  lst = 1:n1_mod;
  chk_sub1_cc = sqrt(diag(sub1.rk(lst,lst)))/2/pi; % Clamped-Clamped
  chk_sub1_cf = eigen_mk(sub1.rm,sub1.rk);         % Clamped-Free
  Blev_freq1  = Blevins_beam_freqs (E,I, A,rho, L/3);
  print_checks ('Sub1', n1_mod, chk_sub1_cc, chk_sub1_cf, Blev_freq1)
end
%--------------------------
% Construct Substructure 2,  Right end fixed
n2_mod = 6;
sub2 = beam_model_reduction (bm_elmt,n2_elt,n2_mod,'Right');
if chk==1
  lst = 1:n2_mod;
  chk_sub2_cc = sqrt(diag(sub2.rk(lst,lst)))/2/pi; % Clamped-Clamped
  chk_sub2_cf = eigen_mk(sub2.rm,sub2.rk);         % Clamped-Free
  Blev_freq2  = Blevins_beam_freqs (E,I, A,rho, 2*L/3);
  print_checks ('Sub2', n2_mod, chk_sub2_cc, chk_sub2_cf, Blev_freq2)
end
% Synthesis WITHOUT MTVs -----------------
% DOFs  g/S1 1:n1_mod,
%       g/S2 n1_mod+(1:n2_mod)
%       2 physical
n_gen = n1_mod+n2_mod;
n_dof = n_gen+2;
lm1 = [1:n1_mod,            0 0 , n_gen+(1:2)];
lm2 = [n1_mod+(1:n2_mod), 0 0 , n_gen+(1:2)];
M1 = zeros(n_dof,n_dof);  K1 = M1;
M1(lm1(lm1>0), lm1(lm1>0)) = M1(lm1(lm1>0), lm1(lm1>0)) ...
   + sub1.rm(lm1>0,lm1>0);
K1(lm1(lm1>0), lm1(lm1>0)) = K1(lm1(lm1>0), lm1(lm1>0)) ...
   + sub1.rk(lm1>0,lm1>0);
M1(lm2(lm2>0), lm2(lm2>0)) = M1(lm2(lm2>0), lm2(lm2>0)) ...
   + sub2.rm(lm2>0,lm2>0);
K1(lm2(lm2>0), lm2(lm2>0)) = K1(lm2(lm2>0), lm2(lm2>0)) ...
   + sub2.rk(lm2>0,lm2>0);
sys_1 = eigen_mk(M1,K1);
% Synthesis WITH  MTVs -----------------
% DOFs  g/S1 1:n1_mod+2,
%       g/S2 n1_mod+2+(1:n2_mod+2)
%       2 physical
n_gen = n1_mod+2+n2_mod+2;
```

```
n_dof = n_gen+2;
lm1 = [1:n1_mod+2 ,             n_gen+(1:2)];
lm2 = [n1_mod+2+(1:n2_mod+2),  n_gen+(1:2)];
M2 = zeros(n_dof,n_dof);  K2 = M2;
M2(lm1,lm1) = M2(lm1,lm1) + sub1.rm;
M2(lm2,lm2) = M2(lm2,lm2) + sub2.rm;
K2(lm1,lm1) = K2(lm1,lm1) + sub1.rk;
K2(lm2,lm2) = K2(lm2,lm2) + sub2.rk;
sys_2 = eigen_mk(M2,K2);
% Comparison of Assemblages ————————————-
n_fs = size(sys_2.freq,1);
n_fs = min(n_fs+2, numel(sys.freq));  % Add a couple more
System Freqs
all_fqs = zeros(n_fs,3);
nn = numel(sys_1.freq);  all_fqs(1:nn,1) = sys_1.freq;
nn = numel(sys_2.freq);  all_fqs(1:nn,2) = sys_2.freq;
all_fqs(:,3) = sys.freq(1:n_fs);
fprintf(['\nSummary of Synthesis Frequecies:\n', ...
        '   No.     2 Substructure   2 Substr with MTV', ...
                                 '            Full Model\n']);
tbl = sprintf('%5d %19.4f %19.4f %19.4f\n',
[(1:n_fs).',all_fqs].');
tbl = strrep(tbl,'0.0000', blanks(6));
disp(tbl);
return
%%%%%%%%%%%%%%%%%%%%%%%%%%%%%eigen_mk.m follows
function [output1, output2,output3] = eigen_mk (m,k)
%    [ ... ] = eigen_mk (m,k)
%
% Solve eigenvalue problem k * phi = lambda*m * phi
% and return values in ascending order.
%     lambda units are (rad/sec)**2
%Returns:
%  2 output arguments:          [ lambda, phi ] = eigen_mk(m,k)
%  Single output argument:      [ out           ] = eigen_mk(m,k)
%    where   "out" consists of:   out.lambda     = lambda
%                                 out.freq       = frequency
%                                 out.phi        = eigenvectors
%  3 output arguments:          [ lambda, phi, freq ] = eigen_mk(m,k)
% Changed sort to algebraic rather than absolute  jmd - 04 July 2007
% Added the structured data set "out" as output option and
cleaned up
%   sort of eigenvalues                          jmd  - 27 July 2008
[nr,nc]=size(m);
%if nr ==1 || nc == 1;  mm=diag(m);
%else                           mm=m;
```

```
%end
if nr'=nc
    mm=diag(m);
else
    mm=m;
end
[vv, ww] =eig (k, mm);
% Organize in ascending order by eigenvalue
w=diag(ww);
%      t=(w.')'.*w;       % Square eigenvalues
%      [t,pointer]=sort(t);
[lambda,pointer]=sort(w);
nn=numel(w);
phi=zeros(nn,nn);
for n=1:nn ;
  if isequal(n,2);  echo off;  end
  temp=vv(:,pointer(n)); scale=temp.'*mm*temp;
  if (scale '= 0.);
    scale=sqrt(1./scale);
  else  msg = sprintf( ...
    'Mode %d has %6.3f generalized mass, no scaling possible', n,
scale);
    disp(' ');  disp(msg);  scale=1.0;
  end
  phi(:,n)=scale*temp;
end;
freqcy =  sqrt(abs(lambda))/(2*pi);
indx = find(lambda<0);   % Mark negative eigenvalues as negative
Freqs
if numel(indx)>0; freqcy(indx) = -freqcy(indx);  end
if    (nargout == 1)
  output1.lambda     = lambda;
  output1.phi        = phi;
  output1.freq       = freqcy;
elseif(nargout == 2)
  output1 = lambda;
  output2 = phi;
else
  output1 = lambda;
  output2 = phi;
  output3 = freqcy;
end
return
%%%%%%%%%%%%%%%%%%%%%%%%%%%%%beam_2d.m follows
function [beam_elemt] = beam_2d (EI, L_e, m_l,Consistent)
%   [beam_elemt] = beam_2d (EI, L_e, m_l, Consistent)
```

```
% Two Dimensional Beam element.  Displacement and Rotation only
% See Craig, 1981, Structural Dynamics, pg 387
%  Input:
%    EI
%    L_e = Length of beam elemet
%    m_l = Mass/unit length =  rho*A
%  Output
%    beam_elemt.stif     4x4 stiffness
%              .mass     4x1 mass, no rotational Mass if only 3 input
%                        arguments, otherwise consistent mass and
%                        4x4 mass
%
EI_Le = EI/L_e;
ke =  EI_Le * [  12/L_e^2     6/L_e    -12/L_e^2     6/L_e
                  6/L_e         4        -6/L_e         2
                -12/L_e^2   -6/L_e     12/L_e^2    -6/L_e
                  6/L_e         2      -6/L_e         4   ];
if nargin <4    % Lateral Lumped mass only
  mL_2 = m_l*L_e/2;
  me =          [  mL_2;      0;   mL_2;     0    ];
else            % Consistent mass, lateral AND rotation,
  rAL_420 = m_l*L_e/420;
  me = rAL_420 * [  156      22*L_e      54            -13*L_e
                  22*L_e     4*L_e^2   13*L_e          -3*L_e^2
                   54       13*L_e     156            -22*L_e
                 -13*L_e   -3*L_e^2  -22*L_e           4*L_e^2];
end
beam_elemt.stif = ke;
beam_elemt.mass = me;
return
%%%%%%%%%%%%%%%%%%%%%%%%%%%boundary_dof_fcn.m follows
function [bnf] = boundary_dof_fcn (stiffness, boundary_dofs)
% [bnf] = boundary_dof_fcn (stiffness, boundary_dofs)
% Input:
%  stiffness      = stiffness matrix
%  boundary_dofs = Degrees of freedom of the model to which the model
%                  is to be reduced using the Guyan procedure
% Output:
%     bnf.rvec  = Guyan boundary function vectors
all_dofs = 1:size(stiffness,1);       % list of all the DOFs of
the model
b_dofs   = boundary_dofs;             % list of the boundary DOFs
num_b    = length(b_dofs);
i_dofs   = setdiff(all_dofs, b_dofs); % list of the interior DOFs
r_vec    = zeros(length(all_dofs), length(b_dofs));    %
Initialize r_vec
```

```
%       formulate -Kii(-1)Kib and insert in r_vec
r_vec(i_dofs,:)  =  - stiffness(i_dofs,i_dofs) \ stiffness(i_dofs,
b_dofs);
r_vec(b_dofs,:)  =  eye(num_b,num_b); % Set the Identity matrix
into
%                                         boundary DOFs
bnf = r_vec;                            %  Return Guyan boundary
functions
                                        %  to calling program
return
%%%%%%%%%%%%%%%%%%%%%%%%%%mtv_vectors.m follows
function [mtv] = mtv_vectors (sub1, sys1, bnf, b_dof)
%  mtv = mtv_vectors(sub1, sys_s1, bnf, b_dof);
% Calculate MTV vectors for beam string
%  Input:
%      sub1.stiff,  sub1.mass,  stiffness and mass of model, both full
%      sys1.phi                 fixed base eigenvectors
%      bnf                      boundary node functions
%      b_dof                    boundary dofs
%
%  MTVs are calculated from:  F_residual = F - M* Phi * Phi.' F
%  Note that Phi * Phi.' * M is an Identity if and only if Phi is
all the
%  modes of the system.  In that case (all the modes have been kept,
%  F_residual or u_residual are = 0), their is no need for MTVs
since all
%  the Force is included in the modes (residuals are = 0).
%
lst_dof = 1:size(sub1.mass,1);                   % All DOFs
i_dof = setdiff(lst_dof,b_dof);                  % Interior DOFs
force = sub1.mass(i_dof,:)*bnf;                  % Force Vector to
Get MTVs
mass_i = sub1.mass(i_dof,i_dof);
f_res = force - mass_i*sys1.phi*sys1.phi.'*force;
u_res = sub1.stif(i_dof,i_dof)\f_res;            % Kii(-1) * F_res
% Normalization and orthgonalization to make MTVs "look like modes"
mm    = u_res.'*mass_i*u_res;
kk    = u_res.'*f_res;
%kk   = u_res.'*sub1.stif(i_dof,i_dof)*u_res;  % same values
as above
sys_mt= eigen_mk(mm,kk);
mtv   = u_res * sys_mt.phi;
return
%%%%%%%%%%%%%%%%%%%%%%%%%%%beam_string.m follows
function [model] = beam_string (beam_stif, beam_mass, n_elt)
% [model] = beam_string (beam_stif, beam_mass)
```

```
% Input:
%    beam_stif  = beam element stiffness, 4x4,
u1,theta1,u2,theta2
%    beam_ass   = beam element mass, 4x4 or (4x1).
%    n_elt       = number of 2D beam elements to attach end-to end
(to
%               string together)
% Output:
%   model        = structured array with fields of: mass, stiffness
%   model.stiff  = stiffness, (n_elt+2) x (n_elt+2)
%   model.mass   = mass, either same size as Stiffness if mass
matrix is
%                  same size as stiffness or a column vector if
the beam
%                  element mass, "beam_mass", is a column vector.
n_dof = 2*n_elt+2;
k = zeros(n_dof,n_dof);
lm1 = (1:4);  lm = lm1;
for n = 1:n_elt
  k(lm,lm) = k(lm,lm) + beam_stif;
  lm = lm+2;
end;
lm = lm1;
if numel(beam_mass)==4;
  m = zeros(n_dof,1);
  beam_mass = reshape(beam_mass,numel(beam_mass),1);
  for n = 1:n_elt
    m(lm)     = m(lm)     + beam_mass;
    lm = lm+2;
  end;
else
  m = zeros(n_dof,n_dof);
  for n = 1:n_elt
    m(lm,lm) = m(lm,lm) + beam_mass;
    lm = lm+2;
  end
end
model.mass = m;
model.stif = k;
return
%%%%%%%%%%%%%%%%%%%%%%%%%%print_checks.m follows
function print_checks (c_sub2, n2_mod, chk_sub2_cc, chk_sub2_cf,
Blev_freq2)
  nn = min(5,n2_mod);
  fprintf('\n%s Frequency Comparisons:\n\n', c_sub2);
  fprintf(' %s    Clamped-Clamped:  %f  %f  %f  %f  %f', ...
```

```
          c_sub2, chk_sub2_cc(1:nn) );
   fprintf('\n Blevins Clamped-Clamped:  %f  %f  %f  %f  %f\n', ...
       Blev_freq2.clamped_clamped);
   nn = min(5,numel(chk_sub2_cf.freq));
   fprintf('\n %s       Clamped-Free:  %f  %f  %f  %f  %f', ...
          c_sub2, chk_sub2_cf.freq(1:nn) );
   fprintf('\n Blevins    Clamped-Free:  %f  %f  %f  %f  %f\n', ...
       Blev_freq2.clamped_free);
   return
%%%%%%%%%%%%%%%%%%%%%%%%%%%Blevins_beam_freqs.m follows
function [Blev_freq] = Blevins_beam_freqs (E,I, A,rho, L)
%   [Blev_freq] = Blevins_beams_freqs (E,I, A,rho, L)
%  List the Beam Frequencies from Blevins [8] for beam:
%     a. Free-Free,  b. Clamped-Free,  c.Clamped-Clamped
EI  = E*I;
m_l = rho*A;   % mass/unit length
const = (EI/m_l)^.5/(2*pi*L^2);
% Free-Free       Blevins page 108, Table 8-1,  case 1
lam_i = [4.73004  7.85320  10.99561 14.13717 17.27876];
Blev_freq.free_free       = const*lam_i.^2;
% Clamped-Free    Blevins page 108, Table 8-1,  case 3
lam_i = [1.87510 4.69409 7.85476 10.99554 14.13717];
Blev_freq.clamped_free    = const*lam_i.^2;
% Clamped-Clamped Blevins page 108, Table 8-1,  case 7
lam_i = [ 4.73004  7.85320  10.99561  14.13717  17.27876];
Blev_freq.clamped_clamped = const*lam_i.^2;
return
%%%%%%%%%%%%%%%%%%%%%%%%%%%beam_model_reduction.m follows
function sub1 = beam_model_reduction
(bm_elmt,num_elt,md_ret,clamped_end)
%
% Create a string of beam elements and reduce to HCB (Hurty-Craig-
Bampton) model.
% Used in the Example Problem in the book
% Clamped End = 'Left' or 'Right'.  If Left end Clamped then the
Right end
% is the boundary for the Model reduction and Vice-versa.
% Construct Substructure Left end fixed, "num_elt" elements,
%   retain "md_ret" modes with 2 MTV vectors
sub1   = beam_string(bm_elmt.stif, bm_elmt.mass, num_elt);
nr     = size(sub1.mass,1);
if strcmpi(clamped_end(1),'L');  lst   = (3:nr);     % Fix Left end
else                             lst   = (1:nr-2);   % Fix Right end
end
sub1.stif   = sub1.stif(lst,lst);
sub1.mass   = sub1.mass(lst,lst);
```

```
sub1.model = sub1;
%  Modes and Truncation Vectors
n_dof1     = size(sub1.mass,1);
if strcmpi(clamped_end(1),'L');
  lst_i     = 1:n_dof1-2;          % Interior DOFs
  bdry_dofs= n_dof1-1:n_dof1;  % Boundary DOFs, last 2 DOFs
else
  lst_i     = 3:n_dof1;            % Interior DOFs
  bdry_dofs= 1:2;                  % Boundary DOFs, First 2 DOFs
end
s1_temp       =
eigen_mk(sub1.mass(lst_i,lst_i),sub1.stif(lst_i,lst_i));
s1_temp.phi    = s1_temp.phi(:,1:md_ret);   % Keep only "md_ret" modes
s1_temp.lambda= s1_temp.lambda(1:md_ret);
s1_temp.freq   = s1_temp.freq(1:md_ret);
bnf = boundary_dof_fcn (sub1.stif,bdry_dofs);
mtv = mtv_vectors(sub1, s1_temp, bnf, bdry_dofs);
n1  = md_ret+2+2;
t   = zeros(size(sub1.stif,1),n1);
t(lst_i,1:md_ret+2) = [s1_temp.phi mtv];
t(:, md_ret+3:end)= bnf;
sub1.rt = t;
sub1.rm = t.' * sub1.mass * t;
sub1.rk = t.' * sub1.stif * t;
return
%%%%%%%%%%%%%%%%%%%%%%%%%%%%
```

7.A.2 Results

Results obtained are shown below for synthesis with (1) component frequencies up to twice desired system frequencies; (2) same as (1) except that two modal truncation vectors (MTV) are added; and (3) full model is kept. In substructure 1, sub1, three component modes to 275.23 Hz were retained to which two MTVs were added. Substructure 2, sub2, had six component modes to 237.26 Hz retained to which two MTVs were added also. These results are summarized in Table 7.A.1. Results show the efficiency of the modal truncation vectors approach for component modeling.

Sub1 Frequency Comparisons:

```
 Sub1    Clamped-Clamped:  50.883797  140.294855  275.233706
 Blevins Clamped-Clamped:  50.882019  140.258123  274.962433
454.526981  678.984488
 Sub1         Clamped-Free:  7.996240  50.113198  140.349490
275.221136  465.580178
 Blevins      Clamped-Free:  7.996199  50.111515  140.313852
```

Table 7.A.1 Summary of Synthesis Frequencies.

No.	Substruct Modes to 2*system Freq	Same Substruct Modes+2 MTVs	Full Model
1	5.6536	5.6536	5.6536
2	15.5849	15.5843	15.5843
3	30.5521	30.5518	30.5518
4	50.5180	50.5047	50.5047
5	75.5044	75.4484	75.4484
6	105.4215	105.3859	105.3859
7	140.7556	140.3228	140.3225
8	181.2012	180.2675	180.2666
9	226.2117	225.2393	225.2303
10	293.8879	276.1798	275.2310
11	573.2914	345.5828	330.2916
12		455.2700	390.4424

```
274.958933  454.526981
Sub2 Frequency Comparisons:
 Sub2     Clamped-Clamped:  12.720536  35.065149  68.744912
113.651135  169.810570
 Blevins Clamped-Clamped:  12.720505  35.064531  68.740608
113.631745  169.746122
 Sub2        Clamped-Free:  1.999058  12.527911  35.079014
68.744037  113.650974
 Blevins     Clamped-Free:  1.999050  12.527879  35.078463
68.739733  113.631745
```

7.10 Conclusion

We have reviewed in this chapter the important subjects of component mode selection, giving detailed codes for finite element formation and synthesis, and eigenvalue problem solution. It is known that public domain software is also available, for example [9] for vibration (and buckling) analysis of assembled lattice structures with repetitive geometry and substructuring options. What is noteworthy here is that we have discussed both model reduction, which is necessary in practice, and compensation for model reduction by modal truncation vectors. We have also discussed the role of geometric stiffness due to rigid body inertia effects that are present with large overall motion of flexible bodies.

References

[1] Craig, R.R. Jr. (1983) *Structural Dynamics, An Introduction to Computer Methods*, John Wiley & Sons, Inc.
[2] Wijker, J. (2004) *Mechanical Vibrations in Spacecraft Design*, Springer Verlag.
[3] Guyan, R.J. (1965) Reduction of stiffness and mass matrices. *AIAA Journal*, **3**, 385.
[4] Ghosh, T.K. (2015) Component Model Reduction by Frequency Filtering. Unpublished Memo to NASA Johnson Space Flight Center, L3-Communications Co., Houston, Texas.

[5] Spanos, J.T. and Tsuha, W.S. (1991) Selection of component modes for flexible multibody simulation. *Journal of Guidance, Control, and Dynamics*, **14**(2), 278–286.
[6] Dickens, J.M. and Stroeve, A. (2000) Modal Truncation Vectors for Reduced Dynamic Substructure Models. Paper No. AIAA-2000-1578. 41st AIAA/ASME/AHS/ASC Structures, Structural Dynamics, and Materials Conference and Exhibit, Atlanta, GA, 3–8 April.
[7] Dickens, J.M., Nakagawa, J.M., and Wittbrodt, M.J. (1997) A critique of mode acceleration and modal truncation methods for modal response analysis. *Computers & Structures*, **62**(6), 985–998.
[8] Blevins, R.D. (2001) *Formulas for Natural Frequency and Mode Shape*, Krieger Publishing.
[9] Anderson, M.S., Williams, F.W., Banerjee, J.R., *et al.* (1986) User manual for BUNVIS-RG: An exact buckling and vibration program for lattice structures, with repetitive geometry and substructuring options. NASA Technical Memorandum 87669.
[10] Banerjee, A.K. and Dickens (1990) Dynamics of an arbitrary flexible body in large rotation and translation. *Journal of Guidance, Control, and Dynamics*, **13**(2), 221–227.

8

Block-Diagonal Formulation for a Flexible Multibody System

In Chapter 6 we presented the dynamics of an n-degree of freedom flexible multibody system with equations of motion in the form, $\tilde{M}\dot{U} = G$, in which $\tilde{M}$ is a $n \times n$ dense mass matrix with time-varying elements, $\dot{U}$ and G are $n \times 1$ matrices, of the derivatives of generalized speeds, and remainder terms of generalized inertia and active forces, respectively. This means that an $n \times n$ matrix $\tilde{M}$ has to be evaluated, decomposed, and back-solved to form $\dot{U}$, at least four times per step in a fourth-order Runge-Kutta integration method. That makes a simulation expensive when n is large. The present chapter describes a more computationally efficient algorithm, covered in Ref. [1], for simulating motions of flexible multibody dynamic systems. As we will, see the computational savings come from an algorithm that forms a new set of matrix equations, $M\dot{U} = R$, where M is a sparse, block-diagonal matrix, meaning that one has to invert only small block-matrices to compute $\dot{U}$. We consider large overall motion of the system, accounting for geometric stiffness due to inertia and interbody forces, and allow motion constraints. First, consider a simple example to see the importance of geometric stiffness due to interbody forces and inertia forces, in large overall motion of flexible body systems.

8.1 Example: Role of Geometric Stiffness due to Interbody Load on a Component

Figure 8.1 shows three massless rigid rods, OA, AB, BC, connected at the hinges at A and B by torsional springs of stiffness k. Rod OA of length r is driven at a prescribed angular speed $\omega(t)$. Rods AB and BC are each of length L. Particles of mass m are attached at point B of AB and at C of BC. The angle between OA and AB is q_1 and the angle between AB and BC is q_2.

As has been described before, correctly linearized dynamical equations for small motions in angles q_1, q_2 are obtained in Kane's method by keeping nonlinear terms in the velocities to extract nonlinear partial velocities, and then linearizing these partial velocities, following

Flexible Multibody Dynamics: Efficient Formulations and Applications, First Edition. Arun K. Banerjee.

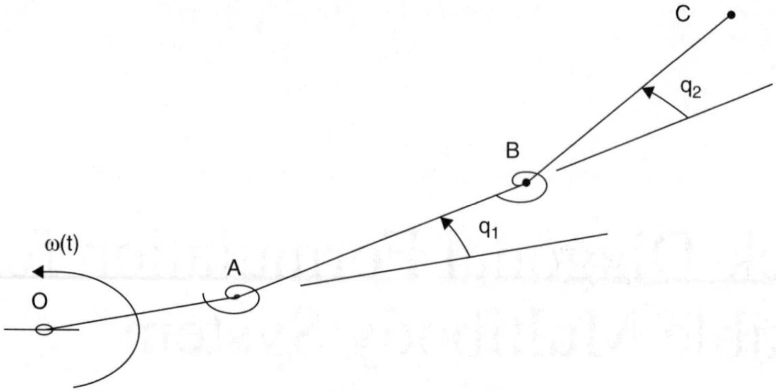

Figure 8.1 A System of Hinged Links Undergoing Large Rotation with Small Vibration. Banerjee [1].

which velocities and accelerations can be linearized, in q_1, q_2. The linear equations for the system in Figure 8.1 can be shown to be, in matrix form, as follows:

$$M\ddot{Q} + KQ = F, \qquad Q = [q_1, q_2]^T \tag{8.1}$$

where

$$\begin{aligned} M &= mL^2 \begin{bmatrix} 5 & 2 \\ 2 & 1 \end{bmatrix}, \quad F = -m\dot{\omega}L \begin{Bmatrix} 5L + 3R \\ 2L + r \end{Bmatrix} Q = \begin{Bmatrix} q_1 \\ q_2 \end{Bmatrix} \\ K &= \begin{bmatrix} k + 3m\omega^2 rL & m\omega^2 rL \\ m\omega^2 rL & k + m\omega^2 L(r + L) \end{bmatrix} \end{aligned} \tag{8.2}$$

It is clear from the stiffness matrix K that effective stiffness increases with any rotation rate ω, due to centrifugal stiffening, as expected. If, on the other hand, premature linearization is done in that velocities are linearized first, and partial velocities and accelerations determined from these linear velocities, then incorrect equations are obtained, where the error shows up in the following incorrect stiffness matrix, K_i, with M remaining the same as in Eq. (8.2):

$$K_i = \begin{bmatrix} k - 5m\omega^2 L^2 & -2m\omega^2 L^2 \\ -2m\omega^2 L^2 & k - m\omega^2 L^2 \end{bmatrix} \tag{8.3}$$

It appears from Eq. (8.3) that the effect of rotation is that, for all rotation rates ω, stiffness decreases. Equation (8.1) with M and F remaining the same, but the stiffness matrix given by Eq. (8.3), then becomes wrong, and this is the well-known effect of premature linearization. By the way, this error has nothing to do with Kane's method used for the derivation; in other words, one would get the same wrong result using the linearized velocity expression in Lagrange's equation. As we have seen in Chapters 1, 5, and 6, that errors of premature linearization can be corrected, after the fact, by adding geometric stiffness due to existing inertia loads [2].

For the hinged links example problem, "existing loads" are just inertia loads due to rotation. Centrifugal force on the rod BC in the existing or undeformed configuration, that is, when $q_1 = q_2 = 0$, is:

$$F^{BC} = m\omega^2(r + 2L) \tag{8.4a}$$

Inertia force on the rod AB consists of the inertia force for the mass on AB plus the interbody force applied by rod BC on AB (which in the absence of any external force on BC is the inertia force of BC), thus yielding the force magnitude on AB in the undeformed configuration as

$$F^{AB} = m\omega^2(r + L) + m\omega^2(r + 2L) \tag{8.4b}$$

Using the well-known form [2] of the geometric stiffness matrix for an existing axial load on a bar, and small transverse displacements from the existing or reference configuration at the ends of the rods AB and BC, the total potential energy associated with motion-induced geometric stiffness is given by:

$$P_g = \frac{F^{AB}}{2L}[0 \quad Lq_1]\begin{bmatrix} 1 & -1 \\ -1 & 1 \end{bmatrix}\begin{Bmatrix} 0 \\ Lq_1 \end{Bmatrix} + \frac{F^{BC}}{2L}[Lq_1 \quad L(2q_1 + q_2)]\begin{bmatrix} 1 & -1 \\ -1 & 1 \end{bmatrix}\begin{Bmatrix} Lq_1 \\ L(2q_1 + q_2) \end{Bmatrix} \tag{8.5}$$

Here the first term is due to force on AB and the second term for force on BC. Generalized force due to this system potential energy is that associated with a geometric stiffness due to loads:

$$-\begin{Bmatrix} \dfrac{\partial P_g}{\partial q_1} \\ \dfrac{\partial P_g}{\partial q_2} \end{Bmatrix} = -[K_g]\begin{Bmatrix} q_1 \\ q_1 \end{Bmatrix}$$
$$[K_g] = m\omega^2 L\begin{bmatrix} 3r + 5L & r + 2L \\ r + 2L & r + 2L \end{bmatrix} \tag{8.6}$$

Now, the sum of the incorrect stiffness, due to premature linearization, and the geometric stiffness due to inertia loads is:

$$\begin{aligned}[K_g] + [K_i] &= m\omega^2 L\begin{bmatrix} 3r + 5L & r + 2L \\ r + 2L & r + 2L \end{bmatrix} + \begin{bmatrix} k - 5m\omega^2L^2 & -2m\omega^2L^2 \\ -2m\omega^2L^2 & k - m\omega^2L^2 \end{bmatrix} \\ &= \begin{bmatrix} k + 3m\omega^2 rL & m\omega^2 rL \\ m\omega^2 rL & k + m\omega^2 L(r + L) \end{bmatrix}\end{aligned} \tag{8.7}$$

This is the same stiffness matrix as in Eq. (8.2)! *This shows clearly that the geometric stiffness matrix K_g in Eq. (8.7), which is based on using the inertia force on BC and inertia force for AB in the reference (undeformed) state, when added to the incorrect stiffness matrix of Eq. (8.3),* *gives precisely the correct stiffness matrix of Eq. (8.2).* It is obvious that the correct equations would not be obtained if the force F^{AB} on AB (see Eq. 8.4b), accounted for only the inertia of B itself, $m\omega^2(r+L)$, and not adding the transmitted inertia force from BC. In other words, the error of premature linearization in velocity can be compensated for interconnected bodies only by the addition of geometric stiffness due to inertia forces on interconnected components. We have seen in Chapters 5 and 6 that errors due to premature linearization are analytically compensated by load-induced stiffness. This example highlights the role of inertia-induced stiffness in determining the motion and loads on a component in a multibody system. We will pursue this idea in the context of a multibody system of arbitrarily general flexible components in large overall motion.

8.2 Multibody System with Rigid and Flexible Components

As a motivating example of a flexible multibody system [3], we revisit in Figure 8.2 the same example as in Figure 6.7, the space station remote manipulator system (SSRMS). It has eight links arranged in a chain with seven active joints. Typically one end of the SSRMS latches on to the space station, while the other end grapples a payload. The manipulator can impart six dof and has an extra dof to overcome singular situations in the control of its motion. The dynamics of the space station, SSRMS, and payload is modeled with 13 rigid body coordinates and n flexible mode coordinates. A dense matrix formulation for (13+n) dof, where n is the total number of modes, will produce a $(13+n)\times(13+n)$ dense, time-varying mass matrix. Even for a reasonable value of the number of modes n, this makes the case for developing a sparse, block-diagonal mass matrix formulation.

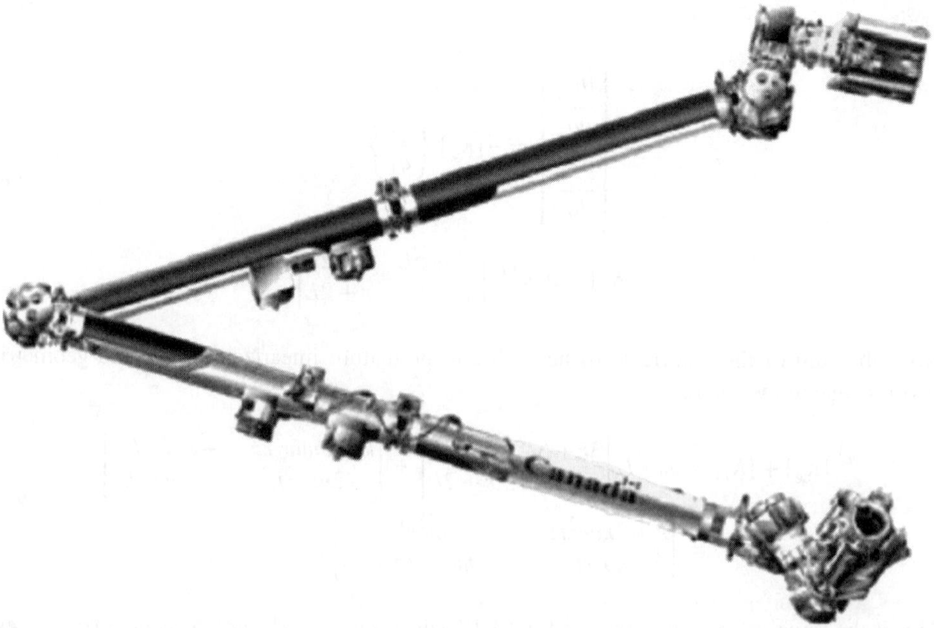

Figure 8.2 Example of a Flexible Multibody System: Space Station Manipulator Arm. Banerjee [3]. Reproduced with permission of the American Institute of Aeronautics and Astronautics, Inc.

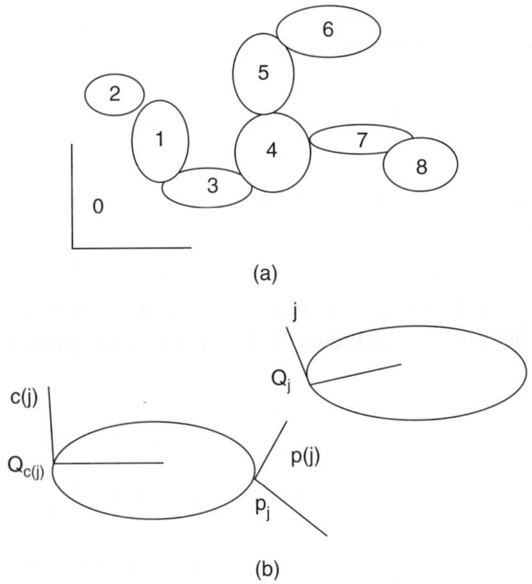

Figure 8.3 (a) System of Rigid and Flexible Bodies Hinge-Connected in a Topological Tree; (b) Bodies $B_j, B_{c(j)}$ Connected at a Hinge Allowing Relative Rotation and Translation for a System of Rigid and Flexible Bodies in a Topological Tree. Banerjee [1].

We review here some of the basic material discussed previously for multibody systems. A system of hinge-connected rigid and flexible bodies, in a topological tree configuration, is shown in Figure 8.3a. The bodies are numbered arbitrarily from 0 to NB (NB equals 8 in Figure 8.3a), with inertial frame labeled as body 0. Any generic body B_j has an inboard body $B_{c(j)}$ in the path going from body j to body *0*. Figure 8.3b shows two adjacent bodies B_j and $B_{c(j)}$ connected at a hinge that allows relative rotation and translation, with Q_j a hinge point on B_j relative to a point P_j on body $B_{c(j)}$. Reference frames j and $p(j)$ are fixed at points Q_j and P_j, respectively. Let τ_j denote a $(T_j \times 1)$ matrix of generalized coordinates representing T_j number of relative translational degree of freedom of Q_j from P_j, and let θ^j denote a $(R_j \times 1)$ matrix of generalized coordinates representing R_j relative rotations of frame j with respect to frame $p(j)$ at P_j; let η^j denote M_j number of modal coordinates of body B_j. This then defines a total of n degree freedom for an NB-body system, where:

$$n = \sum_{j=1}^{NB} (R_j + T_j + M_j) \tag{8.8}$$

8.3 Recurrence Relations for Kinematics

Making the assumption that all elastic displacements and elastic rotations are small, and referring to Figure 8.3b, inertial angular velocity of frame j can be expressed in j-basis by the angular velocity addition theorem, proceeding from the inertial angular velocity of frame $c(j)$ in basis- $c(j)$ by the following matrix expression:

$$\omega^j = C^T_{c(j),j}\left[\omega^{c(j)} + \psi^{c(j)}(P_j)\dot{\eta}^{c(j)} + C_{c(j),p(j)}H^j\dot{\theta}^j\right] \tag{8.9}$$

Here $C_{c(j),j}$ is the coordinate transformation matrix transforming basis vectors in frame j to those in frame $c(j)$; $\psi^{c(j)}(P_j)\eta^{c(j)}$ is the 3×1 vector of small elastic modal rotation at node P_j, $C_{c(j),p(j)}$ is the basis transformation from frame $p(j)$, at the point P_j, to frame $c(j)$ at $Q_{c(j)}$ (see Figure 8.3b), and H^j is the kinematical transformation matrix for rotations between frames j and $p(j)$. Velocity of the point Q_j in Figure 8.3b in the inertial frame is given by the matrix expression,

$$\begin{aligned} v^{Q_j} = C^T_{c(j),j}&\left\{v^{Q_{c(j)}} + \tilde{\omega}^{c(j)}\left[r^{Q_{c(j)}P_j} + \phi^{c(j)}(P_j)\eta^{c(j)}\right] + \phi^{c(j)}(P_j)\dot{\eta}^{c(j)}\right\} \\ &+ C_{c(j),p(j)}L^j\dot{\tau}^j + \left[\tilde{\omega}^{c(j)} + \text{SKEW}\left\{\psi^{c(j)}(P_j)\dot{\eta}^{c(j)}\right\}\right]C_{c(j),p(j)}L^j\tau^j \end{aligned} \tag{8.10}$$

where "SKEW" stands for the skew-symmetric matrix operator representing cross product with a vector, represented by the 3×1 matrix in a brace, in this case $\{\psi^{c(j)}(P_j)\dot{\eta}^{c(j)}\}$, for the small rotation rate vector at P_j, and $\tilde{\omega}^{c(j)}$ is a skew-symmetric matrix formed out of the body components of the angular velocity vector of body $B_{c(j)}$ as:

$$\omega^{c(j)} = \begin{Bmatrix} \omega_1^{c(j)} \\ \omega_2^{c(j)} \\ \omega_3^{c(j)} \end{Bmatrix} \Rightarrow \tilde{\omega}^{c(j)} = \begin{bmatrix} 0 & -\omega_3^{c(j)} & \omega_2^{c(j)} \\ \omega_3^{c(j)} & 0 & -\omega_1^{c(j)} \\ -\omega_2^{c(j)} & \omega_1^{c(j)} & 0 \end{bmatrix} \tag{8.11}$$

In Eq. (8.10), $\phi^{c(j)}(P_j)$, $\psi^{c(j)}(P_j)$ are translational and rotational deformation matrices of size $(3 \times M_j)$, and $L^j\tau^j$ denotes the $p(j)$ frame measure numbers of relative translational displacements of hinge Q_j with respect to the point P_j. Now, obtain the partial velocity of the origin of frame j with respect to the relative translation rates $\dot{\tau}^j$ at hinge Q_j, and the partial angular velocity of frame j with respect to the relative rotation rates, $\dot{\theta}^j$ at hinge j, by inspection from Eqs. (8.9) and (8.10), to define the partial velocity/partial angular velocity $6 \times n_j$ matrix R^j (where n_j is total of the rotation *and* translation hinge degrees of freedom):

$$R^j = \begin{bmatrix} C^T_{p,j}L^j & 0 \\ 0 & C^T_{p,j}H^j \end{bmatrix} \tag{8.12}$$

Note that forming partial velocities and partial angular velocities from Eqs. (8.9) and (8.10), which are linear in the modal coordinates, constitute premature linearization [3,5], and hence to compensate for the concomitant errors, one has to use motion-induced geometric stiffness. Following [6], we separate the inertial (N) frame acceleration of Q_j and the angular acceleration of frame j into two groups, one involving the derivatives of the generalized speeds, denoted subsequently by subscript 0, and the remainder acceleration and angular acceleration terms free of the generalized speed derivatives, denoted by the subscript t. This is always possible because, from [6, 8]:

$$\begin{aligned}
{}^{N}\mathbf{a}^{Q_j} &= \frac{{}^{N}d}{dt}\left(\sum\nolimits_k \mathbf{v}_k^{Q_j} u_k + \mathbf{v}_t^{Q_j}\right) = \sum\nolimits_k \mathbf{v}_k^{Q_j}\dot{u}_k + \left\{\sum\nolimits_k u_k \frac{{}^{N}d}{dt}\left(\mathbf{v}_k^{Q_j}\right) + \frac{{}^{N}d}{dt}\mathbf{v}_t^{Q_j}\right\} \\
&= \mathbf{a}_0^{Q_j} + \left\{\mathbf{a}_t^{Q_j}\right\} \Rightarrow a_0^{Q_j} + a_t^{Q_j} \\
{}^{N}\boldsymbol{\alpha}^{j} &= \frac{{}^{N}d}{dt}\left(\sum\nolimits_k \boldsymbol{\omega}_k^{j} u_k + \boldsymbol{\omega}_t^{j}\right) = \sum\nolimits_k \boldsymbol{\omega}_k^{j}\dot{u}_k + \left\{\sum\nolimits_k u_k \frac{{}^{N}d}{dt}\left(\boldsymbol{\omega}_k^{j}\right) + \frac{{}^{N}d}{dt}\boldsymbol{\omega}_t^{j}\right\} \\
&= \boldsymbol{\alpha}_0^{j} + \left\{\boldsymbol{\alpha}_t^{j}\right\} \Rightarrow \alpha_0^{j} + \alpha_t^{j}
\end{aligned} \tag{8.13}$$

Here the arrows indicate the matrix forms in the j-basis. Separate the groups representing terms with derivatives of generalized speeds further into two groups: one group indicated by a *hat, representing contribution* of second derivative terms pertaining to generalized coordinates of *all inboard bodies* in the path to body 0; and the other group for the degrees of freedom of only hinge j, denoted as follows:

$$\begin{Bmatrix} a_0^{Q_j} \\ \alpha_0^{j} \end{Bmatrix} = \begin{Bmatrix} \hat{a}_0^{Q_j} \\ \hat{\alpha}_0^{j} \end{Bmatrix} + R^j \begin{Bmatrix} \ddot{\tau}^j \\ \ddot{\theta}^j \end{Bmatrix} \tag{8.14}$$

Differentiation of Eq. (8.9) in N and use of the preceding notations in Eq. (8.13) let us produce the formula for recursive generation of α_t^j, the remainder term of the angular acceleration of frame j in the inertial frame:

$$\alpha_t^j = C_{c(j),j}^T \left\langle \begin{aligned} &\alpha_t^{c(j)} + \tilde{\omega}^{c(j)}\{\psi^{c(j)}(P_j)\dot{\eta}^{c(j)} + C_{c(j),p(j)}H^j\dot{\theta}^j\} \\ &+ C_{c(j),p(j)}\dot{H}^j\dot{\theta}^j + \mathrm{SKEW}\{\psi^{c(j)}(P_j)\dot{\eta}^{c(j)}\}C_{c(j),p(j)}H^j\dot{\theta}^j \end{aligned} \right\rangle \tag{8.15}$$

Differentiating Eq. (8.10) in N and using notations in Eq. (8.13) give the remainder acceleration term for Q_j :

$$a_t^{Q_j} = C_{c(j),j}^T \left\langle \begin{aligned} &a_t^{Q_{c(j)}} + \tilde{\alpha}_t^j\{r^{Q_{c(j)}P_j} + \phi^{c(j)}(P_j)\eta^{c(j)} + C_{c(j),p(j)}L^j\tau^j\} \\ &+ \tilde{\omega}^{c(j)}[\tilde{\omega}^{c(j)}\{r^{Q_{c(j)}P_j} + \phi^{c(j)}(P_j)\eta^{c(j)}\} + 2\phi^{c(j)}(P_j)\dot{\eta}^{c(j)}] \\ &+ [\tilde{\omega}^{c(j)} + \mathrm{SKEW}\{\psi^{c(j)}(P_j)\dot{\eta}^{c(j)}\}][[\tilde{\omega}^{c(j)} + \mathrm{SKEW}\{\psi^{c(j)}(P_j)\dot{\eta}^{c(j)}\}]C_{c(j),p(j)}L^j\tau^j \\ &+ 2C_{c(j),p(j)}L^j\dot{\tau}^j] - \mathrm{SKEW}\{C_{c(j),p(j)}L^j\tau^j\}\tilde{\omega}^{c(j)}\psi^{c(j)}(P_j)\dot{\eta}^{c(j)} \end{aligned} \right\rangle \tag{8.16}$$

Equations (8.9), (8.15), and (8.16) are computed in what is the first forward pass going from body 1 to body NB, for use in a backward pass for generating the dynamical equations described in the next section.

8.4 Construction of the Dynamical Equations in a Block-Diagonal Form

It can be seen that an order-n formulation [6], producing an entirely diagonal mass matrix, is impossible for flexible multibody systems because of the presence of component modal coordinates. The objective here is to get the dynamical equations in a form such that the mass-matrix is at least block-diagonal. We will not repeat some of the basic equations. Thus, referring to Chapter 6, and ignoring rotational inertia terms in Eq. (6.11), it can be seen from the latter that the modal equations for a terminal body can be completed by collecting similar terms, in a slightly different matrix form, as:

$$E^j\ddot{\eta}^j + b^{j^T}a^{Q_j} + g^{j^T}\alpha^j - \Omega^j\Delta^j\omega^j + 2\Omega^j\rho^j\dot{\eta}^j + \Gamma_1^j a^{Q_j} + \Gamma_2^j\alpha^j + \Gamma_3^j + \lambda^j\eta^j = \int_j \phi_e^{j^T} df_e^j \tag{8.17}$$

Here E^j, b^j, g^j are, respectively, component generalized modal mass and modal integral matrices defined in the Appendix given at the end of the book; Ω^j is a $M_j \times 3M_j$ banded matrix with body components of the angular velocity ω^j appearing in each row starting with the diagonal, and Δ^j is identifiable as a $(3M_j \times M_j)$ matrix made up of 3x3 matrix form of modal dyadics $D_i^j, (i = 1, \ldots, M_j)$ defined in Appendix 5.A; ρ^j is an $(M_j \times M_j)$ modal integral implicitly defined so as to express in matrix form the term associated with $\dot{\eta}^j$ in Eq. (6.10) of Chapter 6. Nominal generalized structural stiffness is given by the $(M_j \times M_j)$ matrix λ^j in $\lambda^j\eta^j$ of Eq. (8.17). Geometric stiffness due to inertia forces is represented by the terms $\Gamma_1^j, \Gamma_2^j, \Gamma_3^j$ and is reviewed here from Chapter 6, for tailoring it to the block-diagonal formulation. Distributed inertia forces at the node at location x_i, y_i, z_i of a component is written, as before, in terms of 12 acceleration components, with U_3 being the 3×3 unity matrix, in:

$$\begin{Bmatrix} F_1 \\ F_2 \\ F_3 \end{Bmatrix}_{\text{inertia}} = -m_i[U_3 \quad x_iU_3 \quad y_iU_3 \quad z_iU_3]\begin{Bmatrix} d_1^j \\ \vdots \\ \vdots \\ d_{12}^j \end{Bmatrix} \tag{8.18}$$

Here $d_1^j, \ldots\ldots, d_{12}^j$ are defined by splitting acceleration and angular acceleration terms in the groups with and without derivatives of generalized speeds as:

$$\begin{Bmatrix} d_1^j \\ d_2^j \\ d_3^j \end{Bmatrix} = \{a_0^{Q_j} + a_t^{Q_j}\}$$

$$\begin{Bmatrix} d_4^j \\ d_5^j \\ d_6^j \end{Bmatrix} = \begin{Bmatrix} z_1^j \\ z_2^j \\ z_3^j \end{Bmatrix} - \begin{bmatrix} 0 & 0 & 0 \\ 0 & 0 & -1 \\ 0 & 1 & 0 \end{bmatrix}\{\alpha_0^{Q_j} + \alpha_t^{Q_j}\}$$

$$\begin{Bmatrix} a_7^j \\ a_8^j \\ a_9^j \end{Bmatrix} = \begin{Bmatrix} z_2^j \\ z_4^j \\ z_5^j \end{Bmatrix} - \begin{bmatrix} 0 & 0 & 1 \\ 0 & 0 & 0 \\ -1 & 0 & 0 \end{bmatrix} \{\alpha_0^{Q_j} + \alpha_t^{Q_j}\} \tag{8.19}$$

$$\begin{Bmatrix} a_{10}^j \\ a_{11}^j \\ a_{12}^j \end{Bmatrix} = \begin{Bmatrix} z_3^j \\ z_5^j \\ z_6^j \end{Bmatrix} - \begin{bmatrix} 0 & -1 & 0 \\ 1 & 0 & 0 \\ 0 & 0 & 0 \end{bmatrix} \{\alpha_0^{Q_j} + \alpha_t^{Q_j}\}$$

where

$$\begin{bmatrix} z_1^j & z_2^j & z_3^j \\ z_2^j & z_4^j & z_5^j \\ z_3^j & z_5^j & z_6^j \end{bmatrix} = \tilde{\omega}^j \tilde{\omega}^j$$

Motion-induced stiffness terms in Eq. (8.17) are now defined in terms of the geometric stiffness K_{gi}^j for unit values of the rigid body (or frame) acceleration components, $a_i^j, (i = 1, \ldots, 12)$ in Eq. (8.18) and the modal matrix ϕ^j and the modal coordinate η^j for body j as follows:

$$\begin{aligned} G_i^j &= \phi^{j^T} K_{gi}^j \phi^j \quad i = 1, \ldots, 12 \\ \Gamma_1^j &= [G_1^j \eta^j \quad G_2^j \eta^j \quad G_3^j \eta^j] \\ \Gamma_2^j &= [(G_9^j - G_{11}^j)\eta^j \quad (G_{10}^j - G_6^j)\eta^j \quad (G_5^j - G_7^j)\eta^j] \\ \Gamma_3^j &= G_4^j \eta^j z_1^j + G_5^j \eta^j z_2^j + G_6^j \eta^j z_3^j + G_7^j \eta^j z_2^j + G_8^j \eta^j z_4^j + G_9^j \eta^j z_5^j \\ &\quad + G_{10}^j \eta^j z_3^j + G_{11}^j \eta^j z_5^j + G_{12}^j \eta^j z_6^j \end{aligned} \tag{8.20}$$

In terms of the acceleration grouping used in Eq. (8.13), Eq. (8.17) can be written as:

$$E^j \ddot{\eta}^j = A^j \begin{Bmatrix} a_0^{Q_j} \\ \alpha_0^j \end{Bmatrix} + Y_1^j \tag{8.21}$$

where

$$A^j = -[b^{j^T} + \Gamma_1^j \quad g^{j^T} + \Gamma_2^j] \tag{8.22}$$

$$Y_1^j = -\left\{ b^{j^T} a_t^{Q_j} + g^{j^T} \alpha_t^{Q_j} - \Omega^j \Delta^j \omega^j + 2\Omega^j \rho^j \dot{\eta}^j + \Gamma_1^j a_t^{Q_j} + \Gamma_2^j \alpha_t^{Q_j} + \Gamma_3^j + \lambda^j \eta^j \right\} + \int_j \phi_e^{j^T} df_e^j \tag{8.23}$$

By D'Alembert's principle, the sum of inertia and external forces and torques on body j, f^{*j}, f_e^j, t^{*j}, t_e^j, and the total interaction force and torque at the hinge Q_j, from j to $c(j)$, $f_{c(j)}^j$, $t_{c(j)}^j$, is then stated as the (6×1) zero matrix:

$$0 = \begin{Bmatrix} f^{*j} - f_e^j - f_{c(j)}^j \\ t^{*j} - t_e^j - t_{c(j)}^j \end{Bmatrix} = M_1^j \begin{Bmatrix} a_0^{Q_j} \\ \alpha_0^j \end{Bmatrix} + M_2^j \ddot{\eta}^j + X^j - \begin{Bmatrix} f_{c(j)}^j \\ t_{c(j)}^j \end{Bmatrix} \tag{8.24}$$

As before, M_1^j, M_2^j, X^j are expressed in terms of the mass, expressed as a 3×3 diagonal matrix, and the deformation-dependent first and second moments of inertia about Q_j, or $s^{j/Q_j}, I^{j/Q_j}$:

$$M_1^j = \begin{bmatrix} m^j & -\tilde{s}^{j/Q_j} \\ \tilde{s}^{j/Q_j} & I^{j/Q_j} \end{bmatrix} \tag{8.25}$$

$$M_2^j = \begin{bmatrix} b^j \\ g^j \end{bmatrix} \tag{8.26}$$

$$X^j = M_1^j \begin{Bmatrix} a_t^{Q_j} \\ \alpha_t^j \end{Bmatrix} + \begin{Bmatrix} \tilde{\omega}^j \tilde{\omega}^j s^{j/Q_j} + 2\tilde{\omega}^j b^j \dot{\eta}^j - f_e^j \\ \tilde{\omega}^j I^{j/Q_j} \omega^j + 2 \sum_{k=1}^{M_j} D_k^{jT} \dot{\eta}_k^j \omega^j - t_e^j \end{Bmatrix} \tag{8.27}$$

Note in Eq. (8.25), we use the cross-product equivalent matrix $\tilde{s}^{j/Q_j}$ for the (3×1) matrix s^{j/Q_j}, and D_k^{jT} in Eq. (8.27) is, of course, the transpose of the modal matrix D_k^j defined in the Appendix 5.A. Now, using Eq. (8.21) in Eq. (8.24) and defining the following, which requires inversion of the model matrix E^j,

$$\begin{aligned} M_3^j &= M_1^j + M_2^j [E^j]^{-1} A^j \\ Y_2^j &= X^j + M_2^j [E^j]^{-1} Y_1^j \end{aligned} \tag{8.28}$$

one can rewrite Eq. (8.24) by first solving for $\ddot{\eta}^j$ from Eq. (8.21), using Eq. (8.28), and then using Eq. (8.14), as:

$$\begin{aligned} 0 = \begin{Bmatrix} f^{*j} - f_e^j - f_{c(j)}^j \\ t^{*j} - t_e^j - t_{c(j)}^j \end{Bmatrix} &\equiv M_3^j \begin{Bmatrix} a_0^{Q_j} \\ \alpha_0^j \end{Bmatrix} + Y_2^j - \begin{Bmatrix} f_{c(j)}^j \\ t_{c(j)}^j \end{Bmatrix} \\ &= M_3^j \left\{ \begin{pmatrix} \hat{a}_0^{Q_j} \\ \hat{\alpha}_0^j \end{pmatrix} + R^j \begin{pmatrix} \tau^j \\ \theta^j \end{pmatrix} \right\} + Y_2^j - \begin{Bmatrix} f_{c(j)}^j \\ t_{c(j)}^j \end{Bmatrix} \end{aligned} \tag{8.29}$$

Kane's dynamical equations [8] associated with the hinge dof for a terminal body can now be written by pre-multiplying Eq. (8.29) by the transpose of the partial velocity/partial angular velocity matrix, Eq. (8.12), avoiding multiplication by zeros in Eq. (8.12). This operation yields the equations for translational and rotational degrees of freedom at the hinge:

$$\begin{Bmatrix} \ddot{\tau}^j \\ \ddot{\theta}^j \end{Bmatrix} = -[\nu^j]^{-1} R^{j^T} \left[M_3^j \begin{Bmatrix} \hat{a}_0^{Q_j} \\ \hat{\alpha}_0^j \end{Bmatrix} + Y_2^j \right] + [\nu^j]^{-1} \begin{Bmatrix} f_h^j \\ t_h^j \end{Bmatrix}, \quad \text{where} \begin{Bmatrix} f_h^j \\ t_h^j \end{Bmatrix} = R^{j^T} \begin{Bmatrix} f_{c(j)}^j \\ t_{c(j)}^j \end{Bmatrix} \tag{8.30}$$

Here f_h^j, t_h^j are the known $(T_j \times 1)$ hinge force and $(R_j \times 1)$ hinge torque matrices, respectively, between body j and body $c(j)$, and we have inverted the $(R_j + T_j)$ square matrix

$$\nu^j = R^{j^T} M_3^j R^j \tag{8.31}$$

Using Eqs. (8.14) and (8.30) in Eq. (8.28), and the definitions

$$\begin{aligned} M &= M_3^j - M_3^j R^j [\nu^j]^{-1} R^{j^T} M_3^j \\ X &= Y_2^j - M_3^j R^j [\nu^j]^{-1} R^{j^T} Y_2^j + M_3^j R^j [\nu^j]^{-1} \begin{Bmatrix} f_h^j \\ t_h^j \end{Bmatrix} \end{aligned} \tag{8.32}$$

lead to a re-expression of Eq. (8.29) as:

$$\begin{Bmatrix} f_{c(j)}^j \\ t_{c(j)}^j \end{Bmatrix} = \begin{Bmatrix} f^{*j} - f_e^j \\ t^{*j} - t_e^j \end{Bmatrix} = M \begin{Bmatrix} \hat{a}_0^{Q_j} \\ \hat{\alpha}_0^j \end{Bmatrix} + X \tag{8.33}$$

At this stage extend the kinematical propagation equation by a shift operator, which was introduced in Ref. [4], to flexible bodies, relating the terms having to do with second derivatives in generalized coordinates in the acceleration expressions for points $Q_{c(j)}, P_j$ in Figure 8.3b,

$$\begin{Bmatrix} \hat{a}_0^{Q_j} \\ \hat{\alpha}_0^j \end{Bmatrix} = W^j \begin{Bmatrix} a_0^{Q_{c(j)}} \\ \alpha_0^{c(j)} \end{Bmatrix} + N^j \ddot{\eta}^{c(j)} \tag{8.34}$$

where the shift operator W^j and the matrix N^j based on modal matrix of body $c(j)$ for point P_j are as follows:

$$W^j = \begin{bmatrix} C_{c(j),j}^T & -C_{c(j),j}^T \tilde{r}_{c(j)}^{Q_{c(j)} Q_j} \\ 0 & C_{c(j),j}^T \end{bmatrix} \tag{8.35}$$

$$N^j = C^{j^T} \begin{Bmatrix} \phi^{c(j)}(P_j) - \text{SKEW}\{C_{c(j),p(j)} L^j \tau^j\} \psi^{c(j)}(P_j) \\ \psi^{c(j)}(P_j) \end{Bmatrix} \tag{8.36}$$

$$C^j = \begin{bmatrix} C_{c(j),j} & 0 \\ 0 & C_{c(j),j} \end{bmatrix} \tag{8.37}$$

In Eq. (8.35) $\tilde{r}_{c(j)}^{Q_{c(j)}Q_j}$ is a skew-symmetric matrix formed out of the (3×1) matrix whose elements are the components of the position vector from $Q_{c(j)}$ to Q_j, resolved in the $c(j)$ basis:

$$r_{c(j)}^{Q_{c(j)}Q_j} = r_{c(j)}^{Q_{c(j)}P_j} + \phi^{c(j)}(P_j)\eta^{c(j)} + C_{c(j),p(j)}L^j\tau^j \tag{8.38}$$

Dynamic equilibrium, Eq. (8.29), of the system of forces and moments at Q_j, expressed in the j basis, is stated by substituting Eq. (8.34) in Eq. (8.33), as the 6×1 matrix:

$$\begin{Bmatrix} f^{*j} - f_e^j \\ t^{*j} - t_e^j \end{Bmatrix}_{Q_j/j} = M\left\langle W^j \begin{Bmatrix} a_0^{Q_{c(j)}} \\ \alpha_0^{c(j)} \end{Bmatrix} + N^j\ddot{\eta}^{c(j)} \right\rangle + X \equiv \begin{Bmatrix} f_{c(j)}^j \\ t_{c(j)}^j \end{Bmatrix} \tag{8.39}$$

The interbody force applied by body j on body $c(j)$ at P_j is obtained by replacing the force system of Eq. (8.39) from Q_j to P_j. In the $c(j)$ basis this is expressed as follows:

$$d_j^{c(j)} \begin{Bmatrix} f_{c(j)}^j \\ t_{c(j)}^j \end{Bmatrix}_{P_j/c(j)} = d_j^{c(j)} \left[M\left\langle W^j \begin{Bmatrix} a_0^{Q_{c(j)}} \\ \alpha_0^{c(j)} \end{Bmatrix} + N^j\ddot{\eta}^{c(j)} \right\rangle + X \right] \tag{8.40}$$

Here we have defined the 6×6 transformation matrix:

$$d_j^{c(j)} = \begin{bmatrix} C_{c(j),j} & 0 \\ C_{c(j),j}\mathrm{SKEW}\{C_{p(j),j}^T L^j\tau^j\} & C_{c(j),j} \end{bmatrix} \tag{8.41}$$

Geometric stiffness caused by interbody loads on body $c(j)$ contributes to generalized active forces as follows, where only linear terms in $\eta^{c(j)}$ are kept:

$$F_{\text{interbody}}^{c(j)} = -\sum_{i=1}^{6} G_{12+i}^{c(j)}\eta^{c(j)} S_j^{c(j)} d_j^{c(j)} \left\langle MW^j \begin{Bmatrix} a_0^{c(j)} \\ \alpha_0^{c(j)} \end{Bmatrix} + X \right\rangle \tag{8.42}$$

Here G_{12+i}^j, $(i = 1, \ldots, 6)$ are 6 geometric stiffness matrices due to 6 interbody unit forces/moments, and $S_j^{c(j)}$ stands for a selection matrix generated from a 6×6 identity matrix, except for a zero row in the row number corresponding to the rotation axis number in the $p(j)$ frame about which axis moment cannot be transferred. Contributions to the modal equation for body $c(j)$ come from three sources: inertia and active forces on body $c(j)$, structural stiffness, and geometric stiffness due to inertia and interbody forces (see Eq. 8.18), and the equilibrium system of forces and moments acting at Q_j. With the help of Eqs. (8.13), (8.14), and (8.30), this yields:

$$E^{c(j)}\ddot{\eta}^{c(j)} - A^{c(j)} \begin{Bmatrix} a_0^{Q_{c(j)}} \\ \alpha_0^{c(j)} \end{Bmatrix} - Y_1^{c(j)} - F_{\text{interbody}}^{c(j)} + N^{j^T} \begin{Bmatrix} f^{*j} - f_e^j \\ t^{*j} - t_e^j \end{Bmatrix}_{Q_j/j} = 0 \tag{8.43}$$

Substituting Eqs. (8.39) and (8.42) in Eq. (8.43) and collecting terms, one gets the modal equations in the same form as Eq. (8.21),

$$E^{c(j)}\ddot{\eta}^{c(j)} = A^{c(j)} \begin{Bmatrix} a_0^{Q_{c(j)}} \\ \alpha_0^{c(j)} \end{Bmatrix} + Y_1^{c(j)} \tag{8.44}$$

where, in the form of Eq. (8.44), the following "replacements," indicated by arrows replacing the right-hand side expressions by the left-hand terms, have been incorporated:

$$\begin{aligned} E^{c(j)} &\leftarrow E^{c(j)} + N^{j^T} M N^j \\ A^{c(j)} &\leftarrow A^{c(j)} - N^{j^T} M W^j - \sum_{i=1}^{6} G_{12+i}^{c(j)} \eta^{c(j)} S_i^{c(j)} d_i^{c(j)} M W^j \\ Y_1^{c(j)} &\leftarrow Y_1^{c(j)} - N^{j^T} X - \sum_{i=1}^{6} G_{12+i}^{c(j)} \eta^{c(j)} S_i^{c(j)} d_i^{c(j)} X \end{aligned} \tag{8.45}$$

Now, forces and moments acting at Q_j can be replaced at $Q_{c(j)}$ and expressed in the $c(j)$ basis by using again the *shift operator* of Eq. (8.34):

$$\begin{Bmatrix} f^{*j} - f_e^j \\ t^{*j} - t_e^j \end{Bmatrix}_{Q_{c(j)}/c(j)} = W^{j^T} \begin{Bmatrix} f^{*j} - f_e^j \\ t^{*j} - t_e^j \end{Bmatrix}_{Q_j/j} \tag{8.46}$$

To this set of forces and moments is added the resultant of all the active and inertia forces and moments about $Q_{c(j)}$, yielding the zero-sum (6×1) expression:

$$\sum_{j \epsilon S_0} \begin{Bmatrix} f^{*j} - f_e^j \\ t^{*j} - t_e^j \end{Bmatrix} = M_1^{c(j)} \begin{Bmatrix} a_0^{Q_{c(j)}} \\ \alpha_0^{c(j)} \end{Bmatrix} + M_2^{c(j)} \ddot{\eta}^{c(j)} + X^{c(j)} \tag{8.47}$$

Here S_0 is the set of all bodies outward of the hinge $Q_{c(j)}$, and the following replacements state what the terms in (8.47) stand for:

$$\begin{aligned} M_1^{c(j)} &\leftarrow M_1^{c(j)} + W^{j^T} M W^j \\ M_2^{c(j)} &\leftarrow M_2^{c(j)} + W^{j^T} M N^j \\ X^{c(j)} &\leftarrow X^{c(j)} + W^{j^T} X \end{aligned} \tag{8.48}$$

Now, the solution for $\ddot{\eta}^{c(j)}$ is used from Eq. (8.44) in Eq. (8.47), and Eq. (8.14) is invoked with j replaced by $c(j)$, and Kane's dynamical equations associated with the translations and rotations at $Q_{c(j)}$ are written as before ($R^{c(j)}$ being the partial velocity matrix in the sense of Eq. 8.14):

$$\begin{Bmatrix} \ddot{\tau}^{c(j)} \\ \ddot{\theta}^{c(j)j} \end{Bmatrix} = -[\nu^{c(j)}]^{-1} R^{c(j)^T} \left[M_3^{c(j)} \begin{Bmatrix} \hat{a}_0^{Qc(j)} \\ \hat{\alpha}_0^{c(j)} \end{Bmatrix} + Y_2^{c(j)} \right] + [\nu^{c(j)}]^{-1} \begin{Bmatrix} f_h^{c(j)} \\ t_h^{c(j)} \end{Bmatrix} \tag{8.49}$$

This reproduces two cycles of the formulation of dynamical equations, covering the rotation/translation/modal dof of two bodies. This pattern is repeated until the equations for body *1* are written. At this stage, the inboard body linear accelerations/angular accelerations are zero or prescribed. If the zeroth body is the inertial frame (hat quantities are zero), Eq. (8.49) gives for $c(j) = 1$ the hinge translation and rotation,

$$\begin{Bmatrix} \ddot{\tau}^1 \\ \ddot{\theta}^1 \end{Bmatrix} = -[\nu^1]^{-1} R^{1^T} Y_2^1 + [\nu^1]^{-1} \begin{Bmatrix} f_h^1 \\ t_h^1 \end{Bmatrix} \tag{8.50}$$

and concomitantly, Eq. (8.14) reduces to:

$$\begin{Bmatrix} a_0^{Q_1} \\ \alpha_0^1 \end{Bmatrix} = R^j \begin{Bmatrix} \ddot{\tau}^1 \\ \ddot{\theta}^1 \end{Bmatrix} \tag{8.51}$$

This permits the computation of $\ddot{\eta}^j$ from the D'Alembert equation, Eq. (8.21) for $j = 1$. Once the dynamical equations for the base body are formed, the rest of the dynamical equations can be obtained by a second forward pass going up the tree, using the kinematical equations. The algorithm is block-diagonal because it breaks the dynamical equations into sub-blocks for each body j, with the sub-blocks requiring inversion of matrices of order M_j and $(T_j + R_j)$. The structure of the dynamical equations solved for the derivatives of the M_j and $(T_j + R_j)$ generalized speeds is schematically shown in Figure 8.4b for a system of one rigid body hinge-connected to three terminal flexible bodies, shown in Figure 8.4a.

The block-diagonal algorithm given above was used to simulate the Next Generation Space Telescope (NGST) and the Hubble Space Telescope by the author and his colleagues at Lockheed Martin Space Systems Co., shown in Fig. 8.5 as obtained from public domain sources.

8.5 Summary of the Block-Diagonal Algorithm for a Tree Configuration

8.5.1 First Forward Pass

Step 1: For $j = 1, \ldots$, NB (number of bodies) compute ω^j from Eq. (8.9), α_t^j from Eq. (8.15), $a_t^{Q_j}$ from Eq. (8.16), E^j (a constant matrix), A^j from Eq. (8.22), Y_1^j from Eq. (8.23), M_1^j from Eq. (8.25), M_2^j from Eq. (8.26), X^j from Eq. (8.27), W^j from Eq. (8.35), N^j from Eq. (8.36), and specify the selection matrix $S_j^{c(j)}$ in Eq. (8.42).

8.5.2 Backward Pass

Step 2: For $j = \text{NB}, \ldots, 1$ compute M_3^j, Y_2^j from Eq. (8.28), ν^j from Eq. (8.31), and M and X from Eq. (8.32). If $j = 1$, go to Step 5.

Step 3: Compute the updates defined in Eqs. (8.45) and (8.48).

Step 4: Replace j by $j - 1$ and go to step 2.

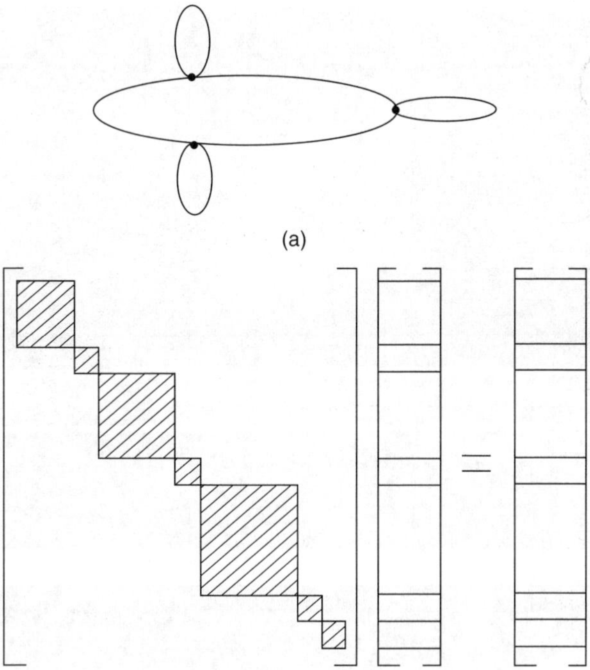

(a)

(b)

Figure 8.4 (a) Rigid Body with Three Articulated Flexible Appendages; (b) Structure of the Block Dynamical Equations for a Tree with a Rigid Body with Three Hinged Flexible Appendages; Base Rigid Body DOF Inertia at Bottom Corner. Banerjee [1].

8.5.3 Second Forward Pass

Step 5: For $j = 1$, evaluate in sequence Eqs. (8.50), (8.51), and (8.44). Set $j = 2$.

Step 6: Compute sequentially, Eqs. (8.34), (8.30), (8.14), and (8.44).

Step 7: If $j = \text{NB}$, stop; otherwise, replace j by $j + 1$ and go to step 6.

8.6 Numerical Results Demonstrating Computational Efficiency

A general purpose flexible multibody dynamics code based on the block-diagonal algorithm given here was developed by the author and his colleagues at the Lockheed Missiles & Space Company. One application involves two wrapped-rib antennas undergoing large deformation during deployment (see Figure 8.8). An antenna is modeled with 160 spring-connected rigid segments attached to an L-shaped elastic cantilever beam, with complex interbody forces and bodies connected to one another at 2 dof hinges by nonlinear springs, representing bending and torsion. Simulation of the deployment dynamics is done by the flexible multibody dynamics code based on the block-diagonal algorithm, and the algorithm involving a dense mass matrix given in Chapter 6. Both codes produce **exactly the same results**, but the **code using the block-diagonal algorithm given here runs 26 times faster** [9]. We present the ground test simulation results, together with the experimental results, at the end of this chapter.

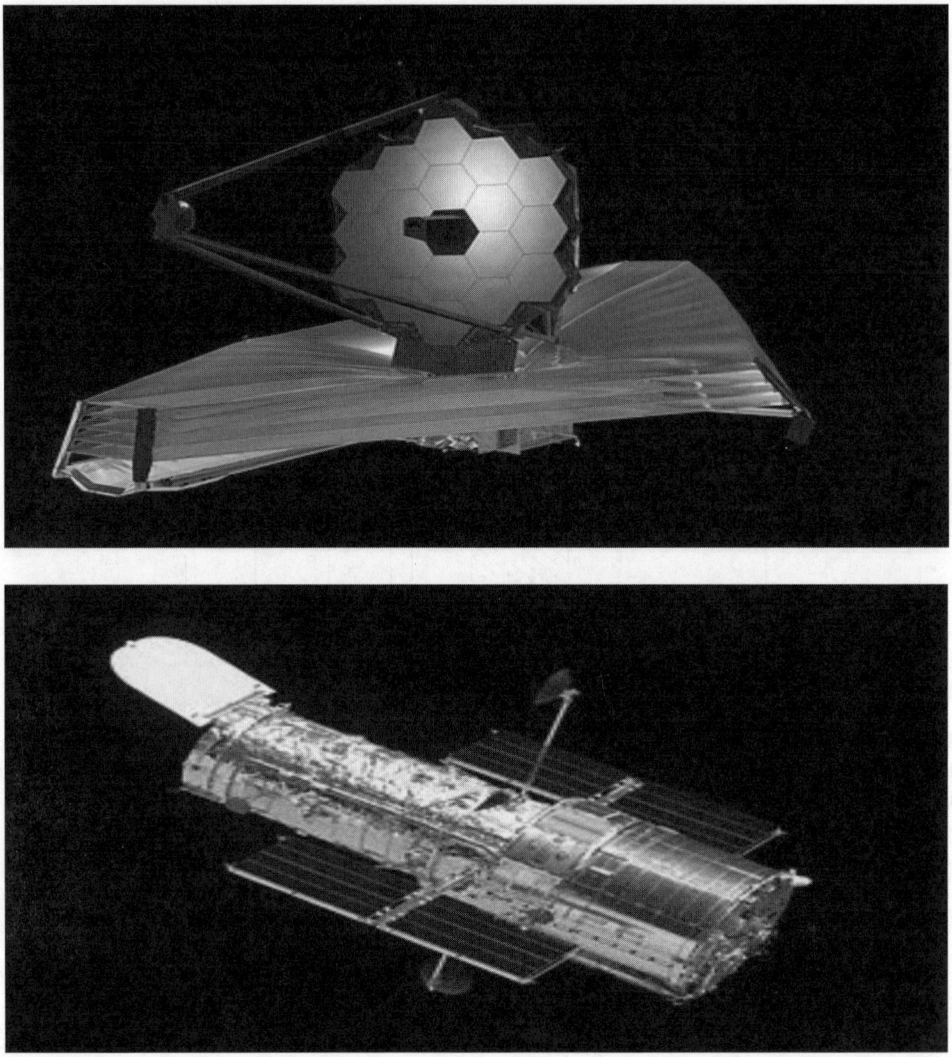

Figure 8.5 The Next Generation Space Telescope and the Hubble Space Telecope Simulated by the Block-Diagonal Algorithm at The Lockheed Martin Space Systems Company. Lemak and Banerjee [9].

8.7 Modification of the Block-Diagonal Formulation to Handle Motion Constraints

To motivate our discussion of systems with motion constraints, we give an example of a four-bar linkage that is commonly used for solar panel deployment in spacecraft. An actual solar panel deployment simulation is reported in Ref. [10] for the INSAT satellite built by the Ford Aerospace Corporation for the Indian Space Research Organization. The solar panel deployment for INSAT from the stowed to the deployed configuration, given in Figure 8.6,

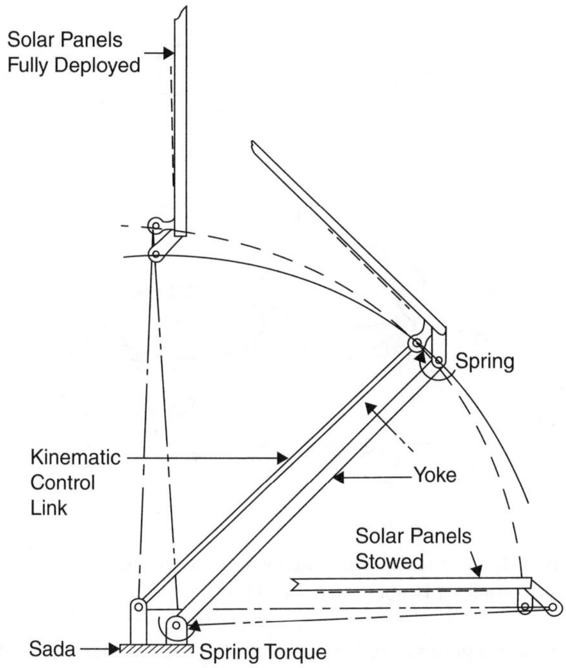

Figure 8.6 INSAT Solar Array Deployment Mechanism (Four-Bar Linkage). Wie, Furumoto, Banerjee, and Barba [10]. Reproduced with permission of the American Institute of Aeronautics and Astronautics, Inc.

shows the closed structural loop represented by the four-bar mechanism: the base, the kinematic control lik, the yoke, and a little bit of the panel.

When there are constraints on the motion of the system such as that represented by a closed loop of bodies, one approach that lends itself to the block-diagonal formulation is to cut the loop, impose equal and opposite unknown forces and torques that originally kept the loop closed, and solve for the unknown forces and torques by the additional consideration of the motion constraints. Figures 8.7a and 8.7b illustrate the closed-loop and cut-loop situations.

Let F_c and T_c denote the body measure numbers of the unknown constraint force and torque acting on one of the bodies at the joint that is cut, and introduce the notation:

$$\lambda = \begin{Bmatrix} F_c \\ T_c \end{Bmatrix} \tag{8.52}$$

Recognizing the body on which λ acts now as a "terminal" body j, denote

$$\begin{aligned} H_1^j &= \phi_c^{jT} - \sum_{i=1}^{6} G_{12+i}^j \eta^j S_i^j \\ H_2^j &= \begin{bmatrix} U & 0 \\ \tilde{r}_c^j & U \end{bmatrix} \end{aligned} \tag{8.53}$$

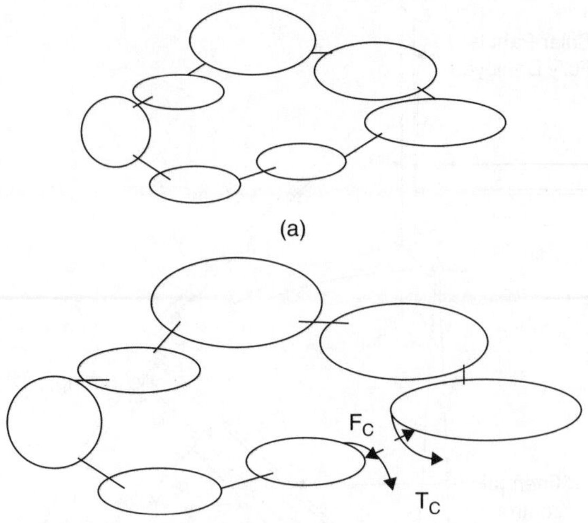

Figure 8.7 (a) System of Bodies in a Closed Kinematic Loop; (b) Same System with Loop Cut, Equal and Opposite Forces and Torques at Hinge. Banerjee [1].

where G^j_{12+i} refers to the ith generalized geometric stiffness matrix due to the interbody force and moment represented by λ; S^j_i is the 6×6 identity matrix except for a row of zeros corresponding to a rotation maxis about which moment cannot be transferred, and ϕ_c, r_c refer to the modal matrix at the cut joint and the position vector from Q_j of this terminal body to the cut, respectively, and U is a 3×3 identity matrix. For the terminal body at the other side of the cut simply reverse the sign of λ. Now with this additional set of forces and moments, Eqs. (8.21) and (8.24) become, respectively:

$$E^j \ddot{\eta}^j = A^j \begin{Bmatrix} a_0^{Q_j} \\ \alpha_0^j \end{Bmatrix} + Y_1^j + H_1^j \lambda \tag{8.54}$$

$$0 = \begin{Bmatrix} f^{*j} - f_e^j - f_{c(j)}^j \\ t^{*j} - f_e^j - t_{c(j)}^j \end{Bmatrix} = M_1^j \begin{Bmatrix} a_0^{Q_j} \\ \alpha_0^j \end{Bmatrix} + M_2^j \ddot{\eta}^j + X^j - \begin{Bmatrix} f_{c(j)}^j \\ t_{c(j)}^j \end{Bmatrix} - H_2^j \lambda \tag{8.55}$$

Using Eq. (8.54) in Eq. (8.55) leads to terms defined in Eq. (8.28) as before, and introducing the new term,

$$H_3^j = H_2^j - M_2^j [E_j]^{-1} H_1^j \tag{8.56}$$

produces a modification of the equations of motion from Eq. (8.30) for the hinge degrees of freedom to the following form:

$$\begin{Bmatrix} \ddot{\tau}^j \\ \ddot{\theta}^j \end{Bmatrix} = -[\nu^j]^{-1} R^{j^T} \left[M_3^j \begin{Bmatrix} \hat{a}_0^{Q_j} \\ \hat{\alpha}_0^j \end{Bmatrix} + Y_2^j - H_3^j \lambda \right] + [\nu^j]^{-1} \begin{Bmatrix} f_h^j \\ t_h^j \end{Bmatrix} \tag{8.57}$$

Note here that f_h^j, t_h^j are explicitly known interaction torques doing work. With this expression, the D'Alembert equation for dynamic equilibrium of external and inertia forces and torques at the hinge Q_j, is modified from Eq. (8.33) to

$$\begin{Bmatrix} f^{*j} - f_e^j \\ t^{*j} - t_e^j \end{Bmatrix} = M \begin{Bmatrix} \hat{a}_0^{Q_j} \\ \hat{\alpha}_0^j \end{Bmatrix} + X - H\lambda = 0 \tag{8.58}$$

where the definitions of M and X are the same as in Eq. (8.32), and we have introduced further:

$$H = H_3^j - M_3^j R^j [v^j]^{-1} R^{j^T} H_3^j \tag{8.59}$$

Equations corresponding to the vibration modal coordinates, Eq. (8.44), for the body inboard to the cut body reduces to:

$$E^{c(j)} \ddot{\eta}^{c(j)} = A^{c(j)} \begin{Bmatrix} a_0^{Q_{c(j)}} \\ \alpha_0^{c(j)} \end{Bmatrix} + Y_1^{c(j)} + H_1^{c(j)} \lambda \tag{8.60}$$

Here the notations of Eq. (8.45) are retained with no change, with geometric stiffness due to interbody forces on body $c(j)$ accounted for in the definition:

$$H_1^{c(j)} = \left[\sum_{i=1}^{6} G_{12+i}^{c(j)} \eta^{c(j)} S_i^{c(j)} d_i^{c(j)} - N^{j^T} \right] H \tag{8.61}$$

Finally, using the shift operator on forces and moments from Q_j to $Q_{c(j)}$ and the updates of Eq. (8.48), and defining further,

$$H_3^{c(j)} = H_2^{c(j)} + W^{j^T} H \tag{8.62}$$

the procedure given earlier for deriving equations of motion for hinge dof, Eq. (8.49), is modified to yield:

$$\begin{Bmatrix} \ddot{\tau}^{c(j)} \\ \ddot{\theta}^{c(j)j} \end{Bmatrix} = -[v^{c(j)}]^{-1} R^{c(j)^T} \left[M_3^{c(j)} \begin{Bmatrix} \hat{a}_0^{Qc(j)} \\ \hat{\alpha}_0^{c(j)} \end{Bmatrix} + Y_2^{c(j)} - H_3^{c(j)} \lambda \right] + [v^{c(j)}]^{-1} \begin{Bmatrix} f_h^{c(j)} \\ t_h^{c(j)} \end{Bmatrix} \tag{8.63}$$

In the two sets, Eqs. (8.54) and (8.57), or Eqs. (8.60) and (8.63), one needs to know the unknown constraint force and moments, whose measure numbers are denoted by λ. These constraint force and moment components can be solved by consideration of the motion constraint equations. However, a direct substitution of the constraint equations will destroy the block-diagonal nature of the coefficient matrix of the second derivatives of the hinge dof generalized coordinates in the dynamical equations. The algorithm given in the following preserves the block-diagonal character of the equations. Defining for $j = 1$,

$$\begin{aligned} f_1^1 &= -[v^1]^{-1} R^{1^T} Y_2^1 + [v^1]^{-1} \begin{Bmatrix} f_h^1 \\ t_h^1 \end{Bmatrix} \\ c_1^1 &= [v^1]^{-1} R^{1^T} H_3^1 \end{aligned} \tag{8.64}$$

This lets us start the sequence of generation of the equations of motion for body $j = 1$ with

$$\begin{Bmatrix} \ddot{\tau}^1 \\ \ddot{\theta}^1 \end{Bmatrix} = f_1^1 + c_1^1 \lambda \tag{8.65}$$

Now, use of Eq. (8.14), where the zeroth body is the inertial frame, yields

$$\begin{Bmatrix} a_0^{Q_1} \\ \alpha_9^1 \end{Bmatrix} = g^1 + h^1 \lambda \tag{8.66}$$

where

$$\begin{aligned} g^1 &= R^1 f_1^1 \\ h^1 &= R^1 c_1^1 \end{aligned} \tag{8.67}$$

Reference to Eq. (8.54) and use of the definitions, again for $j = 1$,

$$\begin{aligned} f_2^1 &= [E^1]^{-1} \left[A^1 g^1 + Y_1^1 \right] \\ c_2^1 &= [E^1]^{-1} \left[A^1 h^1 + H_1^1 \right] \end{aligned} \tag{8.68}$$

yield the equations for the modal coordinates of body 1 in the form:

$$\ddot{\eta}^1 = f_2^1 + c_2^1 \lambda \tag{8.69}$$

Use of Eqs. (8.66) and (8.69) and inserting the definitions

$$\begin{aligned} e^j &= W^j g^{c(j)} + N^j f_2^{c(j)} \\ n^j &= W^j h^{c(j)} + N^j c_2^{c(j)} \end{aligned} \tag{8.70}$$

in Eq. (8.34) lead to the equations:

$$\begin{Bmatrix} \hat{a}_0^{Q_j} \\ \hat{\alpha}_0^{Q_j} \end{Bmatrix} = e^j + n^j \lambda \tag{8.71}$$

When this is substituted in Eq. (8.57), and the following general notations are used,

$$\begin{aligned} f_1^j &= -[\nu^j]^{-1}R^{j^T}\left[M_3^j e^j + Y_2^j\right] + [\nu^j]^{-1}\begin{Bmatrix} f_h^j \\ t_h^j \end{Bmatrix} \\ c_1^j &= -[\nu^j]^{-1}R^{j^T}\left[M_3^j n^j - H_3^j\right] \end{aligned} \tag{8.72}$$

equations for the body *j* hinge degrees of freedom are obtained:

$$\begin{Bmatrix} \hat{\tau}^j \\ \ddot{\theta}^j \end{Bmatrix} = f_1^j + c_1^j \lambda \tag{8.73}$$

Use of Eqs. (8.71) and (8.73) in Eq. (8.14) with the substitutions

$$\begin{aligned} g^j &= e^j + R^j f_1^j \\ h^j &= n^j + R^j c_1^j \end{aligned} \tag{8.74}$$

yield the general relation:

$$\begin{Bmatrix} a_0^{Q_j} \\ \alpha_0^j \end{Bmatrix} = g^j + h^j \lambda \tag{8.75}$$

Finally, substitution of Eq. (8.75) in Eq. (8.54) and use of the notations

$$\begin{aligned} f_2^j &= [E^j]^{-1}\left[A^j g^j + Y_1^j\right] \\ c_2^j &= [E^j]^{-1}\left[A^j h^j + H_1^j\right] \end{aligned} \tag{8.76}$$

produce the equations for the modal coordinates for the *j*th flexible body:

$$\ddot{\eta}^j = f_2^j + c_2^j \lambda \tag{8.77}$$

With generic equations, Eqs. (8.73) and (8.77), a recursive formulation of the equations of motion in the block-diagonal form is completed for the closed-loop system with the loop cut. Now, λ involved in these equations can be solved for by the consideration of the constraints. We consider holonomic constraints, where the constraint conditions are expressed in the form of m *algebraic* equations of loop closure as:

$$f_j(q_1, \ldots, q_n) = 0, \quad j = 1, .., m; \quad m < n \tag{8.78}$$

One way of using these is to differentiate Eq. (8.78) twice with respect to time and use the resulting equations along with the dynamical equations for (n + m) unknowns, $\ddot{q}_1, \ldots, \ddot{q}_n, \lambda_1, \ldots, \lambda_m$. However this approach is known [11] to give rise to a drift with time in the constraint satisfaction at the position/orientation level. A procedure that satisfies the position, velocity, and acceleration constraints simultaneously is given below.

Two differentiations of the set of equations, Eq. (8.78), with respect to time yield,

$$J_i\dot{U}_i + J_d\dot{U}_d = -[\dot{J}_iU_i + \dot{J}_dU_d] \tag{8.79}$$

where the following notations for the Jacobian matrices J_i, J_d have been used,

$$[J_i] = \left[\frac{\partial f_j}{\partial q_i}\right]; \quad [J_d] = \left[\frac{\partial f_j}{\partial q_d}\right]; \quad U_i = \dot{q}_i; \quad U_d = \dot{q}_d \tag{8.80}$$

with the set of generalized coordinates partitioned into a subset q_i of independent generalized coordinates and a chosen subset q_d of dependent generalized coordinates, as many in number as the number of constraints, m. Now, the dynamical equations generated sequentially in Eqs. (8.73) and (8.77) for bodies $j = 1, \ldots, N$ can be stacked together and written as follows:

$$\begin{aligned} \dot{U}_i &= F_i + C_i\lambda \\ \dot{U}_d &= F_d + C_d\lambda \end{aligned} \tag{8.81}$$

Substitution of Eq. (8.81) in Eq. (8.79) provides the equations for λ, the column of measure numbers of the unknown constraint forces and moments:

$$[J_iC_i + J_dC_d]\lambda = -\{J_iF_i + J_dF_d + \dot{J}_iU_i + \dot{J}_dU_d\} \tag{8.82}$$

The dynamical differential equations are now explicitly known by using Eq. (8.82) in Eq. (8.81). In situations where the coefficient matrix in Eq. (8.82) is singular (which should not happen for independent constraints), the constraint stabilization method of Ref. [11] may be used, or a low-order integration with recursive error correction [12] can be used. In summary, the following set of differential-algebraic equations is solved for constrained dynamical systems:

$$\dot{U}_i = F_i - C_i[J_iC_i + J_dC_d]^{-1}\{J_iF_i + J_dF_d + \dot{J}_iU_i + \dot{J}_dU_d\} \tag{8.83}$$

$$\dot{q}_i = U_i \tag{8.84}$$

$$J_iU_i + J_dU_d = 0 \tag{8.85}$$

$$f(q_i, q_d) = 0 \tag{8.86}$$

8.8 Validation of Formulation with Ground Test Results

Results reported below of a comparison of simulation with the closed-loop block-diagonal algorithm given above and ground test measurements for a wrapped-rib antenna deployment are taken from the conference proceedings of the American Institute of Aeronautics and Astronautics [9]. Figure 8.8 is an overall picture of a test set up for an antenna with the support system of an L-shaped boom and a loop-closing cable. Figure 8.9 gives the details of the bending-torsion stiffness vs. angle of twist or bending for a rib element, where an antenna rib is segmented by many ribs connected by circumferential webs. Figures 8.10–8.12 show the shoulder reaction torques in the x-, y-, and z-direction as the antenna deploys from being initially wrapped around a spool. Figure 8.13 is for a closed structural loop, where the

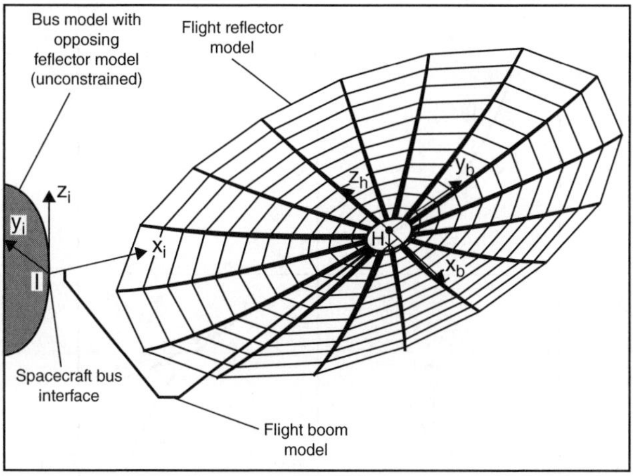

(a)

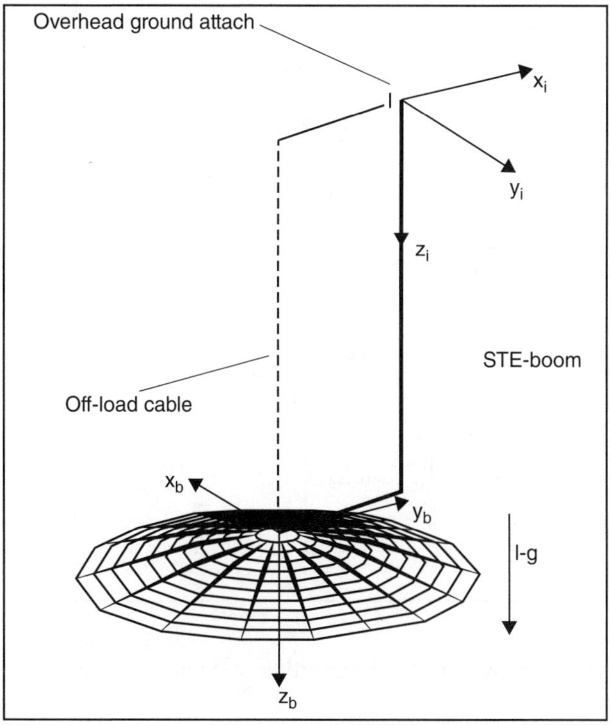

Flight Reflector Test Model

(b)

Figure 8.8 Ground Test Model of Antenna Attached to the Ceiling by an L-shaped Boom and an Off-Load Cable Forming a Closed-Loop Structure. Lemak and Banerjee [9].

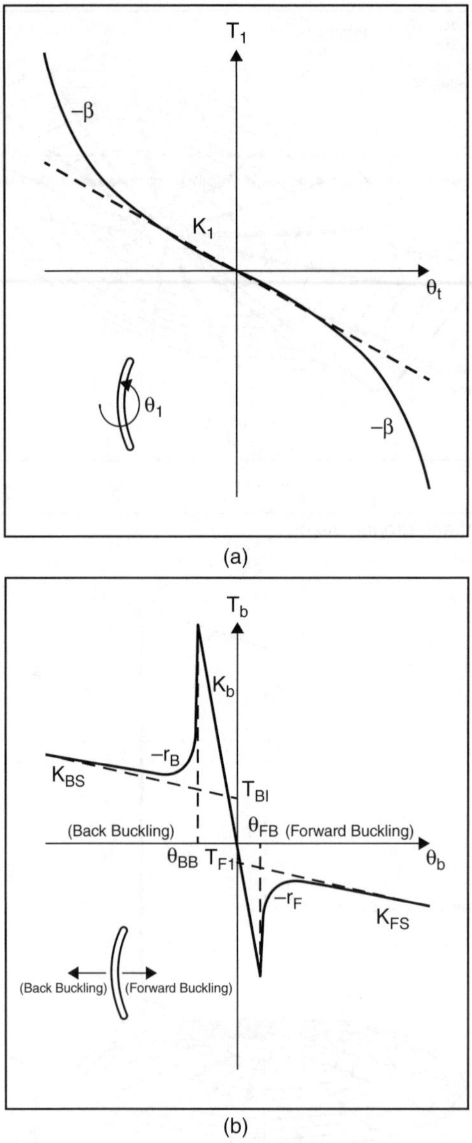

Figure 8.9 (a) Rib Element Torsional Nonlinear Stiffness; (b) Bending Stiffness Leading to Buckling. Lemak and Banerjee [9].

loop closes with the off-loading cable that tries to negate the gravity load on the suspended structure. While simulation results in Figure 8.13 seem to diverge, that may be symptomatic of the facts that a cable is an on-off spring and this fact was not modeled, and no constraint stabilization [11] was implemented in the formulation. All simulation results are obtained by the algorithm reported in this chapter. Considering the fact that many parameters necessary

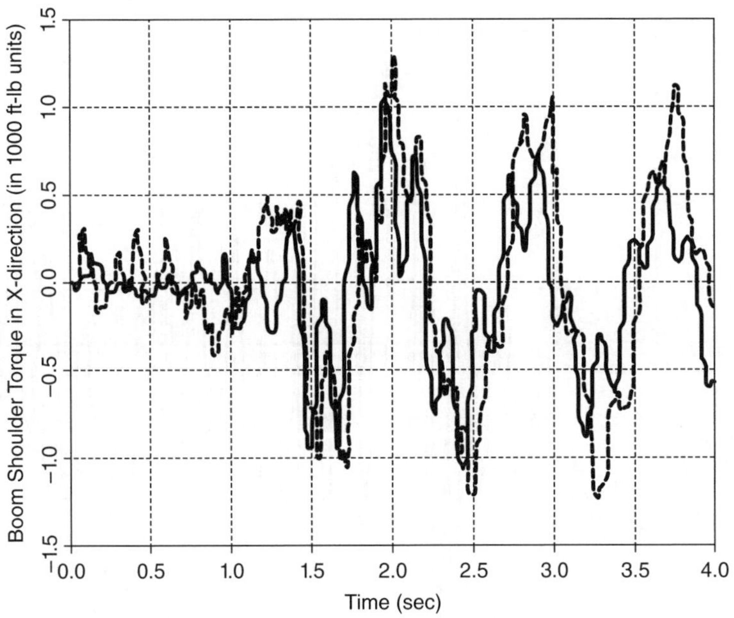

Figure 8.10 Boom Shoulder Torque in X-Direction, X-STQ, Comparison in Figure 8.7 for Deployment of Antenna as it Unfurls from a Spool (Test: Solid Line; Theory: Dashed Line). Lemak and Banerjee [9].

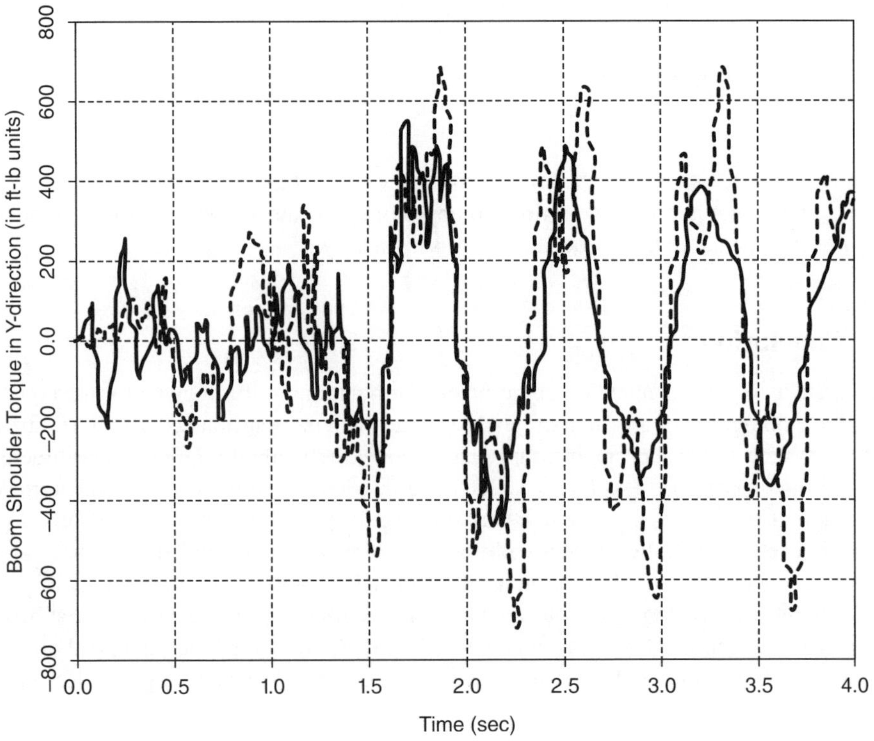

Figure 8.11 Boom Shoulder Torque in Y-Direction, Y-STQ, Comparison in Figure 8.7 due to Deployment of Antenna as it Unfurls from a Spool (Test: solid line; Theory: dashed line).

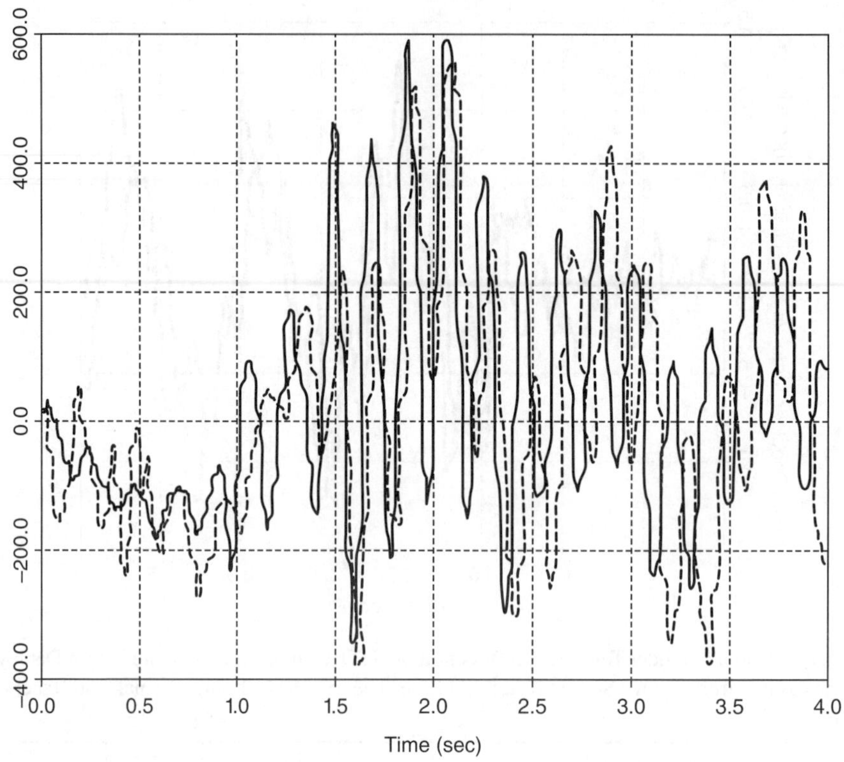

Figure 8.12 Boom Shoulder Torque Comparison in Z-Direction, Z-STQ, in Figure 8.6 due to Deployment of Antenna as It Unfurls from a Spool (Test: solid line; Theory: dashed line).

for the simulation could only be estimated approximately, the validation of the formulation with the experimental data may be deemed as satisfactory.

8.9 Conclusion

A comprehensive and computationally efficient formulation has been given for the equations of large overall motion of a system of flexible bodies with motion-induced geometric stiffness and motion constraints. For an arbitrary elastic body, the method requires a one-time computation of geometric stiffness matrices due to 12 distributed inertia loadings and at most six point-loadings per hinge connection. By accounting for these geometric stiffness effects, the formulation retains its validity even with the use of rotation rates higher than the vibration modes frequencies retained. Use of modes by themselves would have been an act of premature linearization, showing disastrous results when the rotation rate exceeds the first bending frequency of a component. For systems with motion constraints, a block-diagonal algorithm that satisfies the position, velocity, and acceleration constraints has been given. Results of a

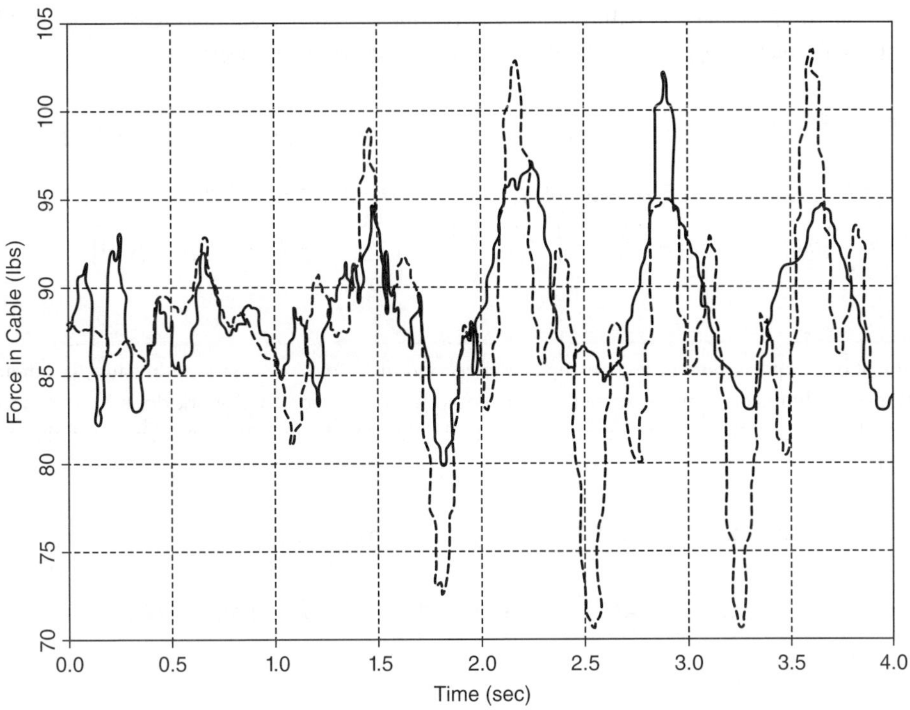

Figure 8.13 Off-Load Cable Tension, Z-STQ, Comparison as the Antenna Unfurls from a Spool (Test–Solid Line; Simulation—Dashed); Cable Closes the Structural Loop.

ground test of an antenna deployment, involving a closed structural loop, show reasonable correlation between the theory and the experiment.

Appendix 8.A An Alternative Derivation of Geometric Stiffness due to Inertia Loads

8.A.1 [Contributed by Dr. Tushar Ghosh of L3 Communications Corporation]

An interesting illustration of geometric stiffness as work done by the nonlinear terms in the acceleration for the existing stress state, as the body goes into further deformation, is shown here for the example system of Figure 8.1.

Let $\mathbf{i}, \mathbf{j}$ be unit vectors along and perpendicular to *OA* in Figure 8.1. The inertia force of link *AB* due to its lumped mass at *B*, and the inertia force of link BC due to its lumped mass at *C* are:

$$\begin{aligned} \mathbf{F}^{AB} &= -m(r+L)[\omega^2\mathbf{i} + \dot{\omega}\mathbf{j}] \\ \mathbf{F}^{BC} &= -m(r+2L)[\omega^2\mathbf{i} + \dot{\omega}\mathbf{j}] \end{aligned} \tag{8.A.1a}$$

Work done by F^{AB} due to small rotation q_1 from the reference state, and work done by F^{BC} due to small rotation $(q_1 + q_2)$ from this same reference state are, respectively:

$$W^{AB} = \mathbf{F}^{AB} \cdot \left[(L\cos q_1 - L)\mathbf{i} + L\sin q_1 \mathbf{j} \approx -\frac{1}{2}m(r+L)(\omega^2\mathbf{i} + \dot{\omega}\mathbf{j})\right] \cdot \left[L\left(-\frac{1}{2}q_1^2\mathbf{i} + Lq_1\mathbf{j}\right)\right] + \cdots$$

$$\begin{aligned} W^{BC} &= \mathbf{F}^{BC} \cdot < \{(L\cos q_1 - L) + [L\cos(q_1+q_2) - L]\}\mathbf{i} + \{L\sin q_1 + L\sin(q_1+q_2)\}\mathbf{j} > \\ &\approx -\frac{1}{2}m(r+2L)(\omega^2\mathbf{i} + \dot{\omega}\mathbf{j}) \cdot \left[L\left\{-\frac{1}{2}q_1^2 - \frac{1}{2}(q_1+q_2)^2\right\}\mathbf{i} + L\{q_1 + (q_1+q_2)\}\mathbf{j}\right] + \cdots \end{aligned} \tag{8.A.1b}$$

In the above expressions only terms up to the second order in q_1, q_2 are kept because geometric stiffness is obtained from the second-order terms in work done. Total work done by the inertia forces, from the reference configuration OAB to the deformed configuration shown in Figure 8.1, is then:

$$\begin{aligned} W^{AB} + W^{BC} &= \frac{1}{2}m\omega^2L\{(r+L)q_1^2 + (r+2L)\left[q_1^2 + (q_1+q_2)^2\right]\} \\ &= \frac{1}{2}m\omega^2L\left[(3r+5L)q_1^2 + 2(r+2L)q_1q_2 + (r+2L)q_2^2\right] \end{aligned} \tag{8.A.1c}$$

This can be rewritten as

$$W^{AB} + W^{BC} = \frac{1}{2}[q_1 \quad q_2]m\omega^2L\begin{bmatrix} 3r+5L & r+2L \\ r+2L & r+2L \end{bmatrix}\begin{Bmatrix} q_1 \\ q_2 \end{Bmatrix} = \frac{1}{2}[q_1 \quad q_2][K_g]\begin{Bmatrix} q_1 \\ q_2 \end{Bmatrix} \tag{8.A.1d}$$

where we have identified the geometric stiffness matrix for the "flexible body" ABC in Figure 8.1:

$$[K_g] = m\omega^2L\begin{bmatrix} 3r+5L & r+2L \\ r+2L & r+2L \end{bmatrix} \tag{8.A.1e}$$

This is the same result for geometric stiffness as in Eq. (8.6)!

Problem Set 8

8.1 **(a)** Show the details of the H^j matrix for a 2 or 3 dof rotational hinge between frames $p(j)$ and j in Figure 8.3b.

(b) Show that the inertia force at a node located by (x_i^n, y_i^n, z_i^n) from the point Q_j in Figure 8.3b is indeed given by Eqs. (8.18) and (8.19).

(c) Show that the geometric stiffness terms in Eq. (8.17) are indeed given by Eq. (8.20).

(d) Show that the dynamic equilibrium equation, Eq. (8.24), is made up of the constituent terms in Eqs. (8.25)–(8.27).

(e) Show that linear and angular acceleration terms involving second derivatives of the generalized coordinates, in Eq. (8.34), are indeed given by Eqs. (8.35)–(8.37).

8.2 Using the algorithm given in this chapter, formulate and code the equations of motion for the two-beam exercise problem, given at the end of Chapter 6, in a block-diagonal form.

8.3 Consider a four-link robotic manipulator on a movable base, used in construction sites for digging soil. Once the end link digs in to scoop up the soil, it is a closed-loop system, while it becomes a free four-link manipulator on the moving base, after the soil is scooped. Write an algorithm for the two stages of the earth-digging manipulator.

8.4 Formulate the block-diagonal equations for planar swinging motion of a triple pendulum made of three hinge-connected beams of uniform mass, with each beam of length L. Code the equations and simulate the motion with your choice parameters and initial conditions. Form the same equations with the dense matrix method of Chapter 6. Code these equations and simulate for the same parameters and initial conditions. Note the time difference in simulation.

References

[1] Banerjee, A.K. (1993) Block-diagonal equations for multibody elastodynamics with geometric stiffness and constraints. State of the Art Survey Lecture in Multibody Dynamics, European Space Agency, Noordwijk, June 1992; republished in the *Journal of Guidance, Control, and Dynamics*, 1992, **16**(6), 1092–1100.

[2] Cook, R.D. (1985) *Concepts and Applications of Finite Element Analysis*, McGraw-Hill, pp. 331–341.

[3] Banerjee, A.K. (2003) Contributions of multibody dynamics to space flight: A brief review. *Journal of Guidance, Control, and Dynamics*, **26**(3), 385–394.

[4] Kane, T.R., Ryan, R.R. Jr., and Banerjee, A.K. (1987) Dynamics of a cantilever beam attached to a moving base. *Journal of Guidance, Control, and Dynamics*, **10**(2), 139–151.

[5] Banerjee, A.K. and Lemak, M.E. (1991) Multi-flexible-body dynamics capturing motion induced stiffness. *Journal of Applied Mechanics*, **5**, 766–775.

[6] Rosenthal, D.E. (1990) An order-*n* formulation for robotic systems. *Journal of Astronautical Systems*, **38**(4), 511–530.

[7] Banerjee, A.K. and Dickens, J.M. (1990) Dynamics of an arbitrary flexible body in large rotation and translation. *Journal of Guidance, Control, and Dynamics*, **13**(2), 221–227.

[8] Kane, T.R. and Levinson, D.A. (1985) *Dynamics: Theory and Applications*, McGraw-Hill, p. 159.

[9] Lemak, M.K. and Banerjee, A.K. (1994) Comparison of Simulation with Test of Deployment of a Wrapped-Rib Antenna. AIAA Conference Paper AIAA-94-3577-CP.

[10] Wie, B., Furumoto, N., Banerjee, A.K., and Barba, P.M. (1986) Modeling and simulation of spacecraft solar array deployment. *Journal of Guidance, Control, and Dynamics*, **9**(5), 593–598.

[11] Park, K.C. and Chiou, J.C. (1988) Stabilization of computational procedures for constrained dynamical systems. *Journal of Guidance, Control, and Dynamics*, **11**(4), 365–370.

[12] Negrut, D., Jay, L.O., and Khude, N. (2009) A discussion of low-order numerical integration formulas for rigid and flexible multibody dynamics. *Journal of Computational and Nonlinear Dynamics*, **4**(2), 021008.

9

Efficient Variables, Recursive Formulation, and Multi-Point Constraints in Flexible Multibody Dynamics

In this chapter we consider efficient formulations of the equations of motion of complex flexible multibody systems with many motion constraints. To simplify equations of motion we use, for a flexible body, efficient generalized speeds, as defined in Appendix A at the end of the book. In the analysis we use these efficient variables while demonstrating their efficiency in reduced simulation times in contrast to more commonly used variables. The kernel of the algorithm given here consists in deriving the equations of motion for a single flexible body described in terms of these so-called efficient variables. As before, these equations are then extended to a multibody system and to closed structural loops by cutting the loops and exposing unknown constraint forces, in a recursive formulation. For constrained systems the final step is to impose the constraint conditions, which together with the dynamical equations form the system of equations to be solved numerically. The development closely follows that given in Refs. [1,2].

9.1 Single Flexible Body Equations in Efficient Variables

Refer to Figure 9.1 for a schematic representation of a flexible body j with a flying reference frame, also labeled j, fixed at the point O of the body. The body deforms with respect to this frame, and a generic point P' (not shown) of j moves to a point P of the frame due to local elastic deformation.

Inertial N-frame velocities ${}^N\mathbf{v}^P, {}^N\mathbf{v}^O$ of points P and O are written in terms of the angular velocity ${}^N\boldsymbol{\omega}^j$ of the flying reference frame j in Figure 9.1, the position vector $\mathbf{r}$ from O to P'

Flexible Multibody Dynamics: Efficient Formulations and Applications, First Edition. Arun K. Banerjee.

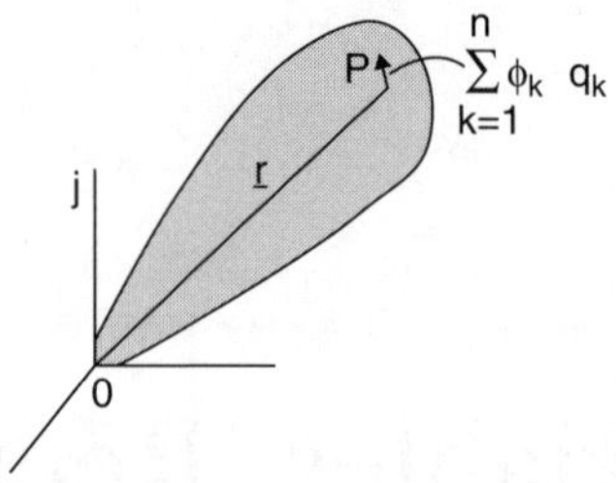

Figure 9.1 Schematic Representation of a Flexible Body with Modal Representation of Deformation. Banerjee and Lemak [1]. Reproduced with permission of Lockheed Martin Space Systems Company.

in the undeformed configuration, and the deformation vector from P' to P given by a linear combination of mode shapes $\boldsymbol{\phi}_k$ multiplied by modal coordinates, $q_k, k = 1, \ldots, n$, as follows:

$$^N\mathbf{v}^P = {}^N\mathbf{v}^O + {}^N\boldsymbol{\omega}^j x \left(\mathbf{r} + \sum_{k=1}^{n} \boldsymbol{\phi}_k q_k \right) + \sum_{k=1}^{n} \boldsymbol{\phi}_k \dot{q}_k \tag{9.1}$$

To make this chapter self-contained, we review here equations for a single flexible body with our choice of "efficient" generalized speeds $u_i, i = 1, \ldots 6 + n$, using vectors expressed in the basis vectors $\mathbf{b}_i, i = 1, 2, 3$ fixed in the frame j, as in Eqs. (9.2)–(9.4):

$$u_i = {}^N\boldsymbol{\omega}^j \cdot \mathbf{b}_i, \quad i = 1, 2, 3 \tag{9.2}$$

$$u_{3+i} = {}^N\mathbf{v}^O \cdot \mathbf{b}_i, \quad i = 1, 2, 3 \tag{9.3}$$

Efficient modal generalized speeds $u_7, \ldots, u_{6+n}$ were introduced [3] indirectly, with vibration modes for body j, indicated by a superscript j, for later convenience, as:

$$\sum_{k=1}^{n} \boldsymbol{\phi}^j_k u_{6+k} = {}^N\boldsymbol{\omega}^j \times \sum_{k=1}^{n} \boldsymbol{\phi}^j_k q_k + \sum_{k=1}^{n} \boldsymbol{\phi}^j_k \dot{q}_k \tag{9.4}$$

Use of Eqs. (9.2)–(9.4) in Eq. (9.1) leads to:

$$^N\mathbf{v}^P = \sum_{i=1}^{3} [u_{3+i}\mathbf{b}_i + u_i\mathbf{b}_i \times \mathbf{r}] + \sum_{k=1}^{n} \boldsymbol{\phi}^j_k u_{6+k} \tag{9.5}$$

Comparing Eq. (9.5) with Eq. (9.1), we see that Eq. (9.5) is free of modal coordinates. As we will see, this has the important consequence of producing equations of motion with a mass matrix that is free of modal coordinates when only a single flexible body, or a terminal body in a recursive formulation of a multibody system, undergoes large overall motion. Having a mass matrix free of modal coordinates is what makes $u_i, i = 1, \ldots 6 + n$ efficient, because customary ways of defining modal generalized speeds as derivatives of modal coordinates give rise to a time-varying mass matrix. We will see later that carrying over the idea to multibody systems also simplifies the equations, making simulations time-efficient. Dot-multiplying the terms in

Eq. (9.4) by ϕ_i^k and integrating over the body, we get the kinematical equations for the modal coordinates of body j in terms of the elastic generalized speeds, $u_7, \ldots, u_{6+n}$:

$$\sum_{k=1}^{n} E_{ki}^{j}(\dot{q}_k - u_{6+k}) = -{}^N\omega^j \cdot \sum_{k=1}^{n} q_k \int_j \phi_k^j \times \phi_i^j \, dm, \quad i = 1, \ldots, n \tag{9.6}$$

Here $E_{ki}^j = \int_j \phi_k^j \cdot \phi_i^j \, dm$ for the (k, i) element of ($n \times n$) "modal mass matrix" E_j of body j. Recall that we are using the index both for the flexible body and its reference frame j. Equation (9.6), describing a coupling between rotational and modal generalized speeds, can be written in the matrix form,

$$\dot{q} = \begin{bmatrix} -E_j^{-1} G_q^j & I \end{bmatrix} \begin{Bmatrix} \omega \\ \sigma \end{Bmatrix}; \quad \dot{q} = [q_1, \ldots.., q_n]^T; \quad \sigma = [u_7, \ldots \ldots, u_{6+n}]^T; \quad \omega = [\omega_1, \omega_1, \omega_1]^T \tag{9.7}$$

where the "centrifugal stiffness" matrix G_q^j, involving the cross-product in Eq. (9.6), also called "gyroscopic stiffness" matrix, is made up of terms containing integrals that are themselves free of modal coordinates (with superscript T denoting a matrix transpose):

$$G_q^j = \begin{bmatrix} \sum_{k=1}^{n} q_k \int_j \left\{ \tilde{\phi}_k^j \phi_1^j \right\}^T dm \\ \cdot \\ \cdot \\ \cdot \\ \sum_{k=1}^{n} q_k \int_j \left\{ \tilde{\phi}_k^j \phi_n^j \right\}^T dm \end{bmatrix} \tag{9.8}$$

Here the sign "~" denotes the usual skew-symmetric matrix (3 × 3) operator for matrix representation of cross products of two vectors, ϕ_k^j, ϕ_i^j. Following Kane's method, the generalized inertia force corresponding to the ith generalized speed u_i, $i = 1, \ldots, 6 + n$ is given by integrating the product of partial velocity with respect to the ith generalized speed and acceleration for a generic particle P:

$$F_i^* = -\int_j \frac{{}^N\partial^N \mathbf{v}^P}{\partial u_i} \cdot \frac{{}^N d^N \mathbf{v}^P}{dt} dm \quad i = 1, \ldots, 6 + n \tag{9.9}$$

Here the integration is carried out over the mass of the entire body j. Referring to Eq. (9.5) for evaluating the partial and total derivatives in Eq. (9.9), it is seen that the integrand in Eq. (9.9) is free of modal coordinates. This is, of course, true for a single flexible body. In fact, the complete set of (6 + n) generalized inertia forces for body j can be shown, as has been done in preceding chapters, to be written from Eq. (9.9) for the rigid and elastic generalized speeds for body j, indicated by subscripts r and e, in the generalized inertia force column matrix:

$$F_r^* = -M_1^j \begin{Bmatrix} \alpha_0^j \\ a_0^j \end{Bmatrix} - M_2^j \dot{\sigma}^j - X_0^j; \qquad X_0^j = M_1^j \begin{Bmatrix} \alpha_t^j \\ a_t^j \end{Bmatrix} + \begin{Bmatrix} \tilde{\omega}^j \tilde{\omega}^j S^{j/o} \\ \tilde{\omega}^j I^{j/o} \omega^j \end{Bmatrix} \tag{9.10}$$

$$F_e^* = -\left[M_2^j\right]^T \begin{Bmatrix} \alpha_0^j \\ a_0^j \end{Bmatrix} - E_j \dot{\sigma}^j - Y_0^j; \qquad Y_0^j = \left[M_2^j\right]^T \begin{Bmatrix} \alpha_t^j \\ a_t^j \end{Bmatrix} \tag{9.11}$$

Here α_0^j, a_0^j in Eqs. (9.10) and (9.11) are, respectively, terms involving derivatives of the generalized speeds in angular acceleration of frame j and linear acceleration of O (in Figure 9.1), as has been shown in Eq. (8.13), and X_0^j, Y_0^j in Eqs. (9.10) and (9.11) are generalized inertia force terms due to the remainder acceleration terms α_t^j, a_t^j in rigid body angular acceleration of frame j and acceleration of point O. Note that in Eqs. (9.10) and (9.11), M_1^j, M_2^j, E^j are **constant** matrices for body j. with

$$M_1^j = \begin{bmatrix} I^{j/o} & -\tilde{S}^{j/o} \\ \tilde{S}^{j/o} & m^j U \end{bmatrix} \tag{9.12}$$

$$M_2^j = \begin{bmatrix} \int_j \tilde{r}\phi dm \\ \int_j \phi dm \end{bmatrix} \tag{9.13}$$

and $I^{j/O}, S^{j/O}$ are, respectively, the matrices of second and first mass moments of inertia about O for body j. In Eq. (9.12), U is a (3×3) unity matrix, and quantities with overhead tilde sign indicate skew-symmetric matrices formed out of corresponding 3×1 matrices. See Chapter 6 for equations similar to Eqs. (9.12) and (9.13). As defined previously, generalized active force due to non-conservative force vectors $d\mathbf{f}^P$ at P for the ith generalized speed for body j is given by:

$$F_i^{nc} = \int_j \frac{{}^N\partial^N \mathbf{v}^P}{\partial u_i} \cdot d\mathbf{f}^P \quad i = 1, \ldots, 6+n \tag{9.14}$$

Again, Kane has shown in Ref. [4] that if the kinematical equations relating the derivatives of the generalized coordinates are related to the generalized speeds in the form,

$$\dot{q} = WU + Y \tag{9.15}$$

where U is a column matrix of generalized speeds and Y is a column matrix of generalized coordinates and time, then generalized force due to a potential function P and a dissipation function D is:

$$F^{cd} = -W^T \left\{ \frac{\partial P}{\partial q} + \frac{\partial D}{\partial \dot{q}} \right\} \tag{9.16}$$

If we take for P the sum of potential energy for mass-normalized modes of frequencies in ω_n, a diagonal matrix, and inertia-load dependent geometric stiffness, K_g, as in Ref. [5], and assume diagonal modal damping, then, for body j,

$$P = \frac{1}{2} q^T \left[\omega_n^2 + \phi^T K_g \phi\right] q \tag{9.17}$$

$$D = \xi \dot{q}^T \omega_n \dot{q} \tag{9.18}$$

and the generalized active force due to stiffness and damping follow from Eqs. (9.2), (9.7), and (9.16). For body j generalized speeds for rigid body rotations they give rise to the (6×1) matrix:

$$F_{\mathbf{r}}^{cd} = -\begin{bmatrix} -\left[G_q^j\right]^T \\ 0 \end{bmatrix} \left\{\left[\omega_n^2 + \phi^T K_g \phi\right] q + 2\xi\omega_n \dot{q}\right\} \tag{9.19}$$

Note that this choice of generalized speeds has yielded nonzero contributions due to stiffness and damping corresponding to rigid body rotation, just as shown in Ref. [3], which is rather peculiar for combining frame rotation and elasticity. Generalized forces due to structural elasticity and damping in the modal generalized speeds are given for body j in the customary manner by:

$$F_e^{cd} = -E_j \left\{\left[\omega_n^2 + \phi^T K_g \phi\right] q + 2\xi\omega_n \dot{q}\right\} \tag{9.20}$$

Generalized active forces due to forces and torques $\mathbf{F}^O, \mathbf{T}^O$ at a hinge at O and $\mathbf{F}^{Q_j}, \mathbf{T}^{Q_j}$ at a point Q_j in contact with an outboard body follow from Kane's method, as the sum of the vector dot-products:

$$F_i = \frac{{}^N\partial^N\mathbf{v}^O}{\partial u_i} \cdot \mathbf{F}^O + \frac{{}^N\partial^N\boldsymbol{\omega}^O}{\partial u_i} \cdot \mathbf{T}^O + \frac{{}^N\partial^N\mathbf{v}^{Q_j}}{\partial u_i} \cdot \mathbf{F}^{Q_j} + \frac{{}^N\partial^N\boldsymbol{\omega}^{Q_j}}{\partial u_i} \cdot \mathbf{T}^{Q_j}, \quad i = 1, \ldots., 6+n \tag{9.21}$$

Use of Eq. (9.7) for body j and the modal matrix $\psi_{Q_j}^j$ for elastic rotation at the node Q_j, the point Q of body j, lead to the inertial angular velocity of a nodal rigid body at Q_j, in matrix form,

$$\omega^{Q_j} = \omega^j + \psi_{Q_j}^j \dot{q} = P_{Q_j}^j \omega^j + \psi_{Q_j}^j \sigma^j \tag{9.22}$$

where, with U defining a 3×3 unity matrix,

$$P_{Q_j}^j = \left[U - \psi_{Q_j}^j E_j^{-1} G_q^j\right] \tag{9.23}$$

Generalized force due to hinge torque and force at points O and Q_j of body j in Figure 9.1 for the rigid body and elastic motion are written, assuming that elastic displacement at O is zero, as:

$$F_r = \begin{Bmatrix} T^O \\ F^O \end{Bmatrix} + Z_{Q_j} \begin{Bmatrix} T^{Q_j} \\ F^{Q_j} \end{Bmatrix}; \quad \text{where } Z_{Q_j} = \begin{bmatrix} P_{Q_j}^{jT} & \tilde{r}^{OQ_j} \\ 0 & U \end{bmatrix} \tag{9.24}$$

$$F_e = \Phi_{Q_j}^T \begin{Bmatrix} T^{Q_j} \\ F^{Q_j} \end{Bmatrix}; \quad \text{where } \Phi_{Q_j} = \begin{bmatrix} \psi^{Q_j} \\ \phi^{Q_j} \end{bmatrix} \tag{9.25}$$

Kane's dynamical equations [4], $F_i + F_i^* = 0, i = 1, \ldots 6 + n$, for a single free-flying flexible body j can now be revisited for rigid and flexible motion variables as two sets of equations, after collecting terms from the above development:

$$M_1^j \begin{Bmatrix} \alpha_0^j \\ a_0^j \end{Bmatrix} + M_2^j \dot{\sigma}^j + X_1^j = \begin{Bmatrix} T^O \\ F^O \end{Bmatrix} + Z_{Q_j} \begin{Bmatrix} T^{Q_j} \\ F^{Q_j} \end{Bmatrix} \tag{9.26}$$

$$[M_2^j]^T \begin{Bmatrix} \alpha_0^j \\ a_0^j \end{Bmatrix} + E_j \dot{\sigma}^j + Y_1^j = \Phi_{Q_j}^T \begin{Bmatrix} T^{Q_j} \\ F^{Q_j} \end{Bmatrix} \tag{9.27}$$

Here the contributions from body forces, remainder acceleration terms, and nominal and geometric stiffness and damping terms in the sequence of rotation, translation, and elastic vibration are lumped as:

$$X_1^j = X_0^j + \begin{Bmatrix} -\left(G_q^j\right)^T \\ 0 \end{Bmatrix} \{[\omega_n^2 + \phi^T K_g \phi]\, q + 2\xi\omega_n \dot{q}\} \tag{9.28}$$

$$Y_1^j = Y_0^j - E_j \{[\omega_n^2 + \phi^T K_g \phi]\, q + 2\xi\omega_n \dot{q}\} \tag{9.29}$$

9.2 Multibody Hinge Kinematics for Efficient Generalized Speeds

Mitiguy and Kane [6] proposes generalized speeds that are efficient in the sense that they lead to simpler equations of motion than are obtained otherwise for a system of rigid bodies undergoing rotations about revolute joints, 2 dof Hooke's joints, and spherical joints, as shown in Figure 9.2. They show that hinge relative rotations are not the best choice, and that equations of motion become simpler when the generalized speeds are chosen as the projections of the Newtonian angular velocity of the rotating body about the hinge axis, as shown in the following figure, taken from Ref. [6]. We now apply the modifications necessary for elastic joints. Consider two bodies, with reference frame j and $c(j)$, connected by rotational and translational joints, with body $c(j)$ inboard to body j along the path to body 1. Consider in Figure 9.3 first that body j is connected at P_j on $c(j)$ by a revolute joint

$u_1 = {}^N\omega^B \cdot a_1$

$u_1 = {}^N\omega^C \cdot a_1$

$u_2 = {}^N\omega^B \cdot b_2$

$u_i = {}^N\omega^B \cdot b_i \ (i = 1, 2, 3)$

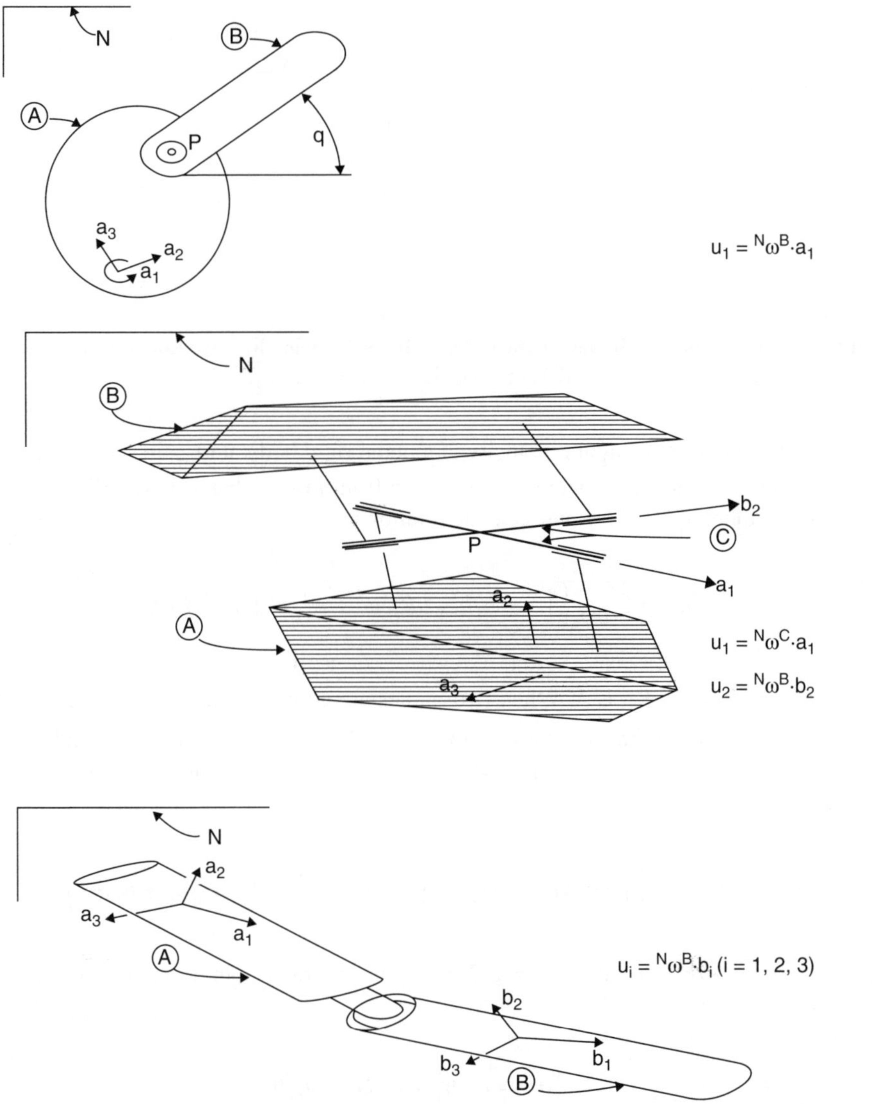

Figure 9.2   Efficient Generalized Speeds for Revolute Joint, Hooke's Joint, Spherical Joint. Mitiguy and Kane [6]. Reproduced with permission of SAGE Publications.

whose axis is given by a vector expressed by the (3×1) column matrix h_1 in the $c(j)$ basis. Reference [6] shows that an efficient rotational generalized speed u^j_{r1} for a revolute joint is such that:

$$\omega^j = C^T_{c(j),j} \left\{ \left[U - h_1 h_1^T \right] \omega^{Q_{c(j)}} + u^j_{r1} h_1 \right\} \tag{9.30}$$

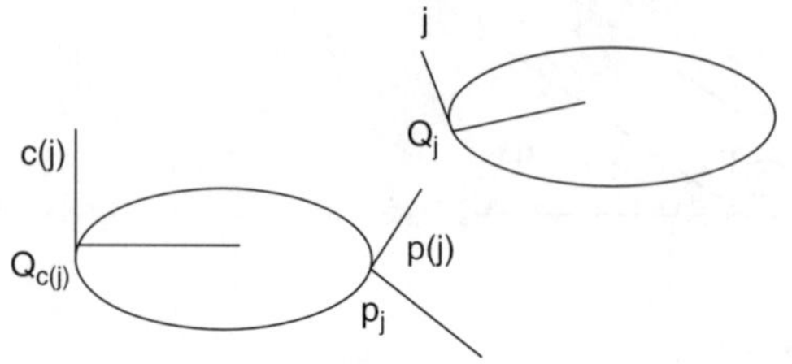

Figure 9.3 Two Adjacent Bodies Connected at a Hinge Allowing Relative Rotation and Translation in a System of Rigid and Flexible Bodies in a Topological Tree. Banerjee [2].

Here $\omega^{Q_{c(j)}}$ is the inertial angular velocity of a nodal rigid body at $Q_{c(j)}$ in $c(j)$ basis, and $C^T_{c(j),j}$ is the transpose of the coordinate transformation from j to $c(j)$ basis. Use of Eq. (9.22) in the above for an inboard body $c(j)$ with elastic rotation at Q yields:

$$\omega^j = C^T_{c(j),j}\left\{\left[U - h_1 h_1^T\right]\left(P^j_{Q_j}\omega^j + \psi^j_{Q_j}\sigma^j\right) + u^j_{r1} h_1\right\}; \tag{9.31}$$

Recall $P^j_{Q_j} = [U - \psi^j_{Q_j} E_j^{-1} G^j_q]$. When two bodies are connected by a 2 dof rotation joint, as shown in Figure 9.2, with the first rotation about the unit vector h_1 in $c(j)$ and the second rotation about an axis vector h_2 expressed in $c(j)$, angular velocity of j is formed by extension of Eq. (9.31) as:

$$\omega^j = C^T_{c(j),j}\left\{\left[U - h_1 h_1^T - h_2 h_2^T\right]\left(P^j_{Q_j}\omega^j + \psi^j_{Q_j}\sigma^j\right) + u^j_{r1} h_1 + u^j_{r2} h_2\right\} \tag{9.32}$$

Efficient generalized speeds $u_{ri} i = 1, 2, 3$ for a spherical joint are, as for the angular velocity:

$$^N\boldsymbol{\omega}^j = u_{r1}\mathbf{b}^j_1 + u_{r1}\mathbf{b}^j_2 + u_{r1}\mathbf{b}^j_3 \tag{9.33}$$

If relative translation is allowed between point P_j on body $c(j)$ and the point Q_j of body frame j, in a slider joint, as shown in Figure 9.3, then depending on the number of relative translational *dof*, one can introduce one, two, or three generalized speeds as follows,

$$\delta^j = \begin{Bmatrix} u^j_{t_1} \\ 0 \\ 0 \end{Bmatrix}; \quad \text{or} \quad \delta^j = \begin{Bmatrix} u^j_{t_1} \\ u^j_{t_2} \\ 0 \end{Bmatrix}; \quad \text{or} \quad \delta^j = \begin{Bmatrix} u^j_{t_1} \\ u^j_{t_2} \\ u^j_{t_3} \end{Bmatrix} \tag{9.34a}$$

where it is defined that

$$u_{t_i} = {}^{t(j)}\mathbf{v}^{Q_j/P_j} \cdot \mathbf{t}_i^j \quad i = 1, 2, 3 \tag{9.34b}$$

Here $\mathbf{t}_i^j$ are the basis vectors of a frame fixed on the flexible body $c(j)$ at node P_j, see Figure 9.3. Velocity of point Q_j is obtained by accounting for deformation at P_j of $c(j)$ and translation from P_j to Q_j:

$$v^{Q_j} = C^T_{c(j),j} \left\{ v^{Q_{c(j)}} + \tilde{\omega}^{c(j)} r^{Q_{c(J)}P_j} + \phi^{c(j)}(P_j)\sigma^{c(j)} + \tilde{\omega}^j C_{c(j),t(j)}\delta^j + C_{c(j),t(j)}\dot{\delta}^j \right\} \tag{9.35}$$

The expressions for angular accelerations and linear accelerations are obtained by differentiations in the Newtonian frame of the angular velocity and the velocity. For the purpose of developing a common notation for all types of joints, we define the following relationships as in Ref. [2]. As we have done before, acceleration terms involving derivatives of generalized speeds are indicated by subscript 0 and the remainder acceleration terms are denoted by subscript t; terms with an overhead caret refer to entities for the inboard link nodes, and R_j denotes the $6 \times (NR + NT)$ partial velocity matrix for the generalized speeds of the joint:

$$\begin{Bmatrix} \alpha^j \\ a^{Q_j} \end{Bmatrix} = \begin{Bmatrix} \alpha_0^j \\ a_0^{Q_j} \end{Bmatrix} + \begin{Bmatrix} \alpha_t^j \\ a_t^{Q_j} \end{Bmatrix} \tag{9.36}$$

$$\begin{Bmatrix} \alpha_0^j \\ a_0^{Q_j} \end{Bmatrix} = \begin{Bmatrix} \hat{\alpha}_0^j \\ \hat{a}_0^{Q_j} \end{Bmatrix} + R_j \begin{Bmatrix} \dot{u}_r^j \\ \dot{u}_t^j \end{Bmatrix} \tag{9.37}$$

$$\begin{Bmatrix} \hat{\alpha}_0^j \\ \hat{a}_0^{Q_j} \end{Bmatrix} = W_j \begin{Bmatrix} \alpha_0^{c(j)} \\ a_0^{Q_{c(j)}} \end{Bmatrix} + N_j \dot{\sigma}^{c(j)} \tag{9.38}$$

The matrices W_j, N_j encapsulate the details of the forward transfer of kinematical information from an inboard body to its outboard body. It can be shown that, for a revolute joint, together with a translational joint from P_j to Q_j, Eqs. (9.37) and (9.38) require the following matrix forms, where $\tilde{\delta}^j$ is a skew-symmetric matrix formed out of the elements in Eq. (9.34a).

$$W_j = \begin{bmatrix} C^T_{c(j),j}\left[U - h_1 h_1^T\right] P^j_{Q_j} & 0 \\ -C^T_{c(j),j}\left[\tilde{r}^{Q_{c(j)}P_j} + C_{c(j),t(j)}\tilde{\delta}^j C^T_{c(j),t(j)} P^j_{Q_j}\right] & C^T_{c(j),j} \end{bmatrix} \tag{9.39}$$

$$N_j = \begin{bmatrix} C^T_{c(j),j}\left[U - h_1 h_1^T\right] \psi^j_{P_j} \\ -C^T_{c(j),j}\left[\phi^{c(j)}_{P_j} - C_{c(j),t(j)}\tilde{\delta}^j C^T_{c(j),t(j)} \psi^{c(j)}_{P_j}\right] \end{bmatrix} \tag{9.40}$$

$$R_j = \begin{bmatrix} C^T_{c(j),j} h_1 & 0 \\ 0 & C^T_{t(j),j} \end{bmatrix} \tag{9.41}$$

In the context of Eq. (9.36), following Eq. (9.31) we also need:

$$\alpha_t^j = C_{c(j),j}^T \left\{ \left[U - h_1 h_1^T\right] \left[P_Q^j \alpha_t^{c(j)} - \rho^j\right] + \dot{q}_{r1}^j \tilde{\omega}^{P_j} h_1 \right\} \tag{9.42}$$

$$\rho^j = \psi_{Q_{c(j)}}^j E_{c(j)}^{-1} \dot{G}_q^j \omega^{c(j)} - \tilde{\omega}^{c(j)} \psi_{Q_{c(j)}}^j \dot{q}^{c(j)} \tag{9.43}$$

$$\begin{aligned} a_t^{Q_j} = C_{c(j),j}^T & \left\{ a_t^{Q_{c(j)}} - \tilde{r}^{Q_{c(j)}P_j} \alpha_t^{c(j)} + \tilde{\omega}^{c(j)} \left[\tilde{\omega}^{c(j)} \mathbf{r}^{Q_{c(j)}P_j} + \phi^{c(j)}(P_j)\sigma^{c(j)}\right] \right. \\ & \left. - C_{c(j),t(j)} \tilde{\delta}^j C_{c(j),t(j)}^T \left\{ P_Q^j \alpha_t^j - \rho^j \right\} + \tilde{\omega}^j \left[\tilde{\omega}^j C_{c(j),t(j)} \delta^j + 2C_{c(j),t(j)} \dot{\delta}^j\right] \right\} \end{aligned} \tag{9.44}$$

The expressions for angular acceleration for a 2 dof rotation joint with a translation joint, corresponding to Eqs. (9.39)–(9.41), follow from Eqs. (9.32)–(9.35):

$$W_j = \begin{bmatrix} C_{c(j),j}^T \left[U - h_1 h_1^T - h_2 h_2^T\right] P_{Q_j}^j & 0 \\ -C_{c(j),j}^T \left[\tilde{r}^{Q_{c(j)}P_j} + C_{c(j),t(j)} \tilde{\delta}^j C_{c(j),t(j)}^T P_Q^j\right] & C_{c(j),j}^T \end{bmatrix} \tag{9.45}$$

$$N_j = \begin{bmatrix} C_{c(j),j}^T \left[U - h_1 h_1^T - h_2 h_2^T\right] \psi_Q^j \\ C_{c(j),j}^T \left[\phi_{P_j}^{c(j)} - C_{c(j),t(j)} \tilde{\delta}^j C_{c(j),t(j)}^T \psi_{P_j}^{c(j)}\right] \end{bmatrix} \tag{9.46}$$

$$R_j = \begin{bmatrix} C_{c(j),j}^T [h_1 \quad h_2] & 0 \\ 0 & C_{t(j),j}^T \end{bmatrix} \tag{9.47}$$

and the equation similar to Eq. (9.42) takes the specific form for two rotations q_{r1}, q_{r2}.

$$\alpha_t^j = C_{c(j),j} \left\{ \left[U - h_1 h_1^T - h_2 h_2^T\right] \left(P_{Q_j}^j \alpha_t^{c(j)} - \rho^j\right) + \left[U - h_2 h_2^T\right] \dot{q}_{r1}^j \tilde{\omega}^Q h_1 + \dot{q}_{r2}^j \tilde{\omega}^j C_{c(j),j}^T h_2 \right\} \tag{9.48}$$

Similarly for body j connected to $c(j)$ by a spherical joint and, say, a 2 *dof* translational joint, the angular acceleration being

$$\alpha^j = \dot{u}_{r1} j_1 + \dot{u}_{r2} j_2 + \dot{u}_{r3} j_3 \tag{9.49}$$

the kinematical transfer matrices become:

$$W_j = \begin{bmatrix} 0 & 0 \\ -C_{c(j),j}^T \left[\tilde{r}^{Q_{c(j)}P_j} + C_{c(j),t(j)} \tilde{\delta}^j C_{c(j),t(j)}^T P_{Q_j}^j\right] & C_{c(j),j}^T \end{bmatrix} \tag{9.50}$$

$$N_j = \begin{bmatrix} 0 \\ C_{c(j),j}^T \left[\phi_{P_j}^{c(j)} - C_{c(j),t(j)} \tilde{\delta}^j C_{c(j),t(j)}^T \psi_{P_j}^{c(j)}\right] \end{bmatrix} \tag{9.51}$$

$$R_j = \begin{bmatrix} C_{c(j),j}^T & 0 \\ 0 & C_{t(j),j}^T \end{bmatrix} \tag{9.52}$$

In coding these equations, multiplications with the zero blocks are to be avoided, of course.

9.3 Recursive Algorithm for Flexible Multibody Dynamics with Multiple Structural Loops

The algorithm given in Ref. [1] for a system of hinge-connected flexible bodies in a tree configuration can now be extended to systems with many closed structural loops by the artifice of cutting the loops and imposing unknown constraint forces [7], while using the efficient variables and their associated kinematical transfer matrices. The general algorithm for the case of n bodies with m structural loops can be deduced by following a specific example. Consider the system in Figure 9.4a that shows five hinge-connected bodies with one closed loop, and with bodies 1 and 4 hinged to the ground. Figure 9.4b shows the same system after cutting the hinges connecting body 4 to the ground and the hinge connecting bodies 2 and 5 and replacing the action of the hinges with forces and torques; note that we have shown in Figure 9.4b, only for generality of notation, the constraint force $\boldsymbol{\lambda}_2$, which equals $(-\boldsymbol{\lambda}_1)$ for the structural loop. We will now form the dynamical equations of the system obtained by cutting the structural loops. The idea is to solve for the constraint forces by the additional equations of motion constraints.

9.3.1 Backward Pass

First, we do the kinematics on the system with loops cut, by recursively forming the acceleration of the hinge points and angular accelerations of all body frames, in a forward pass. To illustrate with a specific system, then, going backward starting with body 5, we assume a joint cut at outboard link node $L(5)$ of body 5 (note the change in notations for hinges, used here). From

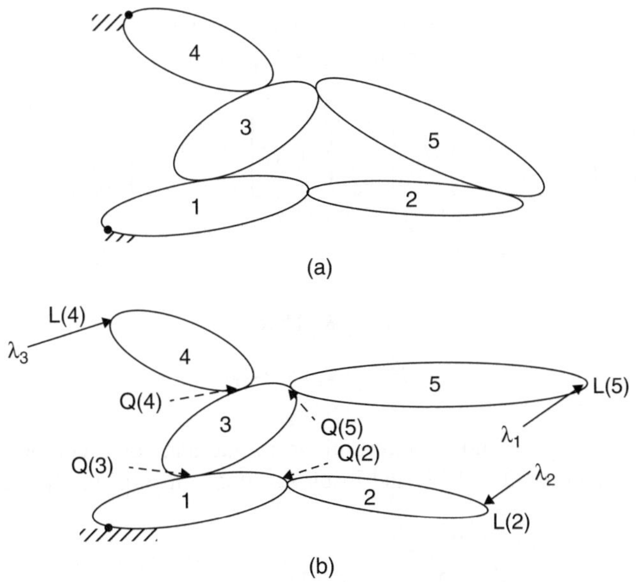

Figure 9.4 (a) Original Constrained System; (b) System Obtained by Cutting Structural Loops and Showing General (actually Equal and Opposite) Forces and Torques. Banerjee and Lemak [1]. Reproduced with permission of Lockheed Martin Space Systems Company.

Eq. (9.27) (recall that superscripts refer to a body number, not a power, except for inverse or transpose operations):

$$\dot{\sigma}^5 = E_5^{-1}\left[\left[M_2^5\right]^T \begin{Bmatrix} \alpha_0^5 \\ a_0^5 \end{Bmatrix} + Y_1^5 \right] + \hat{H}_{1e}^5 \lambda_1 \tag{9.53}$$

$$\hat{H}_{1e}^5 = E_5^{-1}\Phi_{L(5)}^T \tag{9.54}$$

Note here that λ_1 stands for the column matrix of the outboard hinge torque and force components exposed at $L(5)$, in Figure 9.4b. Putting Eq. (9.53) in Eq. (9.26) for $j = 5$ yields:

$$M_1^5 \begin{Bmatrix} \alpha_0^5 \\ a_0^5 \end{Bmatrix} + X_1^5 = \begin{Bmatrix} \tau_h^5 \\ f_h^5 \end{Bmatrix} + G_{51}\lambda_1 \tag{9.55}$$

Here τ_h^5, f_h^5 are working components of the total hinge torque and force, at the inboard hinge O (refer to Eq. 9.26) of body 5, and the following replacements have been made:

$$\begin{aligned} M_1^5 &\leftarrow M_1^5 + M_2^5 E_5^{-1} \left[M_2^5\right]^T \\ X_1^5 &\leftarrow X_1^5 + M_2^5 E_5^{-1} Y_1^5 \\ G_{51} &= Z_{L(5)} - M_2^5 E_5^{-1} \Phi_{L(5)}^T \end{aligned} \tag{9.56}$$

Now, using Eq. (9.37) for $j = 5$ and pre-multiplying the resulting equation by R_5^T yield the rotation and translation equations of motion from Eq. (9.55),

$$\begin{Bmatrix} \dot{u}_r^5 \\ \dot{u}_t^5 \end{Bmatrix} = -\nu_5^{-1} R_5^T \left[M_1^5 \begin{Bmatrix} \hat{\alpha}_0^5 \\ \hat{a}_0^5 \end{Bmatrix} + X_1^5 \right] + \nu_5^{-1} R_5^T \begin{Bmatrix} \tau_h^5 \\ f_h^5 \end{Bmatrix} + \hat{H}_{1r}^5 \lambda_1 \tag{9.57}$$

where

$$\nu_5 = R_5^T M_1^5 R_5 \tag{9.58}$$

$$\hat{H}_{1r}^5 = \nu_5^{-1} R_5^T G_{51} \tag{9.59}$$

The full set of working and non-working hinge torque and force components on body 5 are obtained by revisiting Eq. (9.55) and using Eq. (9.57) in Eq. (9.37) (see Figure 9.3 for connectivity),

$$\begin{Bmatrix} T_h^5 \\ F_h^5 \end{Bmatrix} = M_5 \begin{Bmatrix} \hat{\alpha}_0^5 \\ \hat{a}_0^5 \end{Bmatrix} + X_5 - B_{31}\lambda_1 \tag{9.60}$$

after the following replacements have been made:

$$S_5 = U - M_1^5 R_5 \nu_5^{-1} R_5^T \tag{9.61}$$

$$M_5 = S_5 M_1^5 \tag{9.62}$$

$$X_5 = S_5 X_1^5 + M_1^5 R_5 \nu_5^{-1} R_5^T \begin{Bmatrix} \tau_h^5 \\ f_h^5 \end{Bmatrix} \tag{9.63}$$

$$B_{31} = S_5 G_{51} \tag{9.64}$$

Bodies 2 and 4 in Figure 9.4b, being terminal bodies like body 5, contribute equations similar to Eqs. (9.53)–(9.64) with two important differences in the index for the subscripts: body index changes from 5 to 2 or 4, and terms associated with λ_1 such as B_{31} (first subscript for inboard body number and second subscript constraint force number) become B_{12} for λ_2 and B_{33} for λ_3 (see Figure 9.4b). The starting point for a body with outboard bodies, such as body 3 in Figure 9.4b, follows from Eq. (9.27), with $Q(4)$, $Q(5)$ being points on bodies 3 at which bodies 4 and 5 are connected, imposing working and non-working torques and forces $T_h^4, F_h^4, T_h^5, F_h^5$, respectively:

$$E_3 \dot{\sigma}^3 = M_2^3 \begin{Bmatrix} \alpha_0^3 \\ a_0^3 \end{Bmatrix} + Y_1^3 + \Phi_{Q(4)}^T \begin{Bmatrix} T_h^4 \\ F_h^4 \end{Bmatrix} + \Phi_{Q(5)}^T \begin{Bmatrix} T_h^5 \\ F_h^5 \end{Bmatrix} \tag{9.65}$$

Here $\Phi_{Q(4)}^T, \Phi_{Q(5)}^T$ are the transpose of the modal rotation-translation matrices at $Q(4)$, $Q(5)$. Using Eq. (9.60) for body 5 and its analogous equation for body 4 in Eq. (9.65) with subsequent use of Eq. (9.38) yields

$$\dot{\sigma}^3 = E_3^{-1} \left[M_2^3 \begin{Bmatrix} \alpha_0^3 \\ a_0^3 \end{Bmatrix} + Y_1^3 \right] + \hat{H}_{1e}^3 \lambda_1 + \hat{H}_{3e}^3 \lambda_3 \tag{9.66}$$

$$\hat{H}_{1e}^3 = E_3^{-1} \Phi_{Q(5)}^T d_3^5 B_{31} \tag{9.67}$$

$$\hat{H}_{3e}^3 = E_3^{-1} \Phi_{Q(4)}^T d_3^4 B_{33} \tag{9.68}$$

which incorporate the following replacements, lumping right-hand-side terms by symbols at left:

$$\begin{aligned} E_3 &\leftarrow E_3 + \Phi_{Q(4)}^T d_3^4 M_4 N_4 + \Phi_{Q(5)}^T d_3^5 M_5 N_5 \\ M_2^3 &\leftarrow M_2^3 - \Phi_{Q(4)}^T d_3^4 M_4 W_4 - \Phi_{Q(5)}^T d_3^5 M_5 W_5 \\ Y_1^3 &\leftarrow Y_1^3 - \Phi_{Q(4)}^T d_3^4 X_4 - \Phi_{Q(5)}^T d_3^5 X_5 \end{aligned} \tag{9.69}$$

Equations (9.67)–(9.69) have used the following matrix to transfer hinge force and torque from body j to body $c(j)$ for bodies 4 and 5 accounting for any simultaneous translation

δ^j at the joint, where $\tilde{\delta}^j$ is the skew-symmetric matrix corresponding to cross product with vector δ^j:

$$d^j_{c(j)} = -\begin{bmatrix} C_{c(j),j} & C_{c(j),t(j)}\tilde{\delta}^j C_{t(j),j} \\ 0 & C_{c(j),j} \end{bmatrix} \tag{9.70}$$

The rotation and translation equation for body 3, modeled after Eq. (9.26), becomes:

$$M_1^3 \begin{Bmatrix} \alpha_0^3 \\ a_0^3 \end{Bmatrix} + M_2^3 \dot{\sigma}^3 + X_1^3 = \begin{Bmatrix} T_h^3 \\ F_h^3 \end{Bmatrix} - Z_{Q(4)} d_3^4 \begin{Bmatrix} T_h^4 \\ F_h^4 \end{Bmatrix} - Z_{Q(5)} d_3^5 \begin{Bmatrix} T_h^5 \\ F_h^5 \end{Bmatrix} \tag{9.71}$$

Substitution from Eq. (9.60) and its counterpart for body 4 and use of the replacements,

$$\begin{aligned} M_1^3 &\leftarrow M_1^3 + Z_{Q(4)} d_3^4 M_4 W_4 + Z_{Q(5)} d_3^5 M_5 W_5 \\ M_2^3 &\leftarrow M_2^3 + Z_{Q(4)} d_3^4 M_4 N_4 + Z_{Q(5)} d_3^5 M_5 N_5 \\ X_1^3 &\leftarrow X_1^3 + Z_{Q(4)} d_3^4 X_4 + Z_{Q(5)} d_3^5 X_5 \end{aligned} \tag{9.72}$$

give rise to the equation:

$$M_1^3 \begin{Bmatrix} \alpha_0^3 \\ a_0^3 \end{Bmatrix} + M_2^3 \dot{\sigma}^3 + X_1^3 = \begin{Bmatrix} T_h^3 \\ F_h^3 \end{Bmatrix} + Z_{Q(4)} d_3^4 B_{33} \lambda_3 + Z_{Q(5)} d_3^5 B_{11} \lambda_1 \tag{9.73}$$

Now putting Eq. (9.66) in Eq. (9.73) and making the following replacements,

$$\begin{aligned} M_1^3 &\leftarrow M_1^3 + M_2^3 E_3^{-1} A_3 \\ X_1^3 &\leftarrow X_1^3 + M_2^3 E_3^{-1} Y_1^3 \\ G_{33} &\leftarrow Z_3^4 - M_2^3 E_3^{-1} \Phi^T_{Q(4)} \\ G_{31} &\leftarrow Z_3^5 - M_2^3 E_3^{-1} \Phi^T_{Q(5)} \end{aligned} \tag{9.74}$$

yield the following equations for body 3:

$$M_1^3 \begin{Bmatrix} \alpha_0^3 \\ a_0^3 \end{Bmatrix} + X_1^3 = \begin{Bmatrix} T_h^3 \\ F_h^3 \end{Bmatrix} + G_{33} d_3^4 B_{33} \lambda_3 + G_{31} d_3^5 B_{11} \lambda_1 \tag{9.75}$$

Use of Eq. (9.37) in the above and subsequent pre-multiplication by R_3^t yields:

$$\begin{Bmatrix} \dot{u}_r^3 \\ \dot{u}_t^3 \end{Bmatrix} = -\nu_3^{-1} R_3^T \left[M_1^3 \begin{Bmatrix} \hat{\alpha}_0^3 \\ \hat{a}_0^3 \end{Bmatrix} + X_1^3 \right] + \nu_3^{-1} \begin{Bmatrix} \tau_h^3 \\ f_h^3 \end{Bmatrix} + \hat{H}_{3r}^3 \lambda_3 + \hat{H}_{1r}^3 \lambda_1 \tag{9.76}$$

$$\hat{H}_{3r}^3 = \nu_3^{-1} R_3^T G_{33} d_3^4 B_{33} \tag{9.77}$$

$$\hat{H}_{1r}^3 = \nu_3^{-1} R_3^T G_{31} d_3^5 B_{31} \tag{9.78}$$

Derivatives of the generalized speeds for body 3 having been expressed as in Eqs. (9.66) and (9.76), total working and non-working hinge torque and force on body 3 are obtained from Eq. (9.71) as,

$$\begin{Bmatrix} T_h^3 \\ F_h^3 \end{Bmatrix} = M_3 \begin{Bmatrix} \hat{\alpha}_0^3 \\ \hat{a}_0^3 \end{Bmatrix} + X_1^3 - B_{11}\lambda_1 - B_{13}\lambda_3 \tag{9.79}$$

where, again, the following substitutions have been incorporated in sequence:

$$\begin{aligned} S_3 &= I - M_1^3 R_3 \nu_3^{-1} R_3^T \\ M_3 &= S_3 M_1^3 \\ X_1^3 &= S_3 X_1^3 + M_1^3 R_3 \nu_3^{-1} \begin{Bmatrix} \tau_h^3 \\ f_h^3 \end{Bmatrix} \\ B_{13} &= S_3 d_3^4 B_{33} \\ B_{11} &= S_3 d_3^5 B_{31} \end{aligned} \tag{9.80}$$

The dynamics of body 1 in Figure 9.4b is influenced by hinge torques and forces from both body 3 and body 2. With body 2 being a terminal body, derivatives of its generalized speeds are expressed just for body 5, and its hinge loads can be written as in the development of Eq. (9.60) as:

$$\begin{Bmatrix} T_h^2 \\ F_h^2 \end{Bmatrix} = M_2 \begin{Bmatrix} \hat{\alpha}_0^2 \\ \hat{a}_0^2 \end{Bmatrix} + X_2 - B_{12}\lambda_2 \tag{9.81}$$

With body 1 being an interior body, development of its dynamical equations is similarly identical to that of body 3. Thus we have the sequence of replacements,

$$\begin{aligned} E_1 &\leftarrow E_1 + \Phi_{Q(2)}^T d_1^2 M_2 N_2 + \Phi_{Q(3)}^T d_1^3 M_3 N_3 \\ M_2^1 &\leftarrow M_2^1 - \Phi_{Q(2)}^T d_1^2 M_2 W_2 - \Phi_{Q(3)}^T d_1^3 M_3 W_3 \\ Y_1^1 &\leftarrow Y_1^1 - \Phi_{Q(2)}^T d_1^2 X_2 - \Phi_{Q(3)}^T d_1^3 X_3 \end{aligned} \tag{9.82}$$

where $Q(2)$ and $Q(3)$ are points on body 1 to which body 2 and body 3 are connected. Dynamic equations for the modal coordinates of body 1 follow from the pattern as before:

$$\dot{\sigma}^1 = E_1^{-1}\left[M_2^1\begin{Bmatrix}\alpha_0^1\\ a_0^1\end{Bmatrix} + Y_1^1\right] + \hat{H}_{1e}^1\lambda_1 + \hat{H}_{2e}^1\lambda_2 + \hat{H}_{3e}^1\lambda_3 \tag{9.83}$$

$$\hat{H}_{1e}^1 = E_1^{-1}\Phi_{Q(3)}^T d_1^3 B_{13} \tag{9.84}$$

$$\hat{H}_{2e}^1 = E_1^{-1}\Phi_{Q(2)}^T d_1^2 B_{12} \tag{9.85}$$

$$\hat{H}_{3e}^1 = E_1^{-1}\Phi_{Q(3)}^T d_1^3 B_{13} \tag{9.86}$$

Again making the replacements similar to Eq. (9.72), that is,

$$\begin{aligned} M_1^1 &\leftarrow M_1^1 + Z_{Q(2)}d_1^2 M_2 W_2 + Z_{Q(3)}d_1^3 M_3 W_2 \\ M_2^1 &\leftarrow M_2^1 + Z_{Q(2)}d_1^2 M_2 N_2 + Z_{Q(3)}d_1^3 M_3 N_2 \\ X^1 &\leftarrow X^1 + Z_{Q(2)}d_1^2 X_2 + Z_{Q(3)}d_1^3 X_3 \end{aligned} \tag{9.87}$$

and replacements similar to Eq. (9.74), namely

$$\begin{aligned} M_1^1 &\leftarrow M_1^1 + M_2^1 E_1^{-1}[M_2^1]^T \\ X^1 &\leftarrow X^1 + M_2^1 E_1^{-1} Y_1^1 \\ G_{11} &\leftarrow Z_1^3 - M_2^1 E_1^{-1}\Phi_{Q(3)}^T \\ G_{12} &\leftarrow Z_1^2 - M_2^1 E_1^{-1}\Phi_{Q(2)}^T \\ G_{13} &\leftarrow G_{11} \end{aligned} \tag{9.88}$$

one obtains the equations for the rigid body rotation and translation of body 1:

$$\begin{Bmatrix}\dot{u}_r^1\\ \dot{u}_t^1\end{Bmatrix} = -\nu_1^{-1}R_1^t X^1 + \nu_1^{-1}\begin{Bmatrix}\tau_h^1\\ f_h^1\end{Bmatrix} + \hat{H}_{1r}^1\lambda_1 + \hat{H}_{2r}^1\lambda_2 + \hat{H}_{3r}^1\lambda_3 \tag{9.89}$$

Here we have used the notations:

$$\nu_1 = R_1^T M_1^1 R_1 \tag{9.90}$$

$$\hat{H}_{1r}^1 = \nu_1^{-1}R_1^T G_{11} d_1^3 B_{11} \tag{9.91}$$

$$\hat{H}_{2r}^1 = \nu_1^{-1}R_1^T G_{12} d_1^2 B_{12} \tag{9.92}$$

$$\hat{H}_{3r}^1 = \nu_1^{-1}R_1^T G_{13} d_1^3 B_{13} \tag{9.93}$$

9.3.2 Forward Pass

To produce the dynamical equations for the rest of the bodies, we rewrite Eq. (9.89) as

$$\begin{Bmatrix} \dot{u}_r^1 \\ \dot{u}_t^1 \end{Bmatrix} = F_{rf}^1 + H_{1r}^1\lambda_1 + H_{2r}^1\lambda_2 + H_{3r}^1\lambda_3 \tag{9.94}$$

where

$$F_{rf}^1 = -v_1^{-1}R_1^T Y_2^1 + v_1^{-1}\begin{Bmatrix} \tau_h^1 \\ f_h^1 \end{Bmatrix} \tag{9.95}$$

$$H_{ir}^1 = \hat{H}_{ir}^1 \quad i = 1, 2, 3 \tag{9.96}$$

If body 1 is connected to the ground, then kinematical equations in Eq. (9.37) reduce to:

$$\begin{Bmatrix} \alpha_0^1 \\ a_0^1 \end{Bmatrix} = R_1^T \begin{Bmatrix} \dot{u}_r^1 \\ \dot{u}_t^1 \end{Bmatrix} \tag{9.97}$$

Now with the use of the following defining relations,

$$S_f^1 = R_1 F_{rf}^1 \tag{9.98}$$

$$S_{ci}^1 = R_1 H_{ir}^1 \quad i = 1, 2, 3 \tag{9.99}$$

$$F_{ef}^1 = E_1^{-1}\left\{ Y_1^1 + A_1 S_f^1 \right\} \tag{9.100}$$

Equation (9.83) is written in terms of the elastic generalized speeds for body 1 as

$$\dot{u}_e^1 = F_{ef}^1 + H_{1e}^1\lambda_1 + H_{2e}^1\lambda_2 + H_{3e}^1\lambda_3 \tag{9.101}$$

where the following update of the constraint force coefficients has been made:

$$H_{ie}^1 = \hat{H}_{ie}^1 + E_1^{-1}A_1 S_{ci}^1 \quad i = 1, 2, 3 \tag{9.102}$$

Dynamical equations for body 1, as Eqs. (9.94) and (9.101), can be used to form the equations for the rest of the bodies with the help of the kinematical transfer matrices of Section 9.2. Thus appealing to Eq. (9.38) for body 2, and defining

$$\gamma_f^2 = W_2 S_f^1 + N_2 F_{ef}^1 \tag{9.103}$$

$$\gamma_{ci}^2 = W_2 S_{ci}^1 + N_2 H_{ie}^1, \quad 1, 2, 3 \tag{9.104}$$

one gets:

$$\begin{Bmatrix} \hat{\alpha}_0^2 \\ \hat{a}_0^2 \end{Bmatrix} = \gamma_f^2 + \gamma_{c1}^2\lambda_1 + \gamma_{c2}^2\lambda_2 + \gamma_{c3}^2\lambda_3 \tag{9.105}$$

Specifying $\hat{H}_{ir}^2 = 0, i = 1, 2, 3$, unless a nonzero value is prescribed, as is the case here of the example Figure 9.2 for $\hat{H}_{ir}^2$, we have:

$$H_{ir}^2 = \hat{H}_{ir}^2 - \nu_2^{-1} R_2^T M_1^2 \gamma_{ci}^2 \quad i = 1, 2, 3 \tag{9.106}$$

Derivatives of the rigid body generalized speeds for body 2 then follow with:

$$F_{rf}^2 = -\nu_2^{-1} R_2^T (X^2 + M_2^1 \gamma_f^2) + \nu_2^{-1} \begin{Bmatrix} \tau_2 \\ f_2 \end{Bmatrix} \tag{9.107}$$

$$\begin{Bmatrix} \dot{u}_r^2 \\ \dot{u}_t^2 \end{Bmatrix} = F_{rf}^2 + H_{1r}^2\lambda_1 + H_{2r}^2\lambda_2 + H_{3r}^2\lambda_3 \tag{9.108}$$

Now appealing to Eq. (9.37) for body 2 in the example and forming

$$S_f^2 = \gamma_f^2 + R_2 F_{rf}^2 \tag{9.109}$$

$$S_{ci}^2 = \gamma_{ci}^2 + R_2 H_{ir}^2 \quad i = 1, 2, 3 \tag{9.110}$$

$$\begin{Bmatrix} \alpha_0^2 \\ a_0^2 \end{Bmatrix} = S_f^2 + S_{c1}^2\lambda_1 + S_{c2}^2\lambda_2 + S_{c3}^2\lambda_3 \tag{9.111}$$

allow one to revisit the dynamical equations for the modal generalized speeds for body 2, specifying as before,

$$\hat{H}_{ir}^2 = 0 \quad i = 1, 2, 3 \tag{9.112}$$

unless a nonzero value exists, and updating

$$H_{ie}^2 = \hat{H}_{ie}^2 + E_2^{-1} M_1^2 S_{ci}^2 \quad i = 1, 2, 3 \tag{9.113}$$

to finally present, for this example system:

$$F_{ef}^2 = E_2^{-1}\left(Y_1^2 + M_1^2 S_f^2\right) \tag{9.114}$$

$$\dot{u}_e^2 = F_{ef}^2 + H_{1e}^2\lambda_1 + H_{2e}^2\lambda_2 + H_{3e}^2\lambda_3 \tag{9.115}$$

Equations for rigid and elastic generalized speeds for bodies 3, 4, and 5 in Figure 9.4b follow by continuing the recursion in the forward pass. An algorithm for the general case, of *nb* bodies with m loops, now emerge by induction from this representative example. A pseudo-code for the general case is given in the appendix at the end of this chapter.

9.4 Explicit Solution of Dynamical Equations Using Motion Constraints

So far the dynamical equations have been formed by retaining non-working constraint forces that are as yet unknown. Motion constraint conditions are now stated, in their acceleration forms, as many in number as there are components of unknown constraint forces, and these two sets of equations – that is, the dynamical equations and the constraint equations – are solved together. Of course, the constraint conditions are specific to a problem. For a loop formed by a chain of articulated flexible bodies with end-points of body 1 and body n pinned, cutting the loop at $L(n)$ of body n and using the dynamic equations to construct the acceleration and angular acceleration of $L(n)$ yield the strikingly simple form of the acceleration constraint:

$$\left[Z^T_{L(n)}S^n_c + \Phi_{L(n)}H^n_e\right]\lambda = \left\{\begin{array}{c}\alpha^{L(n)}\\ a^{L(n)}\end{array}\right\} - Z^T_{L(n)}S^n_f + \Phi_{L(n)}F^n_{ef} - \left\{\begin{array}{c}\alpha^{L(n)}_t\\ a^{L(n)}_t\end{array}\right\} \tag{9.116}$$

The bottom three of six rows of this matrix equation corresponding to acceleration constraint can be used to solve for λ. This is because the accelerations, and not the angular accelerations at the joint which may be a revolute, or other type of rotational joint, must be the same. For purposes of assigning initial conditions with the constraints, one must also write the position and velocity constraints.

Position constraints are generally of the nonlinear functional form, $f(q) = 0$, where q are the generalized coordinates, as shown previously, and must be solved by Newton's method, for dependent coordinates in terms of the independent coordinates [8]. Velocity constraints for a loop are linear, and can be written as the matrix equation, $[A_c(q)]\{U\} = 0$, where U is the column matrix of system generalized speeds. The constraint matrix $A_c(q)$ can be filled by recursively forming its entries corresponding to the generalized speeds U_j that represent the rotational, translational, and modal motion of body j, as follows, when one writes the angular velocity and velocity of the link node $L(n)$ of body n that is constrained in rotation and translation. Note the use of kinematical quantities defined in generating the dynamical equations:

$$\begin{aligned}\left\{\begin{array}{c}\omega_{L(n)}\\ v_{L(n)}\end{array}\right\} &= [Z_{L(n)}R_n\Phi_{L(n)}]\{U_n\} + Z_{L(n)}[W_nR_{c(n)}N_n]\{U_{c(n)}\}\\ &\quad + Z_{L(n)}W_n[W_{c(n)}R_{c^2(n)}N_{c(n)}]\{U_{c^2(n)}\} + \dots\dots;\quad c^2(n) \equiv c(c(n))\end{aligned} \tag{9.117}$$

Recall that in contrast to the block-diagonal nature [2] of the mass matrices involved in the recursive algorithm given here, a non-recursive formulation with augmented generalized speeds available in Refs. [7, 8] produce the dynamical and constraint equations:

$$\begin{aligned}[M(q)]\{\ddot{q}\} &= \{C(q,\dot{q})\} + [A_c(q)]^T\{\Lambda\}\\ [A_c(q)]\{\ddot{q}\} &= \{D(q,\dot{q})\}\end{aligned} \tag{9.118}$$

Here $M(q)$ in Eq. (9.118) is a dense, time-varying, square matrix of size equal to the augmented dof of the system, and hence getting $\ddot{q}$ is computationally more expensive compared to the algorithm given here. Of course, constraints in both these approaches can be "stabilized," with no drift, by the procedures given in Chapter 8, with as accurate constraint satisfaction as desired. Finally, note that the dynamical equations are to be integrated together with the kinematical equations for large translation and rotation for body 1, as is customary [9, 13].

9.5 Computational Results and Simulation Efficiency for Moving Multi-Loop Structures

Numerical solutions of the differential equations of dynamics with nonlinear algebraic constraints equations are most accurately done by using differential algebraic solvers, such as DASSL [10]. In the results presented next, we use an alternative procedure [7] of solving the dynamical equations together with the acceleration form of the constraint differential equations by the well-known Kutta-Merson integrator, and then overriding the solution of dependent generalized coordinates, by their solution obtained by Newton's method, in terms of the independent coordinates. A word of caution: this approach is not immune to singularity, which happens when two constraint equations are identical, and this needs to be checked [10, 11, 13].

9.5.1 Simulation Results

Large-angle slewing of a *single* flexible body, a solar sail spacecraft with 6 rigid body *dof* and 44 vibration modes, is simulated first for a bang-bang torque application, using *three formulations*. One follows the efficient generalized speeds for a single flexible body, and is called method I here, given in Section 9.1. The latter two use Kane's method: one using the dense mass matrix of Ref. [3], called method II, and the other using the block-diagonal mass matrix feature of Ref. [2], called method III, with both using the generalized speeds, $u_i = \dot{q}_i, i = 1, \dots, 6 + n$ where q_i includes x-, y-, z- coordinates and the three Euler angles. Table 9.1 shows the comparison of simulation times.

Figures 9.5 and 9.6 show actually **three** plots from the three formulations. The results were identical for a 300 sec simulation.

We now consider the same multibody system considered before and reproduced here in Figure 9.7 for a system of flexible bodies connected by a revolute joint, a Hooke's joint

Table 9.1 Simulation Time Comparison of Order-n Algorithm with Efficient Generalized Speeds vs. Extended Kane Theory for a Constrained Whirling Flexible Chain.

Number of modes	Number of gen. speed	CPU sec for 10 s simulation with extended Kane	Ratio of extended Kane over $O(n)$-Eff
4	35	40.75	1.01
12	75	128.61	2.49

Banerjee and Lemak [1]. Reproduced with permission of Lockheed Martin Space Systems Company.

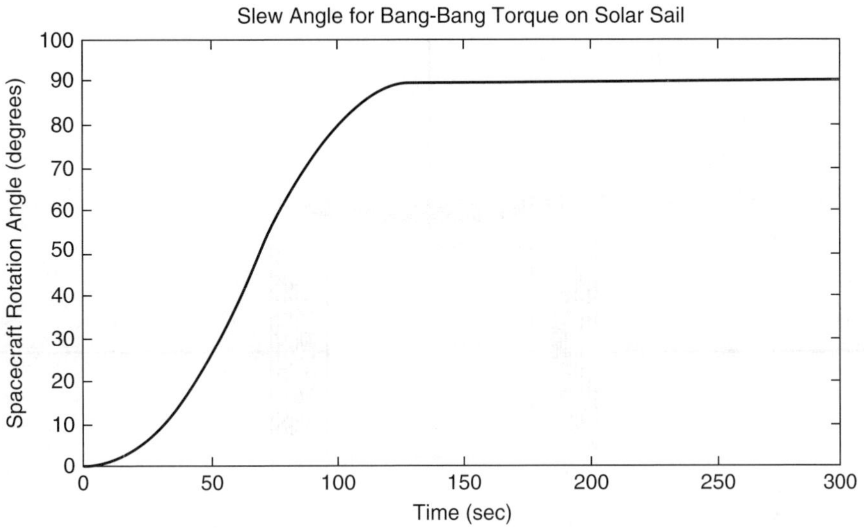

Figure 9.5 Solar Sail Slew Angle vs. Time for a 90 deg Slewing Maneuver due to Bang-Bang Torque, Showing Indistinguishable Results Obtained by Three Simulation Methods. Banerjee and Lemak [1]. Reproduced with permission of Lockheed Martin Space Systems Company.

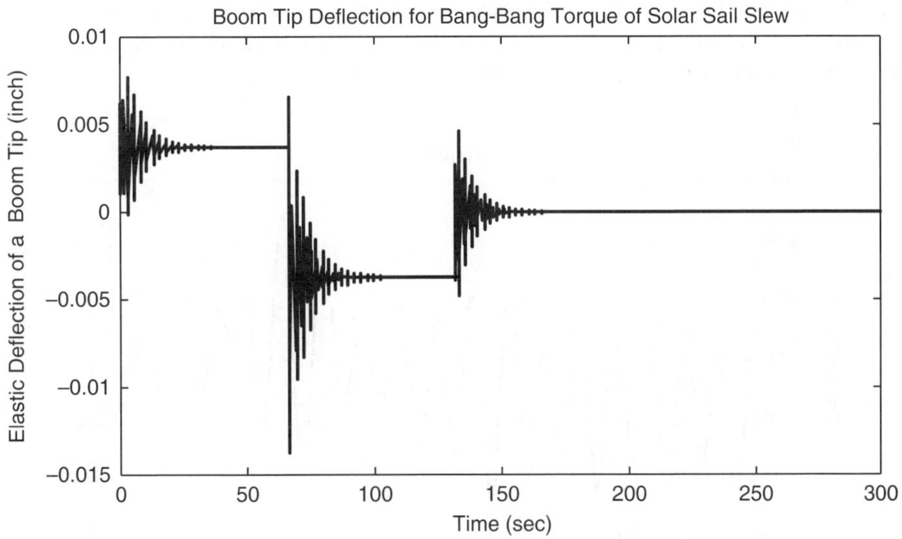

Figure 9.6 Boom Tip Deflection vs. Time for Bang-Bang Torque for Solar Sail Slewing by Three Methods. Banerjee and Lemak [1]. Reproduced with permission of Lockheed Martin Space Systems Company.

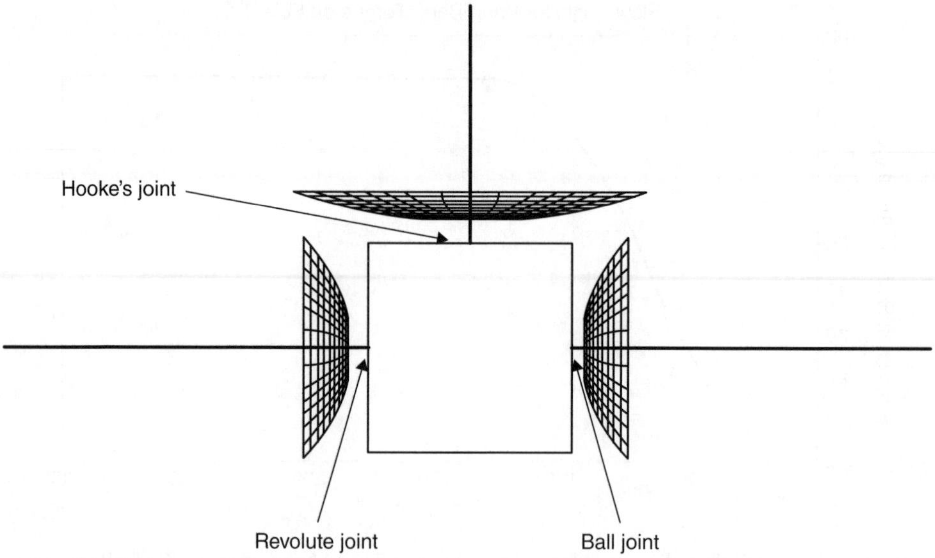

Figure 9.7 Flexible Spacecraft Connected to Three Flexible Antennas via Hooke's Joint, Revolute Joint, Ball Joint. Banerjee and Lemak [1]. Reproduced with permission of Lockheed Martin Space Systems Company.

with two rotational dof, and a spherical ball joint. Each body is modeled by 10 vibration modes. Figure 9.8 shows the internal torque required to prescribe zero rotation for locking one rotational dof of the ball joint, by the same three methods, the analysis requiring a simple prescribed motion modification of the recursive algorithm of Ref. [2].

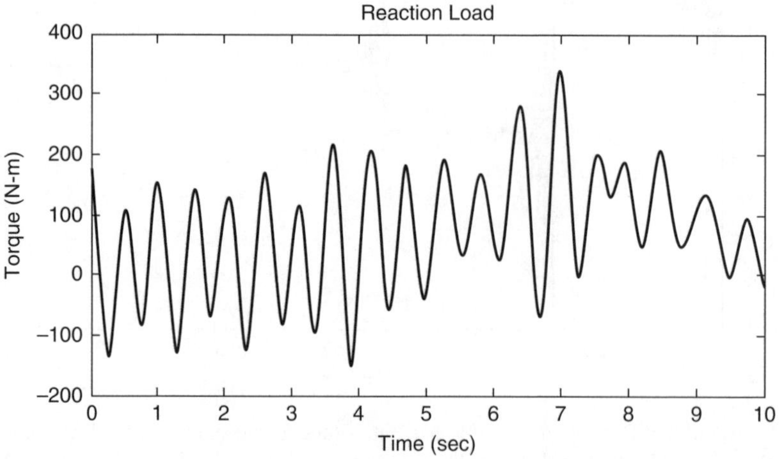

Figure 9.8 Internal Torque vs. Time Required for Locking One Rotational dof of the Ball-and-Socket Joint. Banerjee and Lemak [1]. Reproduced with permission of Lockheed Martin Space Systems Company.

Table 9.2 Simulation Time Comparison of Order-n Algorithm with Efficient Generalized Speeds vs. Extended Kane Theory for a Multibody System with Three Structural Loops.

Number of modes	Number of gen. speeds	CPU sec for 10 s simulation using extended Kane	CPU ratio of extended Kane over order-n efficient method
0	24	6.28	0.969
4	56	57.57	1.516
8	88	909.29	3.053
12	120	2484.3	4.455

Banerjee and Lemak [1]. Reproduced with permission of Lockheed Martin Space Systems Company.

CPU times required for the three different formulations for a 10 sec simulation of this $(6 + 3 + 2 + 1 + 4 \times 10)$ or 52 dof system, with 10 modes for each flexible body, are as shown in Table 9.2.

Thus in this example a recursive method using efficient generalized speed is about two times faster than real time. The next two examples we discuss are for systems with motion constraints. Figure 9.9 gives a stroboscopic plot of a whirling chain of five flexible trusses connected by spherical joints in three-dimensional, constrained motion with both ends in grounded sockets. Each truss has three relative rotational *dof*, and two of them have four modes and the remaining three have 12 modes.

Figure 9.10 shows the plots of large-angle rotation between bodies 4 and 5 from the left end of the truss in Fig. 9.9. We have labeled the plots with indicators such as EKane for the extended-Kane algorithm of Ref. [7] and "ONeff" for the present ("order-n" with efficient variables) method. The plots are identical to the naked eye. Figure 9.11 shows the time-histories of the first five modal coordinate of body 5. The results look identical, with negligible numerical discrepancies.

We show in Table 9.1 the computer times taken by the two methods. The table shows that computational advantage of the present method increases with increased number of modes per truss.

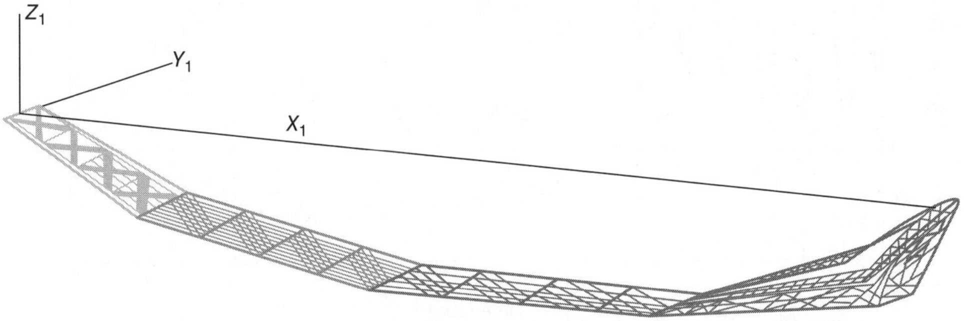

Figure 9.9 Stroboscopic Plot from Simulation of a Whirling Chain of Five Flexible Trusses Each Connected by a Ball Joint, with End Points in Grounded Sockets. Banerjee and Lemak [1]. Reproduced with permission of Lockheed Martin Space Systems Company.

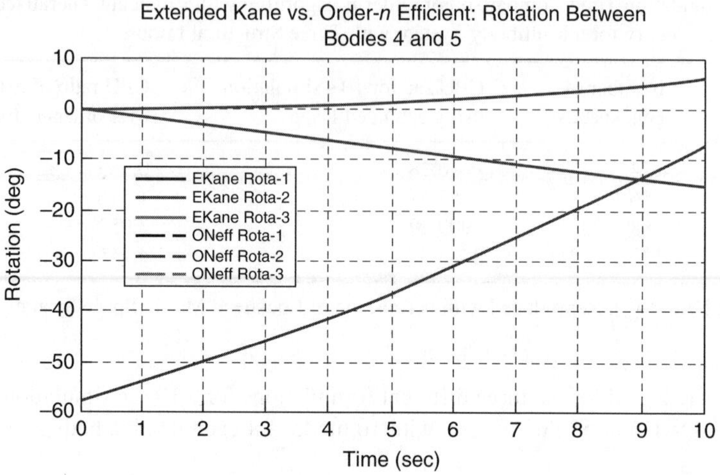

Figure 9.10 Plots of Three Rotations vs. Time, between Bodies 4 and 5, Given by the Extended Kane Method of Ref. [7] and the Present Efficient Order-*n* Method. Banerjee and Lemak [1]. Reproduced with permission of Lockheed Martin Space Systems Company.

Figure 9.12 shows an example of a three-dimensional, multi-loop, flexible structure mechanism, considered to illustrate an application of the main algorithm presented in this chapter. The two parts in Figure 9.13 show the rotations and errors between the two solutions, based on the present theory and the extended Kane method of Ref. [7]. Comparison of modal coordinate solutions for a sample body, body 4, is shown in Figure 9.14.

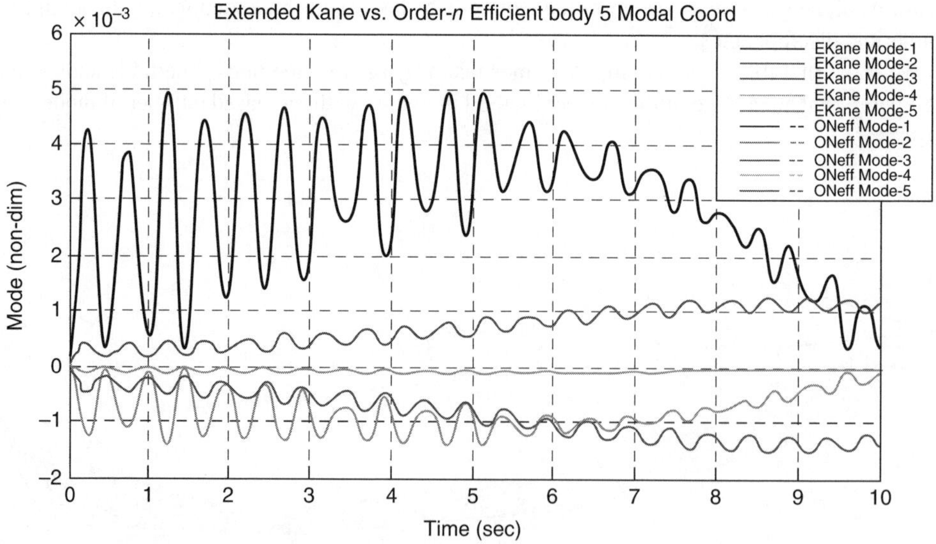

Figure 9.11 Comparison of Solutions for the First Five Modal Coordinates of Body 5 by the Extended Kane Method [7] and the Recursive Method with Efficient Generalized Speeds. Banerjee and Lemak [1]. Reproduced with permission of Lockheed Martin Space Systems Company.

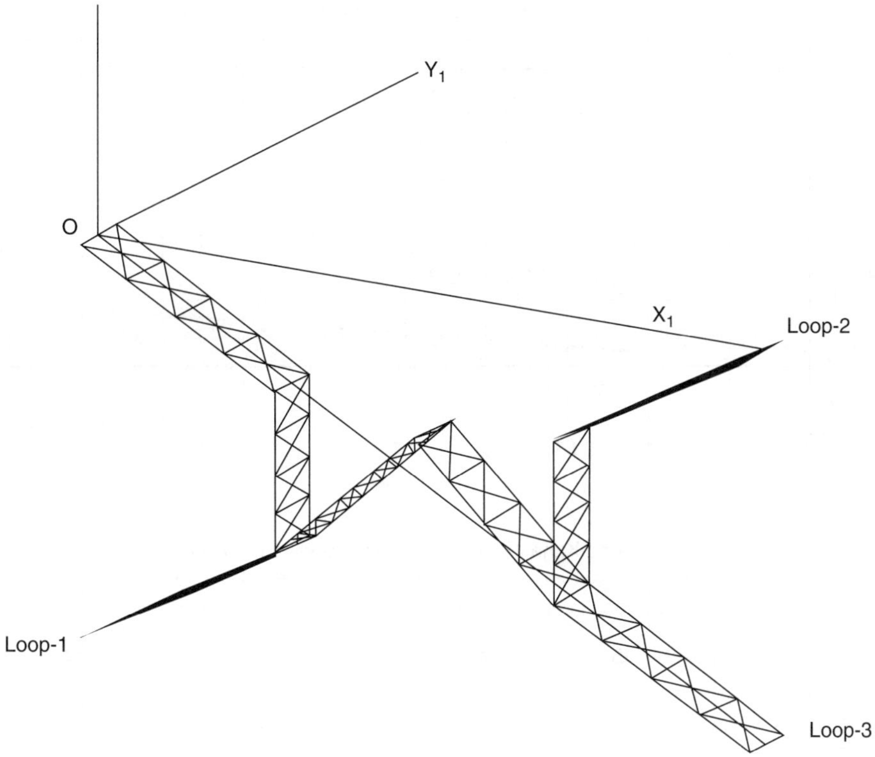

Figure 9.12 Flexible Three-Loops Structure with Eight Ball-Jointed Trusses in Large Motion. Banerjee and Lemak [1]. Reproduced with permission of Lockheed Martin Space Systems Company.

Finally in the two parts of Figure 9.15, we show the computed constraint force and loop closure error for loop 3.

Table 9.2 shows the computational performance of the recursive algorithm with efficient variables for the flexible multi-loop system of Figure 9.11, having three loops and nine constraint equations. In comparison to the algorithm of Ref. [7], it is clear that the overall combination of the block-diagonal algorithm and the choice of efficient motion variables get significantly better performance as the number of modes per body is increased.

It should be noted finally that more recent research [11, 12] on numerical solution of dynamics of multi-loop structures have been devised by solving the differential-algebraic equations [10] of motion and constraint equations in finite difference form recursively. The method may give slightly better results in constraint maintenance than what is shown here in the example.

Acknowledgment

The author is grateful to the Management at the Advanced Technology Center, Lockheed Missiles and Space System Company, where this work [1] was done, for permission to include this material in this book.

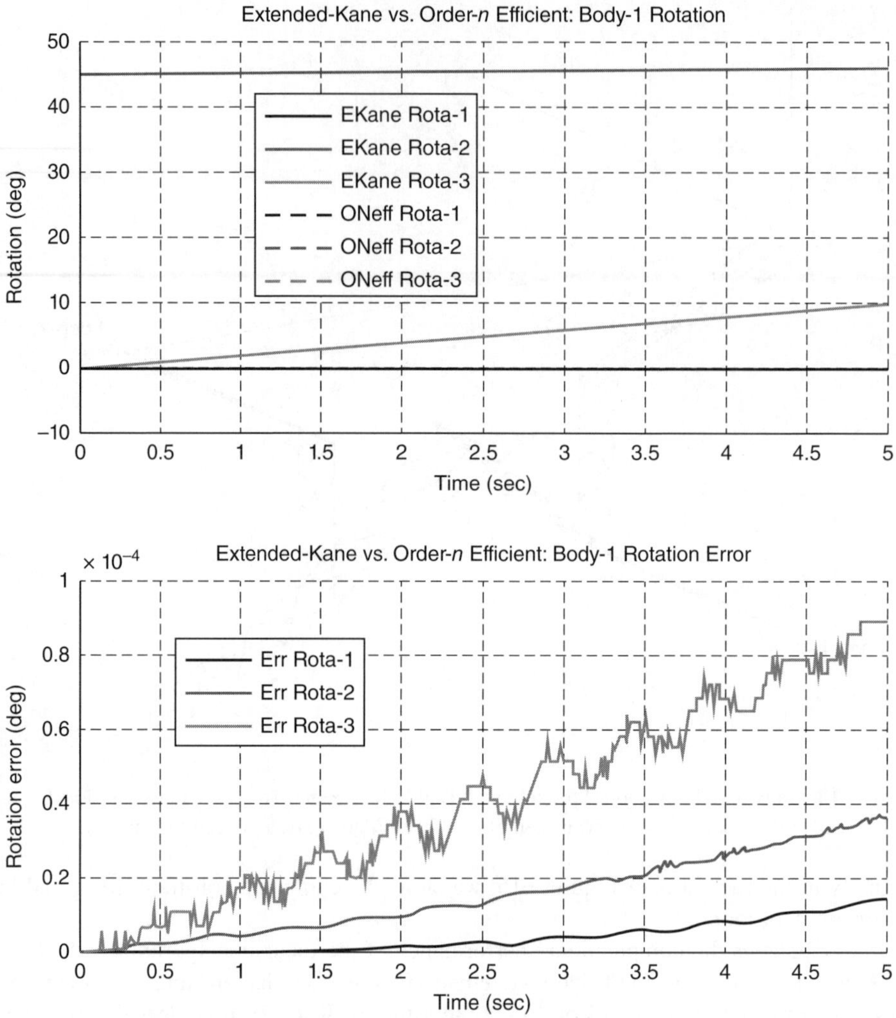

Figure 9.13 Comparison of Rotation Angles and Errors for the Ball-and-Socket Joint for Body 1 by the Efficient Recursive, Order-*n* Formulation and the Extended Kane Method [7]. Banerjee and Lemak [1]. Reproduced with permission of Lockheed Martin Space Systems Company.

Appendix 9.A Pseudo-Code for Constrained *nb-Body* m-Loop Recursive Algorithm in Efficient Variables

9.A.1 Backward Pass

We give here the extensions necessary beyond the backward pass for unconstrained systems that are given in Ref. [2]. The following segment is invoked during the backward pass for each body after all its mass updates are done. All bodies with cut joints become terminal bodies for which the modal mass matrix does not require an update and hence its inverse is computed only once. In the following we use the notion of a direct path as going from a particular cut

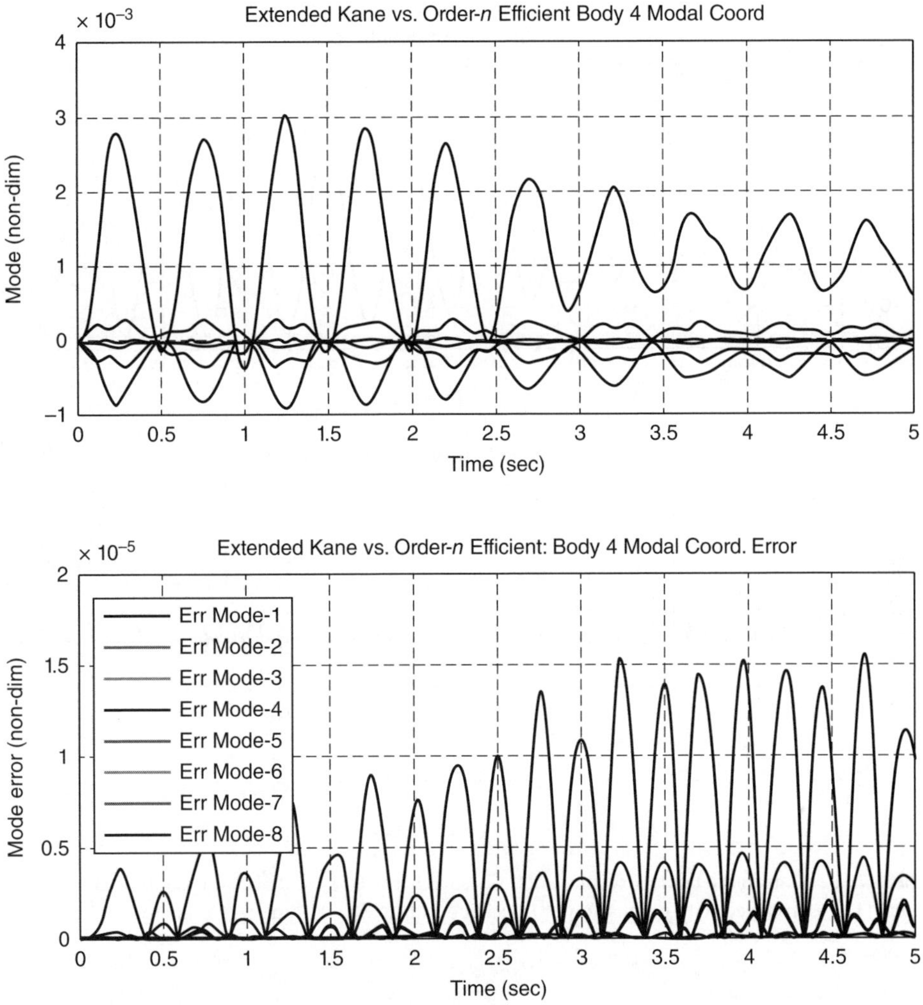

Figure 9.14 Comparison of Modal Coordinates for Body 4 and Error Time History by Two Methods. Banerjee and Lemak [1]. Reproduced with permission of Lockheed Martin Space Systems Company.

joint to body 1. Also $Q(jp)$ is a point on a flexible body j, to which outboard body jp is hinged, and $L(j)$ is the point on body j where the cut occurs. The pseudo-code algorithm description follows.

For j going from nb to 1 **Do** the loop:
If j lies in the direst path for loop k,

$$\hat{H}^j_{ek} = E_j^{-1}\Phi^T_{L(j)} \quad k = 1, \ldots, m \tag{9.A1}$$

$$G_j = Z_{L(j)} - M_2^j \hat{H}^n_{ek} \quad k = 1, \ldots m \tag{9.A2}$$

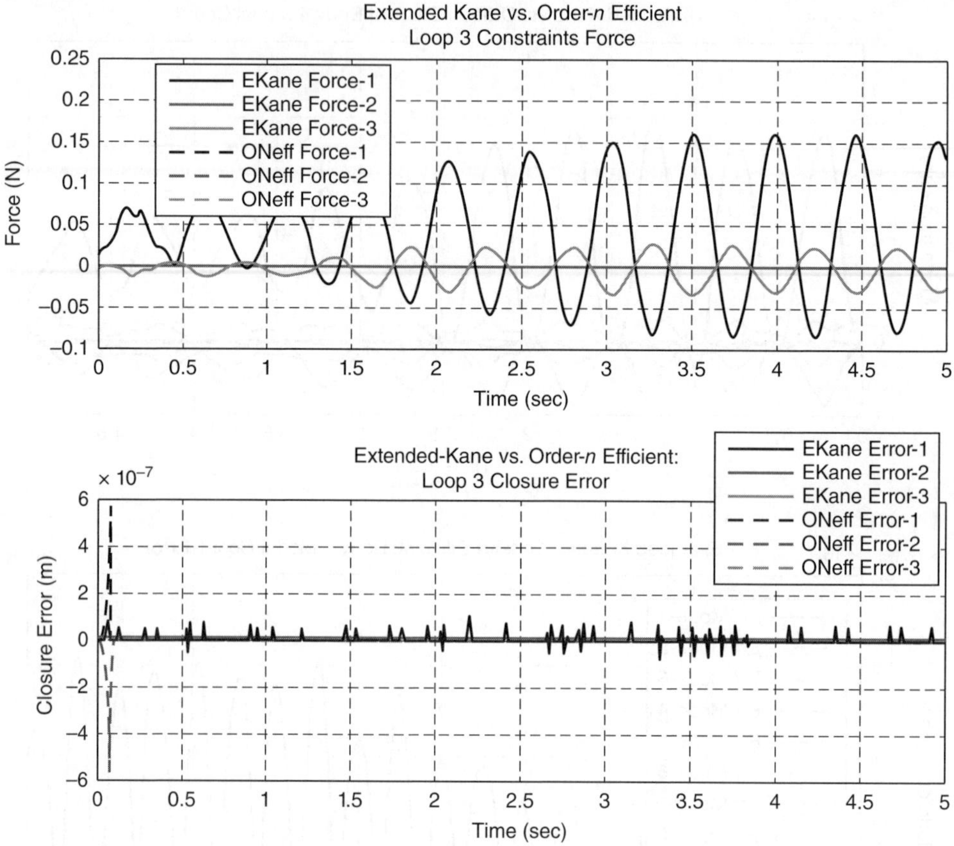

Figure 9.15 Constraint Force (N) and Loop Closure Error (m) vs. Time, for Loop 3. Banerjee and Lemak [1]. Reproduced with permission of Lockheed Martin Space Systems Company.

$$\hat{H}^{n}_{rk} = \nu_j^{-1} R_j^T G_j \tag{9.A3}$$

$$B_{c(j)k} = G_j - M_1^j R_j \hat{H}^j_{rk} \quad k = 1, \ldots, m \tag{9.A4}$$

Else

$jp = o(j)$, outboard body j and denoting

$$d_j^{jp} = \begin{bmatrix} C_{j,jp} & C_{j,t(jp)} \tilde{\delta}^{jp} C^T_{j,t(jp)} \\ 0 & C_{j,jp} \end{bmatrix} \tag{9.A5}$$

$$\hat{H}^j_{ek} = E_j^{-1} \Phi^j_{Q(jP)} d_j^{jP} B_{jk} \quad k = 1, \ldots, m \tag{9.A6}$$

$$G_j = Z_{jP} - M_2^j E_j^{-1} \Phi^j_{Q(jP)} \tag{9.A7}$$

$$\hat{H}^j_{rk} = \nu_j^{-1} R_j^T G_j d_j^{jP} B_{jk} \quad k = 1, \ldots, m \tag{9.A8}$$

$$B_{c(j)k} = G_j d_j^{jP} B_{jk} - M_1^j R_j \hat{H}^j_{rk} \quad k = 1, \ldots, m \tag{9.A9}$$

9.A.2 Continue

9.A.3 Forward Pass

Initialize by setting $H^j_{rk} = H^j_{ek} = 0, j = 1, \dots, nb; k = 1, \dots m.$

Refer to Eqs. (9.95) and (9.100) for F^1_{rf}, F^1_{ef}:

$$H^1_{rk} = \hat{H}^1_{rk} \quad k = 1, \dots, m \tag{9.A10}$$

$$S^1_f = R_1 F^1_{rf} \tag{9.A11}$$

$$S^1_{ck} = R_1 H^1_{rk} \quad k = 1, \dots, m \tag{9.A12}$$

$$H^1_{ek} = \hat{H}^1_{ek} + E^{-1}_j M^j_2 S^1_{ck} \quad k = 1, \dots, m \tag{9.A13}$$

Dynamical equations for the rigid and elastic dof for body 1 then become

$$\dot{u}^1_r = F^1_{rf} + \sum_{k=1}^{m} H^1_{rk}\lambda_k \tag{9.A14}$$

$$\dot{u}^1_e = F^1_{ef} + \sum_{k=1}^{m} H^1_{ek}\lambda_k \tag{9.A15}$$

For $j = 2 \longrightarrow nb$

$$\gamma^j_f = W_j S^{c(j)}_f + N_j F^{c(j)}_{ef} \tag{9.A16}$$

$$\gamma^j_{ck} = W_j S^{c(j)}_{ck} + N_j H^{c(j)}_{ek} \quad k = 1, \dots., m \tag{9.A17}$$

$$F^j_{rf} = -\nu^{-1}_j R^T_j \{Y^j_1 + M^1_j \gamma^j_f\} + \nu^{-1}_j \begin{Bmatrix} \tau_j \\ f_j \end{Bmatrix} \tag{9.A18}$$

$$H^j_{rk} = \hat{H}^j_{rk} - \nu^{-1}_j R^T_j M^j_1 \gamma^j_{ck} \quad k = 1, \dots., m \tag{9.A19}$$

$$S^j_f = \gamma^j_f + R_j F^j_{rf} \tag{9.A20}$$

$$S^j_{ck} = \gamma^j_{ck} + R_j H^j_{rk} \quad k = 1, \dots., m \tag{9.A21}$$

$$F^j_{ef} = E^{-1}_j \{Y^j_1 + M^j_2 S^j_f\} \tag{9.A22}$$

$$H^j_{ek} = \hat{H}^j_{ek} + E^{-1}_j A_j S^j_{ck} \quad k = 1, \dots., m \tag{9.A23}$$

Dynamical equations for body j in rigid and elastic degrees of freedom are now given as:

$$\dot{u}_r^j = F_{rf}^j + \sum_{k=1}^{m} H_{rk}^j \lambda_k \tag{9.A24}$$

$$\dot{u}_e^j = F_{ef}^j + \sum_{k=1}^{m} H_{ek}^j \lambda_k \tag{9.A25}$$

Continue until $j = nb$.

Problem Set 9

9.1 Consider the planar motion of three rigid rods connected by revolute joints and acted on only by gravity. Derive the recursive equations of motion of in two ways: (1) using $u_i = \dot{\theta}_i, i = 1, 2, 3$ where θ_1 is angle from the vertical for the first link, and θ_2, θ_3 are the relative angles; and (2) using $u_1 = \dot{\theta}_1, u_2 = \dot{\theta}_1 + \dot{\theta}_2, u_3 = \dot{\theta}_1 + \dot{\theta}_2 + \dot{\theta}_3$. Show that the latter set of definitions is more efficient in the sense that it produces simpler equations.

9.2 Repeat the above exercise with the third link considered flexible, with two flexible cantilever modes, once using $u_1 = \dot{\theta}_1, u_2 = \dot{\theta}_1 + \dot{\theta}_2, u_3 = \dot{\theta}_1 + \dot{\theta}_2 + \dot{\theta}_3 + \sum_{i=1}^{2} \phi_i \dot{q}_i$, ϕ_i being the ith mode, and the other using Eq. (9.4) for an efficient choice of modal generalized speed. Use for ϕ_i the cantilever modes given in Chapter 5.

9.3 Consider the analysis of a slider crank mechanism of Chapter 1 by starting with a model of a double pendulum. Use efficient generalized speeds for the two links and then impose the constraint condition that the end of the second link must move on a horizontal plane. Derive the equations of motion with the constraint, introducing a Lagrange multiplier to represent the force normal to the horizontal plane. Now consider the crank rod and the connecting rod as bending beams and do the analysis of a system with a closed structural loop. Choose your own parameters to represent the length, mass of the links elastic modulus, and area moment of inertia, and simulate the motion given an initial velocity of the links.

9.4 Consider a "four bar" linkage, made up of three beams, with the first beam connected to the ground by a pin joint, the second beam connected to the first beam at its end by a pin, and the third beam connected to the second beam by a pin and closing a structural loop by being connected to the ground by another pin. Consider each beam of a different length L; assign one rotational dof and one elastic mode $\sin(\pi x/L)$ to each beam. Write two constraint equations for the horizontal and vertical directions, describing loop closure. Derive the equations of motion and solve numerically the differential algebraic equations involved.

References

[1] Banerjee, A.K. and Lemak, M.E. (2007) Recursive algorithm with efficient variables for flexible multibody dynamics with multiloop constraints. *Journal of Guidance, Control, and Dynamics*, **30**(3), 780–790.

[2] Banerjee, A.K. (1993) Block-diagonal equations for multibody elastodynamics with geometric stiffness and constraints. State of the Art Survey Lecture in Multibody Dynamics, European Space Agency, Noordwijk, June; republished in *Journal of Guidance, Control, and Dynamics*, **16**(6), 1092–1100.

[3] D'Eleuterio, G.M.T. and Barfoot, T.D. (1999) Just a second, we'd like to go first: A first order discretized formulation for elastic multibody dynamics, in *Proceedings of the 4th International Conference on Dynamics and Control of Structures in Space* (eds. C.L. Kirk and R. Vignjevic), Cranfield University, UK, May 24–28.

[4] Kane, T.R. and Levinson, D.A. (1985) *Dynamics, Theory and Applications*, McGraw-Hill.

[5] Banerjee, A.K. and Dickens, J.M. (1990) Dynamics of an arbitrary flexible body in large rotation and translation. *Journal of Guidance, Control, and Dynamics*, **13**(2), 221–227.

[6] Mitiguy, P.C. and Kane, T.R. (1996) Motion variables leading to efficient equations of motion. *International Journal of Robotics Research*, **15**(5), 522–532.

[7] Wang, J.T. and Huston, R.L. (1987) Kane's equations with undetermined multipliers—application of constrained multibody systems. *Journal of Applied Mechanics*, **54**(2), 424–429.

[8] Amirouche, F.M.L. (1992) *Computational Methods in Multibody Dynamics*, Prentice-Hall, pp. 434–460.

[9] Kane, T.R., Likins, P.W., and Levinson, T.R. (1983) *Spacecraft Dynamics*, McGraw-Hill.

[10] Brennan, K.E., Campbell, S.L., and Petzold, L. (1989) *Numerical Solutions of Initial Value Problems in Differential-Algebraic Equations*, Elsevier Publications.

[11] Negrut, D., Jay, L.O., and Khude, N. (2009) A discussion of low-order numerical integration formulas for rigid and flexible multibody dynamics. *Journal of Computational and Nonlinear Dynamics*, **4**(4), 021008.

[12] Simeon, B. (2013) *Computational Flexible Multibody Dynamics: A Differential Algebraic Approach*. Springer Verlag.

[13] Lemak, M.E. (1992) Enforcement of Configuration Satisfaction Concurrent with the Numerical Solution of Multi-Flexible-Body Equations of Motion. Engineer's Thesis, Stanford University.

10

Efficient Modeling of Beams with Large Deflection and Large Base Motion

This chapter presents a specific simplification of the block-diagonal formulation given in Chapter 8. The basic method used here to describe large overall motion of beams undergoing large deflection is to model the beam as many rigid rods connected by springs at revolute joints. Note that a representation of a beam deflection by its vibration modes is not commensurate with large deflection. We demonstrate here only the case of planar bending, but three-dimensional bending and torsion can be handled by changing the orientation of the rotation axes at joints. The associated simulation is made time-efficient by the so-called order-n formulation, where the computational effort increases with n, the degree of freedom, and not as n^3 as in a non-recursive multibody formulation, given in Chapter 6. A side benefit of the order-n formulation is a new way of computing internal loads in a system of bodies connected in a topological tree. Two numerical integration schemes are considered: Kutta-Merson and Newmark. Newmark integration scheme is linked here with the order-n equations. To summarize, we show that the order-n formulation with discrete rigid segments is more efficient timewise than a nonlinear finite element formulation, even with a fourth-order integrator to get the same level of accuracy as a finite element formulation with a second-order integrator. The material reported is based on Refs. [1, 2].

10.1 Discrete Modeling for Large Deflection of Beams

The example system under consideration is the Shuttle-Antenna system shown in Figure 10.1, where, in an experimental study of waves in space plasma (WISP), the shuttle rotates slowly carrying two 150 m long antennas. We treat the shuttle as a rigid body, while a flexible antenna with bending rigidity of 1676 N-m^2 is modeled so as to bring out possibly very large deflections.

Flexible Multibody Dynamics: Efficient Formulations and Applications, First Edition. Arun K. Banerjee.

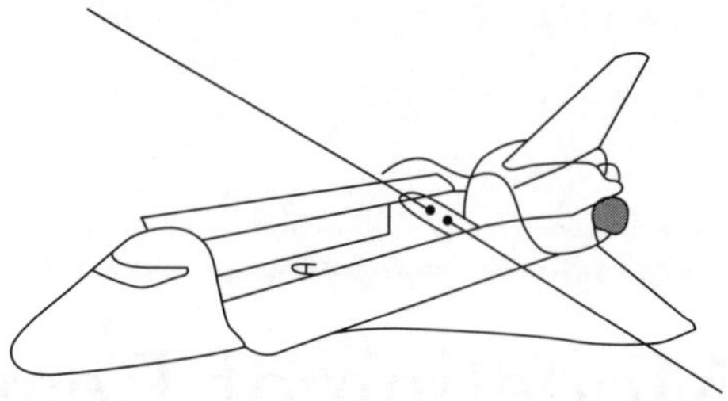

Figure 10.1 Shuttle WISP Antenna System. Banerjee [1]. Reproduced with permission of the American Astronautical Society.

Figure 10.2 shows a particular way of discretizing beams, with rigid rods connected by torsional springs. This rigid finite segment method has been shown in Ref. [3] to represent adequately the bending of beams. We use the same model here for large deflections. Kruszewski *et al.* [3] use *EI/L* as the stiffness coefficient for all the springs, where *E* is the elastic modulus, *I* is the sectional area moment of inertia, and *L* is the length of a segment. We derive here a set of spring coefficients consistent with Bernoulli-Euler beam theory. Required spring coefficient

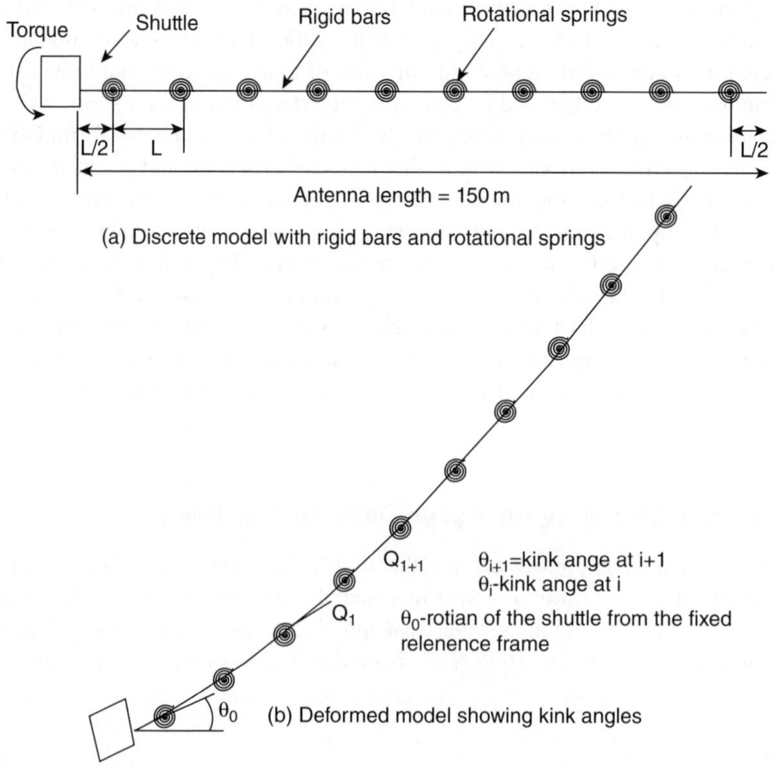

Figure 10.2 Beam Modeled with Rigid Rods Connected by Rotational Springs. Banerjee [1]. Reproduced with permission of the American Astronautical Society.

σ at any joint can be computed by equating the bending moment at that location due to a tip load P to the restoring moment of the spring at that joint. Thus, for joint 1 at a distance of $L/2$ from the fixed end of the beam in Figure 10.2, one has:

$$\sigma_1\theta_1 = P\left(nL - \frac{L}{2}\right) \tag{10.1}$$

One computes from the second joint from the fixed end in Fig. 10.2,

$$\theta_1 = \frac{\delta_2}{L} \tag{10.2}$$

In Figure 10.2 δ_2 is the deflection at joint 2 from the fixed end of the beam. Deflection, due to a tip load, δ at a distance x from the fixed end, is given by linear beam theory as:

$$\delta = \frac{P}{6EI}x^2(3nL - x) \tag{10.3}$$

Evaluating the deflection at $x = 3L/2$ (see Figure 10.2) and substituting in Eqs. (10.1) and (10.2) yield:

$$\sigma_1 = \frac{8}{9}\frac{EI}{L} \tag{10.4}$$

For joint 2, the relation between the relative angle θ_2 and deflections at joints 2 and 3 from the fixed end is:

$$\theta_2 = \frac{(\delta_3 - 2\delta_2)}{L} \tag{10.5}$$

and the bending moment equation is:

$$\sigma_2\theta_2 = P\left(nL - \frac{3L}{2}\right) \tag{10.6}$$

Using Eq. (10.3) for the distances to joints 2 and 3 to evaluate δ_2 , δ_3 and substitution in Eqs. (10.5) and (10.6) lead to:

$$\sigma_2 = \frac{48n - 72}{42n - 71}\frac{EI}{L} \tag{10.7}$$

For a general joint i, the relative angle is given by the central difference relation

$$\theta_i = \frac{(\delta_{i+1} - 2\delta_i + \delta_{i-1})}{L} \tag{10.8}$$

and the moment equilibrium requires:

$$\sigma_i\theta_i = P\left[nL - \left(i - \frac{1}{2}\right)L\right] \tag{10.9}$$

Evaluation of $\delta_{i-1}, \delta_i, \delta_{i+1}$ from Eq. (10.3) and substitution in Eqs. (10.8) and (10.9) give:

$$\sigma_i = \frac{EI}{L}, \qquad i = 3, \ldots, n \tag{10.10}$$

Equations (10.4), (10.7), and (10.10) thus list all the necessary spring constants.

10.2 Motion and Loads Analysis by the Order-*n* Formulation

Motion analysis by order-n formulation is well known [6,7] but is reviewed here only to bring out the features needed for the determination of loads. A system of bodies connected in a topological tree configuration by single dof rotational joints is considered, with the bodies numbered as $j = 1, \ldots, n$. With a slight change of notion here, we call a generic body j as connected to its inboard body i at the hinge Q_j. To demonstrate the process, and restricting ourselves to planar motion, we choose as generalized speed u_j,

$$u_j = {}^i\boldsymbol{\omega}^j . \boldsymbol{\lambda}_j \qquad j = 1, \cdots, n \tag{10.11}$$

where ${}^i\boldsymbol{\omega}^j$ stands for the angular velocity vector of body j with respect to the inboard body i, and $\boldsymbol{\lambda}_j$ is the unit vector along the axis of rotation of the revolute joint. The angular velocity of j in the inertial frame N is related to the angular velocity of i by the angular velocity addition theorem:

$${}^N\boldsymbol{\omega}^j = {}^N\boldsymbol{\omega}^i + u_j\boldsymbol{\lambda}_j \tag{10.12}$$

The kinematical equation defining generalized speeds here u_j is

$$u_j = \dot{q}_j \qquad j = 1, \cdots, n \tag{10.13}$$

where q_j is the angle of rotation of a line fixed in body j with respect to a line fixed in body i. The velocity of the hinge point Q_j, is related to the velocity Q_i, inboard anchor point of body i as

$${}^N\mathbf{v}^{Q_j} = {}^N\mathbf{v}^{Q_i} + {}^N\boldsymbol{\omega}^i \times r^j \tag{10.14}$$

where $\mathbf{r}^j$ is the position vector from Q_i to Q_j. Equations (10.2) and (10.11) give rise to the following expression for Kane's partial angular velocity of body j and the partial velocity of Q_i with respect to the jth generalized speed:

$${}^N\omega^j_j = \boldsymbol{\lambda}_j \tag{10.15}$$

$${}^N\mathbf{v}^{Q_j}_j = 0 \tag{10.16}$$

As is done in previous chapters, the angular acceleration for body j is split into two groups, one involving derivatives of the generalized speeds and the other not:

$$ {}^N\boldsymbol{\alpha}^j = {}^N\boldsymbol{\alpha}_0^j + {}^N\boldsymbol{\alpha}_t^j \tag{10.17} $$

Here the first group, involving derivatives of generalized speeds, is further split as

$$ {}^N\boldsymbol{\alpha}_0^j = {}^N\hat{\boldsymbol{\alpha}}_0^j + {}^N\boldsymbol{\omega}_j^j\,\dot{u}_j \tag{10.18} $$

and the hat-quantities refer to angular acceleration terms containing derivatives of generalized speeds for inboard body i:

$$ {}^N\hat{\boldsymbol{\alpha}}_0^j = {}^N\boldsymbol{\alpha}_0^i \tag{10.19} $$

The second group of terms in Eq. (10.17), the remainder acceleration terms, is zero only for planar motion, that is:

$$ {}^N\boldsymbol{\alpha}_t^j = 0 \tag{10.20} $$

Similarly, all acceleration terms are split into terms involving derivatives of generalized speeds and the remainder terms in the acceleration:

$$ {}^N\mathbf{a}^{Q_j} = {}^N\mathbf{a}_0^{Q_j} + {}^N\mathbf{a}_t^{Q_j} \tag{10.21} $$

$$ {}^N\mathbf{a}_0^{Q_j} = {}^N\hat{\mathbf{a}}_0^{Q_j} \tag{10.22} $$

$$ {}^N\hat{\mathbf{a}}_0^{Q_j} = {}^N\mathbf{a}_0^{Q_i} + {}^N\boldsymbol{\alpha}_0^i \times \mathbf{r}^j \tag{10.23} $$

$$ {}^N\mathbf{a}_t^{Q_j} = {}^N\mathbf{a}_t^{Q_i} + {}^N\boldsymbol{\alpha}_t^i \times \mathbf{r}^j + {}^N\boldsymbol{\omega}^i \times \left({}^N\boldsymbol{\omega}^i \times \mathbf{r}^j\right) \tag{10.24} $$

Now all vectors pertaining to body j are expressed in the vector basis of body j, and their measure numbers are expressed in matrix form. Thus, a matrix R_j of partial velocity and partial angular velocity in the j-basis for the joint frame at Q_j is obtained by inspection of Eqs. (10.15) and (10.16) as the 6×1 matrix of scalar elements,

$$ R_j = \begin{Bmatrix} 0 \\ \lambda_j \end{Bmatrix} \tag{10.25} $$

so that Eqs. (10.22) and (10.18) are written in matrix form with scalars as:

$$ \begin{Bmatrix} {}^N a_0^{Q_j} \\ {}^N \alpha_0^j \end{Bmatrix} = \begin{Bmatrix} {}^N \hat{a}_0^{Q_j} \\ {}^N \hat{\alpha}_0^j \end{Bmatrix} + R_j \dot{u}_j \tag{10.26} $$

As in previous chapters, Eqs. (10.19) and (10.23) are rewritten in terms of a matrix shift transformation,

$$\begin{Bmatrix} {}^N\hat{\mathbf{a}}_0^{Q_j} \\ {}^N\hat{\alpha}_0^j \end{Bmatrix} = W^j \begin{Bmatrix} {}^N\mathbf{a}_0^{Q_i} \\ {}^N\alpha_0^i \end{Bmatrix} \tag{10.27}$$

where, using the skew-symmetric matrix $\tilde{r}^j$ for cross product corresponding to a (3×1) matrix r^j,

$$W^j = \begin{bmatrix} C_{ij}^T & -C_{ij}^T \tilde{r}^j \\ 0 & C_{ij}^T \end{bmatrix} \tag{10.28}$$

Here C_{ij} is the direction cosine matrix transforming basis vectors of body j to basis vectors in body i. At this stage, we can form all the kinematical variables we need for all the bodies, going in a forward kinematical pass from body *1* to body n.

Dynamical equations are generated in two steps: a backward pass starting from body n and ending in body 1, for which its dynamical equations are explicitly written, after which a second forward pass is undertaken, where the rest of the dynamical equations are unraveled. For a generic body j, we write the D'Alembert equations of equilibrium in terms of the inertia forces and torques and the resultant of the external forces and moments about the hinge Q_j as follows,

$$\begin{Bmatrix} f^{*j} - f_e^j \\ t^{*j} - t_e^j \end{Bmatrix} \equiv M^j \begin{Bmatrix} {}^N a_0^{Q_j} \\ {}^N \alpha_0^j \end{Bmatrix} + X^j \tag{10.29}$$

where the matrices M^j, X^j are defined in terms of the mass m^j, the first and second moments of inertia about the hinge Q_j, that is, $s^{j/Q_j}, I^{j/Q_j}$ (with $\tilde{s}^{j/Q_j}$ the skew-symmetric form of cross-product with s^{j/Q_j} as before):

$$M^j = \begin{bmatrix} m^j & -\tilde{s}^{j/Q_j} \\ \tilde{s}^{j/Q_j} & I^{j/Q_j} \end{bmatrix} \tag{10.30}$$

$$X^j = M^j \begin{Bmatrix} {}^N a_t^{Q_j} \\ {}^N \alpha_t^j \end{Bmatrix} + \begin{Bmatrix} \tilde{\omega}^j \tilde{\omega}^j s^{j/Q_j} - f_e^j \\ \tilde{\omega}^j I^{j/Q_j} \omega^j - t_e^j \end{Bmatrix} \tag{10.31}$$

Again $\tilde{\omega}^j$ is the skew-symmetric matrix formed for cross product out of the three elements in the vector ω^j. Kane's Equations [8] associated with the hinge dof for a terminal body j are obtained by pre-multiplying the force and moment terms in Eq. (10.29) augmented by any hinge spring torque t_h^j, by the transpose of the partial velocity/partial angular velocity matrix in Eq. (10.25),

$$\ddot{\theta}^j = -[v^j]^{-1}[R^j]^T \left[M^j \begin{Bmatrix} {}^N\hat{a}_0^{Q_j} \\ {}^N\hat{\alpha}_0^j \end{Bmatrix} + X^j \right] + [v^j]^{-1} \begin{Bmatrix} 0 \\ t_h^j \end{Bmatrix} \tag{10.32}$$

where we have introduced the matrix:

$$\nu^j = [R^j]^T M^j R^j \tag{10.33}$$

Using Eqs. (10.26) and (10.32) in Eq. (10.29), and making the substitutions,

$$\begin{aligned} M &= M^j - M^j R^j [\nu^j]^{-1} [R^j]^T M^j \\ X &= X^j - M^j R^j [\nu^j]^{-1} [R^j]^T X^j + M^j R^j [\nu^j]^{-1} \begin{Bmatrix} 0 \\ t_h^j \end{Bmatrix} \end{aligned} \tag{10.34}$$

lead to a re-expression of Eq. (10.29) as:

$$\begin{Bmatrix} f^{*j} - f_e^j \\ t^{*j} - t_e^j \end{Bmatrix} = M \begin{Bmatrix} \hat{a}^{Q_j} \\ \hat{\alpha}_0^j \end{Bmatrix} + X \tag{10.35}$$

The system of inertia and external forces and their moments at and about Q_j expressed in the j-basis is obtained by substituting Eq. (10.27) in Eq. (10.35) to get the dynamic equilibrium terms represented by the left-hand side of the following equation:

$$\begin{Bmatrix} f^{*j} - f_e^j \\ t^{*j} - t_e^j \end{Bmatrix}_{Q_j/j} \equiv M W^j \begin{Bmatrix} a_0^{Q_{c(j)}} \\ \alpha_0^{c(j)} \end{Bmatrix} + X \tag{10.36}$$

Now, the forces and moments acting at Q_j can be equivalently replaced at Q_i, and expressed in the i-basis by using the shift transform of Eq. (10.27):

$$\begin{Bmatrix} f^{*j} - f_e^j \\ t^{*j} - t_e^j \end{Bmatrix}_{Q_i/i} = [W^j]^T \begin{Bmatrix} f^{*j} - f_e^j \\ t^{*j} - t_e^j \end{Bmatrix}_{Q_j/j} \tag{10.37}$$

To this set of forces and moments is added the resultant of all active and inertia forces and moments of body i about Q_i, yielding the dynamic equilibrium zero-equation,

$$\begin{Bmatrix} f^{*i} - f_e^i \\ t^{*i} - t_e^i \end{Bmatrix} \equiv \sum_{k \in S_i} \begin{Bmatrix} f^{*k} - f_e^k \\ t^{*k} - t_e^k \end{Bmatrix} = M^i \begin{Bmatrix} {}^N a_0^{Q_i} \\ {}^N \alpha_0^i \end{Bmatrix} + X^i \tag{10.38}$$

where S_i is the set of all bodies outboard of hinge Q_i, and we have introduced the following replacements:

$$M^i \to M^i + [W^j]^T M W^j \tag{10.39}$$

$$X^i \to X^i + [W^j]^T X \tag{10.40}$$

Equation (10.38) is of the same form as Eq. (10.29). This pattern is repeated until the equations for body 1 are written. At that stage, the inboard body acceleration and angular acceleration are either zero or prescribed. Assigning the zeroth body as the inertial frame, or of known prescribed motion, Eq. (10.32) reduces for $j = 1$ to

$$\ddot{\theta}^1 = -[v^1]^{-1}[R^1]^T X^1 + [v^1]^{-1} \begin{Bmatrix} 0 \\ t_h^1 \end{Bmatrix} \tag{10.41}$$

and concomitantly, Eq. (10.26) yields:

$$\begin{Bmatrix} a_0^{Q_1} \\ \alpha_0^1 \end{Bmatrix} = R^1 \begin{Bmatrix} 0 \\ \ddot{\theta}^1 \end{Bmatrix} \tag{10.42}$$

Once the dynamical equations for body 1 are formed, the rest of the dynamical equations are completed by using a second forward pass on the kinematical equations.

Computation of internal loads is a bonus feature attendant with the order-n algorithm given above. This can be seen from Eq. (10.38), which shows the result of all external and inertia forces and torques outboard of any joint. To compute, for example, the loads at joint 2 of the system (see Figure 10.2), one proceeds in the *second forward pass* to obtain $\ddot{\theta}_2$ and uses Eq. (10.26):

$$\sum_{j\in S_2} \begin{Bmatrix} f^{*j} - f_e^j \\ t^{*j} - t_e^j \end{Bmatrix} = M_1^2 \begin{Bmatrix} {}^N a_0^{Q_2} \\ {}^N \alpha_0^2 \end{Bmatrix} + X^2 \tag{10.43}$$

Equation (10.43) is precisely what we need for the loads, except that the force system is shifted to the root of the cantilever beam in our example by (recall that the superscript 2, included in $\tilde{r}^2$ below, refers just to quantities associated with body 2):

$$\begin{Bmatrix} s \\ m \end{Bmatrix} = \begin{bmatrix} C_{12} & 0 \\ \tilde{r}^2 C_{12} & C_{12} \end{bmatrix} \left\langle \sum_{j\in S_2} \begin{Bmatrix} f^{*j} - f_e^j \\ t^{*j} - t_e^j \end{Bmatrix} \right\rangle \tag{10.44}$$

to get the shear force, s, and bending moment, m, due to forces to the right of joint 2. Finally, the additional force and moment due to inertia of the part of the beam that represents body 1 (see Figure 10.2) is added to the shear force and moment from Eq. (10.44) to get the total shear and moment at the root of the cantilever.

10.3 Numerical Integration by the Newmark Method

The order-n equations generated in the preceding section for large bending are nonlinear ordinary differential equations of the form:

$$\ddot{q} = f(\dot{q}, q) \tag{10.45}$$

In rigid body dynamics, integration of these equations are customarily done by using the Runge-Kutta-Merson method. In structural dynamics involving mass, stiffness, and damping matrices, the differential equations tend to be stiff due to a mixture of high and low frequencies, and traditionally implicit integration methods such as the Newmark-δ method are used. Here the Newmark method [9] is modified for the order-n formulation as follows. The equations for velocity and displacement are written in Newmark integration for a time-step h as:

$$\dot{q}_{t+h} = \dot{q}_t + [(1-\delta)\,\ddot{q}_t + \delta\,\ddot{q}_{t+h}]h \tag{10.46}$$

$$q_{t+h} = q_t + \dot{q}_t h + [(0.5-\alpha)\ddot{q}_t + \alpha\ddot{q}_{t+h}]h^2 \tag{10.47}$$

The order-n dynamical equations of Eq. (10.45) are discretized, for example, as:

$$\ddot{q}_{t+h} = f(\dot{q}_{t+h}, q_{t+h}) \tag{10.48}$$

Upon substitution of the following notations,

$$\begin{aligned} a_1 &= \delta/(\alpha h); \quad a_2 = (1-\delta/\alpha); \\ a_3 &= [(1-\delta) - (0.5-\alpha)\delta/\alpha]\,h; \\ a_4 &= (0.5-\alpha)h^2; \quad a_5 = \alpha h^2; \quad a_6 = \delta\, h \end{aligned} \tag{10.49}$$

$$G_1 = -a_1 q_t + a_2 \dot{q}_t + a_3 \ddot{q}_t \tag{10.50}$$

$$G_2 = q_t + h\dot{q}_t + a_4 \ddot{q}_t \tag{10.51}$$

equations for (angular) velocity and (angular) displacement, with a choice of $\delta = 2\alpha$, Eqs. (10.46) and (10.47), take the form:

$$\dot{q}_{t+h} = a_1 q_{t+h} + G_1 \tag{10.52}$$

$$q_{t+h} - G_2 - a_5 f\{\dot{q}_{t+h}, q_{t+h}\} = 0 \tag{10.53}$$

Equation (10.53) is an implicit, nonlinear algebraic equation in the angular displacement q_{t+h}, and must be solved by Newton iteration method using the Jacobian matrix:

$$J = \left[I - a_6 \frac{\partial f}{\partial \dot{q}_{t+h}} - a_5 \frac{\partial f}{\partial q_{t+h}} \right] \tag{10.54}$$

Convergence of the Newton iterations is accepted when the norm of the correction vectors between successive iterations for the root of the vector function $f(\dot{q}, q)$ in Eq. (10.45) is less than some small number, like 10^{-6}. This method is similar to that proposed in Ref. [10].

10.4 Nonlinear Elastodynamics via the Finite Element Method

In this section we give a brief outline of the finite method used as an independent, standard approach to solve large deformation problems in nonlinear elastodynamics, to check simulation results of the order-n formulation presented in this chapter. The finite element code used for this purpose is NEPSAP [11] (nonlinear elastic plastic structural analysis program), a comprehensive system of linear/nonlinear, static/dynamic analysis code developed at Lockheed. For the WISP antenna problem, it is the geometric nonlinearity that is of interest because of the large deformation. Two types of nonlinearity are significant here: (a) the use of complete kinematic expression for finite strains, that is, the inclusion of nonlinear terms in the strain-displacement relations; and (b) the use of the deformed configuration of the body to obtain the equations of motion that, as a consequence, become dependent on the total deformation. The nonlinear strain-displacement relation is shown below, with $u_{i,j}$ meaning the partial derivative of deformation u_i with respect to x_j, and so on,

$$\varepsilon_{ij} = \frac{1}{2}\left[\left(u_{i,j} + u_{j,i}\right) + \left(u_{m,i}u^1_{m,j} + u^i_{m,i}u_{m,j}\right) + u_{m,i}u_{m,j}\right] \tag{10.55}$$

where the terms without superscript denote the incremental values involved in the motion of a body from a configuration numbered at state 1 to an infinitesimally close state 2. In this definition, terms with superscript 1 denote the known values at state 1. There are three groups of terms in Eq. (10.55); the first one is the conventional small displacement strain, while the second one represents the effect of large displacements manifested in terms of gradients of known displacements at state 1, and the final term represents the nonlinear part of the strain-displacement relation; repetition in m indicates a tensor sum over m in the range 1,2,3. The first two terms are linearly dependent on the incremental deformations u, whereas the last term varies quadratically as the incremental deformations. For many structural analysis problems it is essential to consider both of these sources of nonlinearities. The effects of both geometric nonlinearities on the behavior of systems are quite easily incorporated in the general formulation, and the equations to be solved in nonlinear elasto-statics are the linearized incremental equations of equilibrium in the form,

$$[K_t]\{\Delta u\} = \{\Delta P\} \tag{10.56}$$

where K_i is the instantaneous stiffness, or "tangent stiffness," including geometric stiffness, and Δu, ΔP are the incremental displacement and incremental load, respectively. The total displacements are computed by adding the incremental solution to the previously computed total values. Particularly relevant to the WISP problem considered here is the large displacement analysis presented by Bisshopp and Drucker [12], exact comparison with their results having been obtained using NEPSAP. The dynamics problem, of course, has inertia (and damping, neglected here) forces in addition to the elastic forces represented by Eq. (10.56), and so, the dynamic incremental equation becomes

$$M\Delta\ddot{u} + K_t\Delta u = \Delta P \tag{10.57}$$

where the incremental displacements, velocities, and accelerations occur during a time interval Δt, from time t to $t + \Delta t$. Newmark integration scheme is invoked, with Eq. (10.57) transformed into a simple set of linear algebraic equations of the form given below, representing Newton iterations required in the integration, where equilibrium is considered at the time step ahead:

$$\tilde{K}_t\Delta u^{(k)} = \Delta P^{(k)} \tag{10.58}$$

$$\tilde{K}_t = K_t + \frac{4}{\Delta t^2}M \tag{10.59}$$

$$\Delta P^{(k)} = R^2 - F^{2,(k-1)} - M\left(\frac{4}{\Delta t^2}\Delta u^{(k-1)} - \frac{4}{\Delta t}\dot{u}^1 - \ddot{u}^1\right) \tag{10.60}$$

Here k is the iteration counter, superscripts 1 and 2 denote variables at t and $t + \Delta t$, and R, F are the total load and the equilibrium load, respectively. For a good discussion of the details of the integration scheme for nonlinear finite element analysis, see Bathe [9].

10.5 Comparison of the Order-*n* Formulation with the Finite Element Method

As we have done several times before, we consider a spin-up problem for the shuttle antenna system for a crucial test. The shuttle has a centroidal moment of inertia of 10^7 kg- m^2 and two 150-m-long antennas are attached at points (0.5 m, −10 m; −0.5 m, −10 m) from the mass center of the shuttle. The antennas are beams of flexural rigidity 1676 N-m^2 and of mass density 0.335 kg/m. In terms of differential equations this is not a "stiff" problem, with the ratio of maximum frequency to the minimum frequency of 20. Order-*n* equations for the proposed spring-connected rigid finite element model are integrated by using both the Newmark and a variable step fourth-order Kutta-Merson integrator. Simulation results are given for a torque pulse of 2687 Newton-meters acting for 20 seconds on the shuttle. Figure 10.3 shows the rotation angle of the shuttle vs. time during spin-up. It is clear that the reference frame for the nonlinearly elastic antenna beams goes through a large angles of rotation. In what follows we use as the truth model the NEPSAP finite element solution obtained by the Newmark integration to assess the accuracy of the rigid segment order-*n* solution. First, we establish the accuracy of the NEPSAP solution itself for a cantilever beam against the classical solution by Bisshopp and Drucker [12] for a cantilever beam with a tip load. Figure 10.4 shows agreement for normalized displacement.

Now we present results for spin-up for both the rigid element order-*n* and NEPSAP models. Figures 10.5 and 10.6 present results for discretizing each beam into 10 elements, each 15 m long. Order-*n* solutions by the Newmark method for step sizes 1.0 sec and 0.5 sec are overlaid on the NEPSAP solution for step size of 1.0 sec. It is seen that while the general behavior of beam tip deflection and bending moment as given by the order-*n* spring-connected rigid element model agrees quite well with the nonlinear finite element solution, the rigid element model predicts the peak deflection 4% higher and the peak moment 8% lower. Slightly better results are expected by tightening the error tolerance in Newmark integration, or reducing the length of the rigid segments; for recent developments with low-order integrators, see Ref. [10].

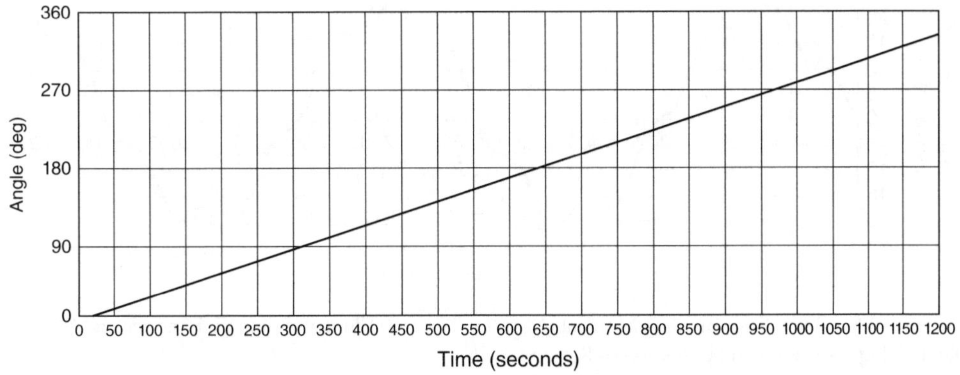

Figure 10.3 Vertical Axis: Shuttle Yaw Rotation Angle in Degrees, vs. Time During Spin-up. Banerjee and Nagarajan [2]. Reproduced with permission of Springer.

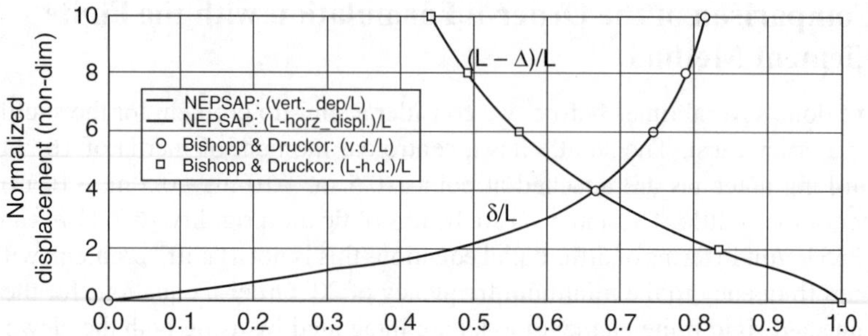

Figure 10.4 Vertical Axis: Normalized Displacement (non-dim) Comparison for the NEPSAP Code, with Classical Bisshopp and Drucker [12] Solution for Tip Displacement. Banerjee and Nagarajan [2]. Reproduced with permission of Springer.

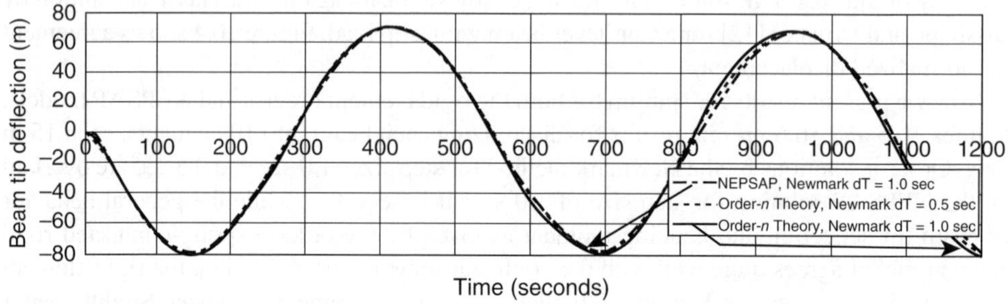

Figure 10.5 Beam Tip Deflection (m) vs. Time, 10 Element Solution. Banerjee and Nagarajan [2]. Reproduced with permission of Springer.

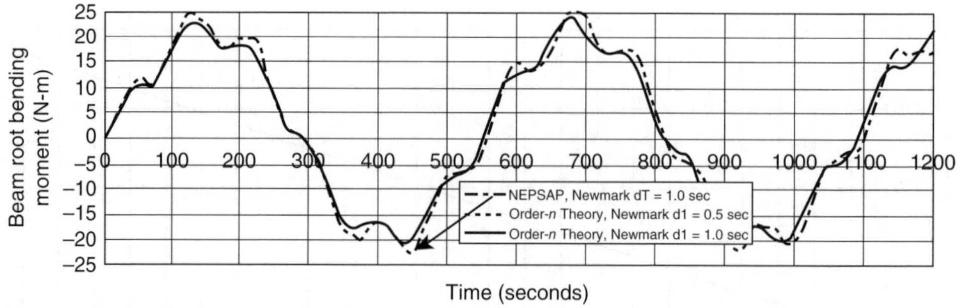

Figure 10.6 Beam Root Bending Moment (N-m) vs. Time, 10 Element Solution. Banerjee and Nagarajan [2]. Reproduced with permission of Springer.

Table 10.1 CPU Times for Order-n and Finite Element Large Deflection Solutions.

Formulation	Number of elements	Step size (sec)	CPU time (sec)
order-n	10	1.0	24
order-n	10	0.5	40.9
finite element	10	1.0	890

Banerjee and Nagarajan [2]. Reproduced with permission of Springer.

Computer time for Newmark integration for a 1200 sec rotational maneuver is in Table 10.1: Newmark integration of the order-n dynamical equations is done with update of the associated Jacobian matrix (that does not change much) every 100 sec. It is clear that while the rigid element formulation with Newmark integration is computationally very efficient, it is not as accurate, as Figure 10.6 shows. To address the accuracy question, the order-n equations are integrated with a fourth-order variable step Kutta-Merson ordinary differential equation solver, with maximum step size of 1.0 sec, and the results compared better with the finite element solution using Newmark integrator for 1.0 step size, for both models using 20 elements. Figure 10.7 shows the almost perfect agreement between the two tip deflection solutions.

Figure 10.8 shows the difference in moment loads computation with the two models is smaller with more elements. Finally, a comparison of the shear force at the fixed end of the cantilever, obtained by the two methods for the 20 element models, are given in Figure 10.9.

Computer solution time for Figures 10.7 and 10.8, for the order-n solution with the rigid element model and the NEPSAP finite element model using a Newmark integrator are compared in Table 10.2, showing, as in Table 10.1, the efficiency of the rigid-segment order-n method over the finite element method.

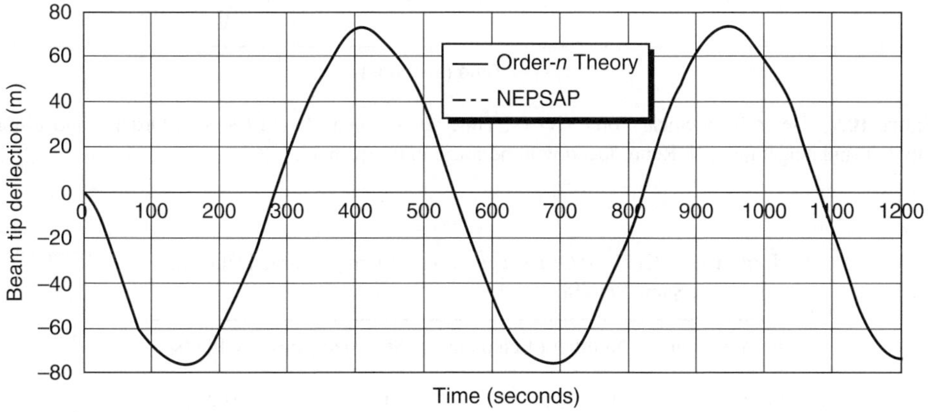

Figure 10.7 Beam Tip Deflection (m) vs. Time: by Rigid Segment Order-n and Runge-Kutta Integrator vs. Finite Element Solution with Newmark Integrator, Both Using 20 Elements. Banerjee and Nagarajan [2]. Reproduced with permission of Springer.

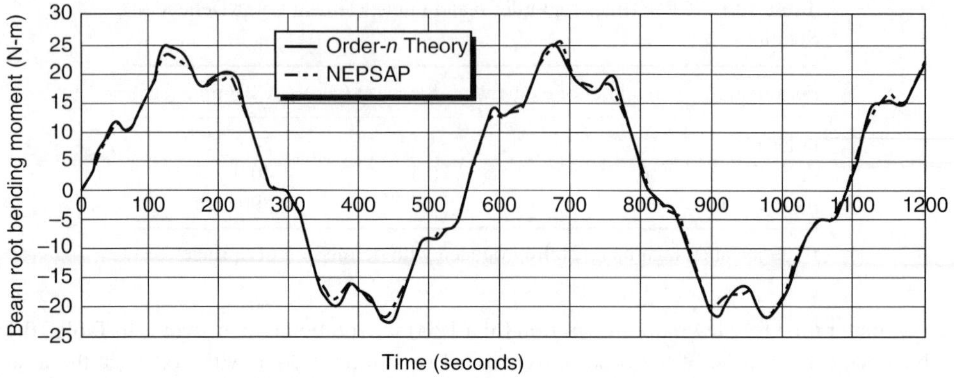

Figure 10.8 Beam Root Bending Moment (N-m) vs. Time, by Rigid Segment Order- and Runge-Kutta Integrator and Finite Element Solutions with Newmark Integrator, Both Using 20 Elements. Banerjee and Nagarajan [2]. Reproduced with permission of Springer.

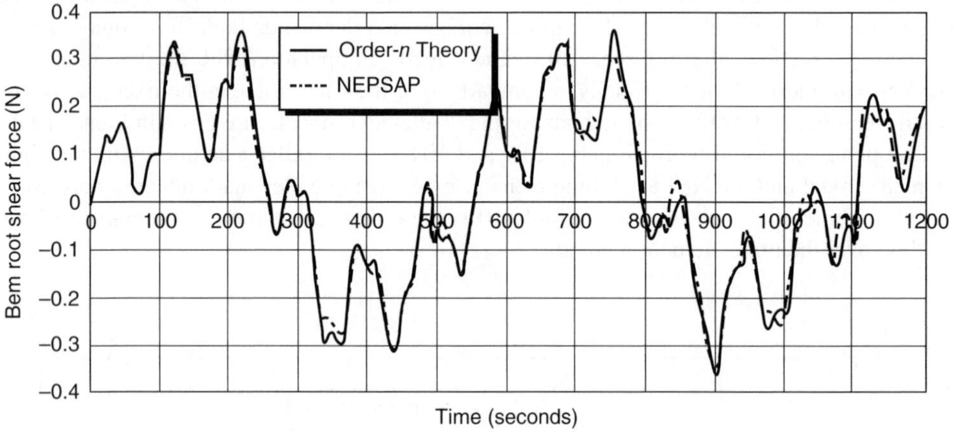

Figure 10.9 Beam Root Shear Force (N) vs. Time, 20 Element, Rigid Element Order-*n* and Finite. Banerjee and Nagarajan [2]. Reproduced with permission of Springer.

Table 10.2 CPU for Order-*n* and Finite Element Large Deflection Solutions, Same Step Size.

Formulation	Number of elements	Step size (sec)	CPU (sec)
order-*n*	20	1.0	182
finite element	20	1.0	1160

Banerjee and Nagarajan [2]. Reproduced with permission of Springer.

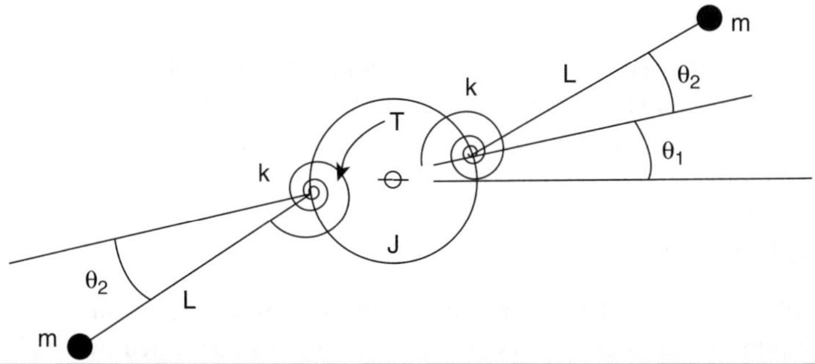

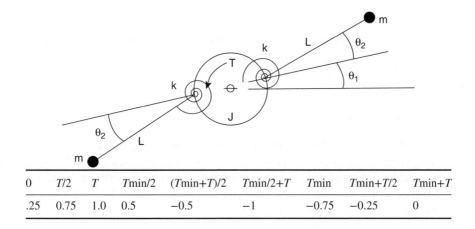

0	*T*/2	*T*	*T*min/2	(*T*min+*T*)/2	*T*min/2+*T*	*T*min	*T*min+*T*/2	*T*min+*T*
.25	0.75	1.0	0.5	−0.5	−1	−0.75	−0.25	0

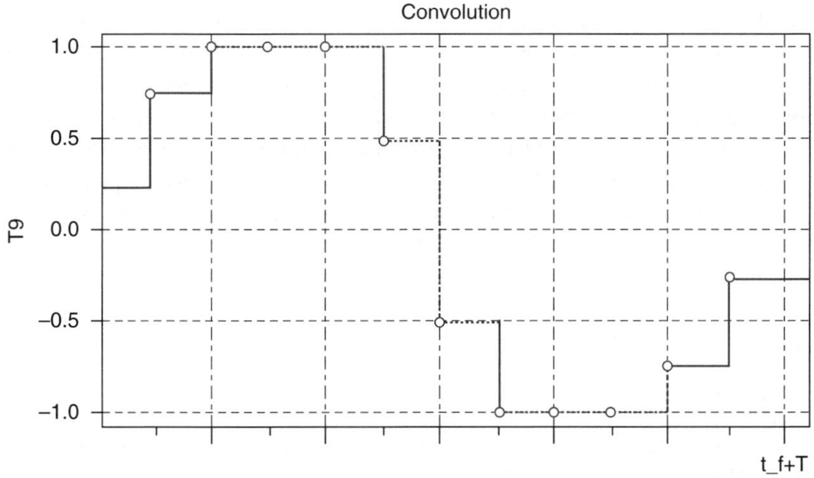

Figure 10.10 Convolution of a Bang-Bang Command with a Three-Impulse Sequence for Suppression of Vibration of Period T for the Mechanical System shown above. Singhose, W., Banerjee, A.K., and Seering [14]. Reproduced with permission of the American Institute of Aeronautics and Astronautics, Inc.

10.6 Conclusion

Results show that use of spring coefficients obtained from linear beam theory for a spring-connected rigid element model gives good results for large deflection, assuming the finite element solution as the truth. The order-*n* formulation given for the rigid segment model provides computation of internal loads as a natural by-product of the algorithm. The solution procedure indicates that while the rigid segment order-*n* formulation is much superior to the finite element method in terms of computing time, it requires a higher-order integrator to give comparable levels of accuracy with a finite element solution. A Fortran code given as Appendix B at the end of the book can be modified for this simulation.

Acknowledgment

Dr. S. Nagarajan performed the numerical simulations based on the nonlinear finite element code, NEPSAP, that he and Dr. P. Sharifi developed for Lockheed Missiles and Space Company and provided the basis of the comparison with results for the order-*n* algorithm given in this chapter. The author is grateful for their help.

Problem Set 10

10.1 Code the recursive order-*n* algorithm given in Section 10.3 of this chapter and use it to represent the motion of a system of massive links connected by revolute joints and hanging in a uniform gravity field. Consider an initial value of the angles to represent a curved swing and consider rate damping due to air on each link. Choose your own parameter values and simulate the ensuing motion.

10.2 Derive the vibration equations of the rotary system model of the WISP shuttle-antenna system shown below, and calculate the time period T of the first vibration mode for your value of J, m, L, and k. For a minimum time slew of the undeformed system through an angle of 90 degrees, the minimum time T_{min} due to a bang-bang torque τ_{max} can be computed as $T_{\text{min}} = \sqrt{\pi J/(2\tau_{\text{max}})}$. Simulate the response of the system for this bang-bang torque. There is a way [13] to suppress the resulting vibrations of period of T by a convolution of the bang-bang torque with a three-impulse sequence of (0.25, 0.5, 0.25) coming at 0, $T/2$, T. This produces the shaped torque scaled to *unit* torque magnitude, shown in Figure 10.10; actual torque has to be multiplied by this scaling. Obtain simulation results with this torque time history.

References

[1] Banerjee, A.K. (1993) Dynamics and control of the WISP shuttle-antenna system. *Journal of the Astronautical Sciences*, **41**, 73–90.

[2] Banerjee, A.K. and Nagarajan, S. (1997) Efficient simulation of large overall motion of beams undergoing large deflection. *Multibody System Dynamics*, **1**, 113–126.

[3] Kruszewski, J., Gawronski, W., Wittbrodt, E., *et al.* (1975) Metoda Sztywnych Elementow Skonczonvch (Rigid Finite Element Method), Arkady Warszawa (in Polish).

[4] Huston, R.L. (1990) *Multibody Dynamics*, Butterworth-Heinemann, USA, pp. 344–360.

[5] Hollerbach, J.M. (1980) A recursive Lagrangian formulation of manipulator dynamics and a comparative study of dynamics formulation complexity. *IEEE Transactions on Systems, Man, and Cybernetics*, **SMC-10**(11), 730–736.

[6] Rosenthal, D.E. (1990) An order-n formulation for robotic systems. *Journal of Astronautical Sciences*, **38**(4), 511–530.

[7] Keat, J.E. (1990) Multibody system order-n formulation based on velocity transform method. *Journal of Guidance, Control, and Dynamics*, **13**(2), 207–212.

[8] Kane, T.R. and Levinson, D.A. (1985) *Dynamics: Theory and Applications*, McGraw-Hill.

[9] Bathe, K.J. (1982) *Finite Element Procedure in Engineering Analysis*, Prentice-Hall.

[10] Negrut, D., Jay, L.O., and Khude, N. (2009) A discussion of low-order numerical integration formulas for rigid and flexible multibody dynamics. *Journal of Computational and Nonlinear Dynamics*, **4**(2), 021008.

[11] Nagarajan, S. and Sharifi, P. (1980) *NEPSAP theory Manual*, Lockheed Missiles & Space Co..

[12] Bisshopp, K.E. and Drucker, D.C. (1945) Large displacement of cantilever beams. *Quarterly of Applied Mathematics*, **3**, 272–275.

[13] Singer, N.C. and Seering, W.P. (1990) Preshaping command inputs to reduce system vibration. *Journal of Dynamic Systems, Measurement and Control*, **112**, 76–82.

[14] Singhose, W., Banerjee, A.K., and Seering, W.E. (1997) Slewing Flexible Spacecraft with Deflection Limiting Input Shaping. *Journal of Guidance, Control, and Dynamics*, **20**(2), 291–298.

11

Variable-n Order-n Formulation for Deployment and Retraction of Beams and Cables with Large Deflection

This chapter describes applications of a variable-n, order-n formulation, where n is the degree of freedom of a system, to deployment or retraction of beams discretized from a moving base such as a spacecraft, and of a cable from a winch on a ship reeling in or out an underwater maneuvering vehicle. Deployment and retraction with large deflection are handled in the order-n algorithm by simply varying the number of discretized bodies out at any given time, with careful state updates to maintain continuity of motion as one goes from one value of n to another. The material is based on the author's papers [1, 2], while [3, 4] are related material. Ref. [5] considers a related problem, deployment of a beam from a vibrating base. We consider large deflections in the main body of the chapter and offer the analysis in Ref. [4] as a guided exercise for the reader. We review first the problem of beam extrusion from the space shuttle, which was proposed as an experiment. Deployment of a cable from a ship to an underwater vehicle, doing seafloor mine search, using its own maneuver control system requires a constrained variable-n order-n algorithm, and will be described in the second half of this chapter.

11.1 Beam Discretization

An extrusion or retraction of a beam process is idealized by the discretization represented in Figure 11.1, where the body B rotates in a prescribed manner in a Newtonian reference frame N, with the beam anchored at a moving point A at distance d, from a point O fixed in B. Specifying deployment or retraction rate of the beam amounts to specifying d as a function of time, with the number of beam segments updated when a whole segment is deployed or

Flexible Multibody Dynamics: Efficient Formulations and Applications, First Edition. Arun K. Banerjee.

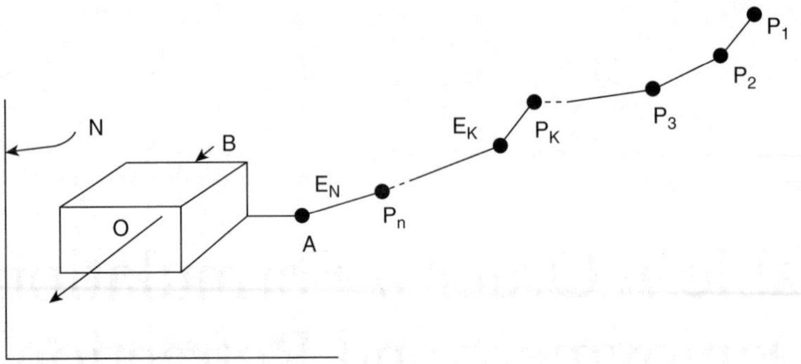

Figure 11.1 Discretized Representation of a Beam Being Extruded from a Rotating Base. Banerjee [1]. Reproduced with permission of the American Institute of Aeronautics and Astronautics, Inc.

retracted. In Figure 11.1, the discrete links are connected by rotational springs, as modeled previously and in Ref. [3].

For convenience, the beam is discretized by massless links with lumped mass at the ends of the links. Torsion springs resist relative rotation between links, and the stiffness coefficients are obtained by equating the deflections due to a tip load with that of a Bernoulli-Euler beam, as in Chapter 10, where the following relations are given, with ρ, EI/L as the linear mass density, and flexural rigidity of the beam, respectively, with i referring to a generic link:

$$m_i = \rho L \qquad (i = 1, \dots, n) \tag{11.1}$$

$$\begin{aligned} k_1 &= EI/L, \qquad (i = 1, \dots, n-1) \\ k_2 &= \frac{6(n-1)EI}{(3n-1)L} \\ k_n &= k_2 + (k_1 - k_2)\, d/L \end{aligned} \tag{11.2}$$

11.2 Deployment/Retraction from a Rotating Base

11.2.1 Initialization Step

Consider only revolute joints for simplicity and specify a starting value of n for the number of bodies, numbering them from the free end as in Figure 11.1. Assign inertia properties and stiffness as per Eqs. (11.1) and (11.2). Initialize generalized coordinates q_k and generalized speeds defined as $u_k = \dot{q}_k, (k = 1, \dots\ n)$ for all the hinge relative rotations. Prescribe the base motion and the extrusion rate – that is, the rate of the change of the distance d from O to A in Figure 11.1 – as a function of time.

11.2.2 Forward Pass

Step 1: Define the coordinate transformation matrix C_{ik} between body k and its inboard body i along the path to B, for all k. The elements of C_{ik} are

$$C_{ik}(l, m) = \mathbf{b}_i(l).\mathbf{b}_k(m) \qquad (k = n, \dots, 1; i, m = 1, 2, 3) \tag{11.3}$$

where $\mathbf{b}_i(m)$ is the mth component of the basis vector of the ith body frame.

Step 2: Form the angular velocity vector ${}^N\boldsymbol{\omega}^k$ of the kth body in the k-basis and using the same notation define corresponding 3×1 matrices ${}^N\omega^k$ $(k = 1, \ldots, n)$ using kth body vector basis components [note body n is nearest to B],

$$\begin{aligned} {}^N\boldsymbol{\omega}^n &= {}^N\boldsymbol{\omega}^B + u_n\boldsymbol{\lambda}_n \\ {}^N\boldsymbol{\omega}^k &= {}^N\boldsymbol{\omega}^i + u_k\boldsymbol{\lambda}_k \qquad : (k = n-1, \ldots\ldots, 1) \end{aligned} \tag{11.4}$$

where $\boldsymbol{\lambda}_k$ is the unit vector along the rotation axis for the kth hinge and B is the reference body to which body n is hinged and whose motion is prescribed. Note that the hinge-axis vectors are not restricted to being all parallel; this feature allows three-dimensional motion.

Step 3: Form the partial angular velocity of the kth body with respect to the ith generalized speeds in the k-basis, from Eq. (11.4),

$$ {}^N\boldsymbol{\omega}^k_k = \boldsymbol{\lambda}_k \qquad (k = n, \ldots, 1) \tag{11.5}$$

forming the 3×1 matrix with kth body measure numbers, ${}^N\omega^k_k$ $\quad (k = n, \ldots, 1)$.

Step 4: From the kinematical relation between the inertial velocity of k^* in Figure 11.2 and the joint between body k and body i, express the partial velocity of the mass center of the kth body with respect to the generalized speed u_k in the k-basis:

$$\begin{aligned} {}^N\mathbf{v}^{k^*} &= {}^N\mathbf{v}^{i^*} + {}^N\boldsymbol{\omega}^i \times \mathbf{r}_1 - {}^N\boldsymbol{\omega}^k \times \mathbf{r}_2 \qquad k = (n-1, \ldots, 1) \\ {}^N\mathbf{v}^{n^*} &= {}^N\mathbf{v}^A + {}^N\boldsymbol{\omega}^n \times \mathbf{r}^{An^*} \end{aligned} \tag{11.6a}$$

The *minus sign arises because* $\mathbf{r}^k_2$ is directed opposite to the outward direction, going from the mass center of body k *inward* to the kth joint (see Figure 11.2). Note $\mathbf{r}^{An*}$ is the position vector from A to n^*. Now form the corresponding 3×1 matrices v^{k*}_k $(k = 1, \ldots, n)$:

$$ v^{k*}_k = -\tilde{\omega}^k_k r^k_2 \qquad (k = n, \ldots., 1) \tag{11.6b}$$

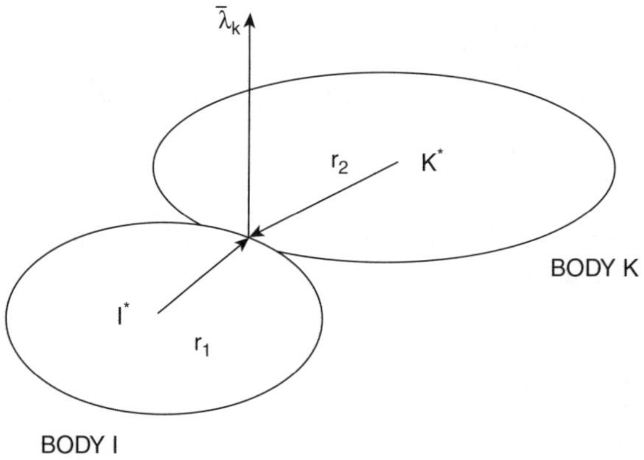

Figure 11.2 Two Hinged Bodies i and k in a Chain, with Body k Outboard of Body i. Banerjee [1]. Reproduced with permission of the American Institute of Aeronautics and Astronautics, Inc.

Step 5: Express the remainder angular acceleration vector $\boldsymbol{\alpha}_t^k$ from the angular acceleration of the kth body, defined indirectly by the following equations:

$$
\begin{aligned}
{}^N\boldsymbol{\alpha}^k &= \sum_{j=1}^{n} {}^N\boldsymbol{\omega}_j^k \dot{u}_j + \boldsymbol{\alpha}_t^k \qquad (k = n, \ldots\ldots, 1) \\
{}^N\boldsymbol{\alpha}_t^n &= {}^N\boldsymbol{\alpha}^B \qquad\qquad (11.7a) \\
{}^N\boldsymbol{\alpha}_t^k &= {}^N\boldsymbol{\alpha}_t^i + {}^N\boldsymbol{\omega}^k \times (u_k\boldsymbol{\lambda}_k) \qquad (k = n-1, \ldots\ldots, 1)
\end{aligned}
$$

Equation (11.7a) defines $\boldsymbol{\alpha}_t^k$ in the k-basis, forming the corresponding 3×1 matrix, ${}^N\boldsymbol{\alpha}_t^k$. Here ${}^N\boldsymbol{\alpha}^B$ is the prescribed angular acceleration of the body B out of which the beam is being deployed. Now form in the k-basis the 3×1 *matrices* α_t^k, $(k = n, \ldots, 1)$,

$$
\begin{aligned}
\alpha^k &= \sum_{j=1}^{n} \omega_j^k u_j + \alpha_t^k \qquad (k = n, \ldots\ldots, 1) \\
\alpha_t^n &= \alpha^n \qquad\qquad (11.7b) \\
\alpha_t^k &= \alpha_t^i + \sum \tilde{\omega}^k \lambda_k u_k \qquad (k = n-1, \ldots\ldots, 1)
\end{aligned}
$$

where ${}^N\tilde{\omega}^k$ is the skew-symmetric matrix, for vector cross-product, formed out of the basis elements of ${}^N\omega^k$.

Step 6: Express the remainder acceleration term, ${}^N\mathbf{a}_t^{k^*}$, of the acceleration of the mass center k^* of k, and those of the inboard body i as follows:

$$
\begin{aligned}
{}^N\mathbf{a}^{k^*} &= \sum_{j=1}^{n} {}^N\mathbf{v}_j^{k^*} \dot{u}_j + {}^N\mathbf{a}_t^{k^*} \qquad (k = n, \ldots\ldots, 1) \\
{}^N\mathbf{a}_t^{n^*} &= {}^N\mathbf{a}^B - {}^N\boldsymbol{\omega}^n({}^N\boldsymbol{\omega}^B \times \mathbf{r}_2^n) \qquad\qquad (11.8a) \\
{}^N\mathbf{a}_t^{k^*} &= {}^N\mathbf{a}_t^{i^*} + {}^N\boldsymbol{\alpha}_t^i \times \mathbf{r}_1^k +{}^N \boldsymbol{\omega}^i \times ({}^N\boldsymbol{\omega}^i \times \mathbf{r}_1^k) - {}^N\mathbf{a}_t^k \times \mathbf{r}_2^k - {}^N\boldsymbol{\omega}^K \times ({}^N\boldsymbol{\omega}^k \times \mathbf{r}_2^k) \\
&\quad (k = n-1, \ldots\ldots, 1)
\end{aligned}
$$

Prescribed linear acceleration of the base of the beam is reflected in the term ${}^N\mathbf{a}^B$. Now form in the k basis the 3×1 *matrix* ${}^N a_t^{k^*}$, $(k = n, \ldots ., 1)$:

$$
\begin{aligned}
a^{k^*} &= \sum_{j=1}^{n} v_j^{k^*} u_j + a_t^{k^*} \qquad (k = n, \ldots\ldots, 1) \\
a_t^{n^*} &= a^A - \tilde{\omega}^n \tilde{\omega} r_2^n \qquad\qquad (11.8b) \\
\alpha_t^{k^*} &= \alpha_t^{i^*} - \tilde{r}_1^k \alpha^i + \tilde{\omega}^i \tilde{\omega}^i r_1^k + \tilde{r}_t^k \alpha^k - + \tilde{\omega}^i \tilde{\omega}^i r_2^k \qquad (k = n-1, \ldots\ldots, 1)
\end{aligned}
$$

Step 7: Represent now the external force F_k and external torque T_k, both 3×1 matrices corresponding to vectors in the k basis, and I^k the moment of inertia matrix of body k, form the following 3×1 matrices:

$$
\begin{aligned}
F_t^{*k} &= m_k \, a_t^{k*} - F^k \qquad (k = n, \cdots, 1) \\
T_t^{*k} &= I^k \, \alpha_t^k + \tilde{\omega}^k \, I^k \quad \omega^k - T^k \qquad (k = n, \ldots, 1)
\end{aligned}
\tag{11.9}
$$

Step 8: Form the 6×6 matrix M_k and the 6×1 matrices $X_k, Y_k, \ (k = n, \ldots, 1)$ given below:

$$
\begin{aligned}
M_k &= \begin{bmatrix} m_k U & 0 \\ 0 & I^k \end{bmatrix} \\
X_k &= \begin{Bmatrix} F_t^{*k} \\ T_t^{*k} \end{Bmatrix} \\
Y_k &= \begin{Bmatrix} {}^N v_k^{k*} \\ {}^N \omega_k^k \end{Bmatrix}
\end{aligned}
\tag{11.10}
$$

Here U is a 3×3 unity matrix, and 0 is a 3×3 null matrix.

11.2.3 Backward Pass

Step 1: *Set* $k = 1, \ M = M_k, \ X = X_k$

Step 2: Form and store the 6×1 matrix Z_k and the scalars m_{kk} and f_k as follows,

$$
\begin{aligned}
Z_k &= M Y_k \\
m_{kk} &= Y_k^T Z_k \\
f_k &= Y_k^T X + \tau_k
\end{aligned}
\tag{11.11}
$$

where τ_k is the hinge torque applied by body i on body k at the kth hinge, and superscript T indicates the transpose of the matrix. Form the 6×6 augmented mass matrix and the 6×1 augmented remainder inertia force column matrix X:

$$
\begin{aligned}
M &= M - [Z_k Z_k^T]/m_{kk} \\
X &= X - Z_k (f_k / m_{kk})
\end{aligned}
\tag{11.12}
$$

Step 3: Define the 6×6 shift transform matrix W between bodies i and k,

$$
W = \begin{bmatrix} C_{ik}^T & -\tilde{r}^{i^* k^*} C_{ik}^T \\ 0 & C_{ik}^T \end{bmatrix}
\tag{11.13}
$$

where $\tilde{r}^{i^* k^*}$ is the skew-symmetric matrix for cross product with the components of the position vector from the mass center i* of body i to the mass center k^* of body k in the k-basis. Next, shift the inertia force and inertia torque of the kth body, represented by the augmented mass and remainder inertia force matrices, M and X of Eq. (11.12), to the ith body and add to the inertia force matrices for body i, thus forming new matrices M , K:

$$\begin{aligned} M &= M_i + W^T M W \\ X &= X_i + W^T X \end{aligned} \tag{11.14}$$

Step 4: **Increase k by 1** and repeat steps 2 and 3 until m_{nn} and f_n are formed.

11.2.4 Forward Pass

Step 5: Set $k = n - 1$, and $i = n$. Form the following:

$$\begin{aligned} \dot{u}_n &= -f_n / m_{nn} \\ \begin{Bmatrix} {}^N a_0^{i*} \\ {}^N \alpha_0^i \end{Bmatrix} &= Y_n \dot{u}_n \end{aligned} \tag{11.15}$$

Step 6: Form two 6×1 matrices and the scalar derivative $\dot{u}_k$ in the sequence shown below, with Z_k, f_k defined in Eq. (11.11) and Y_k in Eq. (11.10):

$$\begin{aligned} \begin{Bmatrix} {}^N \hat{a}_0^k \\ {}^N \hat{\alpha}_0^k \end{Bmatrix} &= W \begin{Bmatrix} {}^N a_0^{i*} \\ {}^N \alpha_0^i \end{Bmatrix} \\ \dot{u}_k &= -\left([Z_k]^T \begin{Bmatrix} {}^N \hat{a}_0^k \\ {}^N \hat{\alpha}_0^k \end{Bmatrix} + f_k \right) / m_{kk} \\ \begin{Bmatrix} {}^N a_0^{k*} \\ {}^N \alpha_0^k \end{Bmatrix} &= \begin{Bmatrix} {}^N \hat{a}_0^{k*} \\ {}^N \hat{\alpha}_0^k \end{Bmatrix} + Y_k \dot{u}_k \end{aligned} \tag{11.16}$$

Step 7: **Decrease the indices** k and i by 1 and repeat step 6 until $\dot{u}_1$ is formed.

11.2.5 Deployment/Retraction Step

If one link has been extruded, *replace n by n + 1* and assume that $q_{n+1} = u_{n+1} = 0$. For retraction, replace n by $n-1$ when one link is retracted and update state as described below to maintain continuity in displacements and velocities. Reset the values of the beam stiffness based on Eq. (11.2) for the current value of n. Go to the step 1 of forward pass. For retraction, an update of dynamical states is needed as one link goes in. The situation is shown in Figure 11.3. The initial values of the generalized coordinates and generalized speeds for a new set of links have to be computed on the basis of continuity of physical displacements and velocities of the new system and the old system. This requires the following two sets of equations for displacement and velocity to be satisfied, going from n to $n-1$ links in retraction, with a superscript (+)

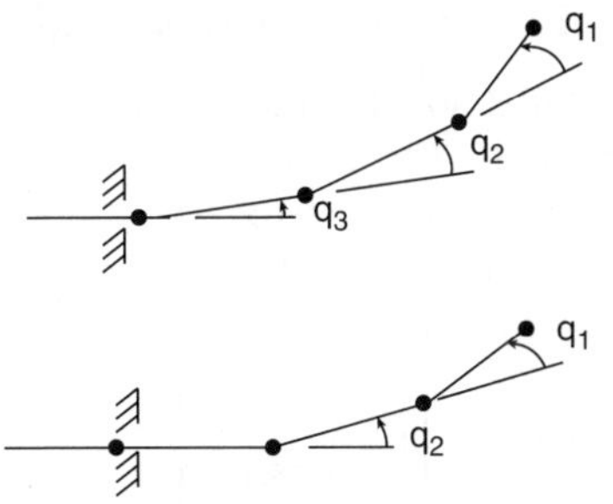

Figure 11.3 Representation of Jump Discontinuity as One Link Goes in During Retraction. Banerjee [1]. Reproduced with permission of the American Institute of Aeronautics and Astronautics, Inc.

indicating unknown initial value in the $n-1$ system, to the n-system with superscript (–) for known end value:

$$\sum_{j=i}^{n-1} L\sin\sum_{k=j}^{n-1} q_k^+ = \sum_{j=i}^{n} L\sin\sum_{k=j}^{n} q_k^- \quad (i=1,\ldots,n-1)$$
$$\sum_{j=i}^{n-1} L\cos\left(\sum_{k=j}^{n-1} q_k^+\right)\sum_{m=j}^{n-1}\dot{q}_m^+ = \sum_{j=i}^{n} L\cos\left(\sum_{k=j}^{n} q_k^-\right)\sum_{m=j}^{n}\dot{q}_m^- \quad (i=1,\ldots,n-1) \tag{11.17}$$

The second set of equations in Eq. (11.17) is linear. The first set, displacement equations, in Eq. (11.17) is nonlinear in the new generalized coordinates, of the form $f_i(q_1^+,\ldots,q_{n-1}^+)=0,\quad i=1,\ldots,n-1$, and they can be solved for the angles by Newton's iterative method for solving nonlinear equations, where the associated Jacobian matrix turns out to be triangular because the deflection at any revolute joint does not depend on the motion of the outboard links:

$$\left[\frac{\partial f_i}{\partial q_j^+}\right]\left(\left\{q_j^+\right\}_{i+1}-\left\{q_j^+\right\}_i\right) = -\{f_i\}, \qquad i=1,\cdots,n-1 \tag{11.18}$$

$$\begin{aligned}\frac{\partial f_i}{\partial q_j^+} &= 0 \qquad j>i\\ &= L\cos\sum_{k=1}^{n-1} q_k^+ \qquad (i=1,\cdots,n-1)\\ &= \frac{\partial f_i}{\partial q_{j-1}^+} + L\cos\sum_{k=j}^{n-1} q_k^+ \qquad (i=1,\cdots,n-1; j=i+1,\cdots,n-1)\end{aligned} \tag{11.19}$$

Note that no initial value update is needed for extrusion, since the two equation sets in Eq. (11.17), with superscripts (+) and (−) interchanged, are identically satisfied as the system grows from $(n-1)$ to n links, with the new link having zero transverse displacement and velocity.

11.3 Numerical Simulation of Deployment and Retraction

The variable-n order-n algorithm is now used to simulate extrusion/retraction of the WISP antenna from a slowly rolling shuttle. The antenna is a tube of 0.0636-m diameter, .00254-m wall thickness and modulus of elasticity $1.96491 \times 10^{10}\ \mathrm{N/m^2}$. Fully deployed, the antenna is 150-m long. The shuttle is assumed to rotate at 1 deg/sec about an axis normal to the direction of extrusion/retraction. Extrusion from 5-m to 150-m in 750m seconds is prescribed at the rate

$$\dot{d} = 0.26666\left[1 - \left(\frac{t}{750}\right)^3\right] \qquad t \leq 750 \tag{11.20}$$

and retraction from 150-m to 5-m in 750m seconds is prescribed at the rate:

$$\dot{d} = 0.26666\left[1 - \left(1 - \frac{t}{750}\right)^3\right] \qquad t \leq 750 \tag{11.21}$$

Figures 11.4a and 11.4b show the transverse beam deflections for extrusion at the beam tip and at a point 50 m from the tip, respectively; the associated length time history is in Figure 13.4c. Note that the beam deflection around 130 m is about 16 m. This is already a large deflection, beyond the range of linear beam theory, so that any modal superposition would have been inapplicable, whereas our formulation does not have any such restriction. Checking up to 750 sec we see that the extrusion from a rotating base makes the beam *lag* its original undeformed configuration; this is because of the nature of Coriolis inertia loading; also note that the lumped mass at 50 m from the beam tip starts late in its deflection because it does not come out until the time that depends on the extrusion rate. Figures 11.5a and 11.5b show the deflection results for retraction, and Figure 11.5c shows the length vs. time. From Figures 11.5a and 11.5c, note that the tip deflection is 23 m when the beam instantaneous length is 133 m; this is a larger deflection than that obtained with extrusion. Here the beam leads the undeformed deformation, again due to the direction of the Coriolis inertia load. The frequency increase as the length gets shorter reminds one of "spaghetti eating"! The oscillation of the point 50 m from the tip stops as the point goes into the containing body, whereas for the tip mass the terminal oscillations Figure 11.5 shows the transverse vibrations of a stubby beam. The results in Figure 11.5 show that the deflections of a retracting beam do not necessarily grow unbounded; this is in contrast to the case of a tether that has a tendency to wrap around the exiting body. Because a tether can be thought of as a beam of zero bending stiffness, the method given in this chapter can be used to analyze the retraction of a tether into a rotating body. Figure 11.6 shows the deflections and instantaneous length of a cable at half the retraction rate used for Figure 11.5. It is seen that the cable has very large deflections with less than 15 m of retraction completed, and the cable tends to fly off. This indicates that it is the bending stiffness of a beam that prevents its deflection from going unbounded. Finally, we show some results of spin-up of the shuttle body in three dimensions as in:

$$\begin{aligned} {}^N\boldsymbol{\omega}^B &= 0.001\mathbf{b}_1 + \frac{0.0174533}{600}\left[t - \frac{300}{\pi}\sin\left(\frac{\pi t}{300}\right)\right]\mathbf{b}_3, t \leq 600 \\ &= 0.001\mathbf{b}_1 + 0.0174533\mathbf{b}_3, t > 600 \end{aligned} \tag{11.22}$$

Three-dimensional response of the beam is obtained by modeling two degrees of rotation at the joints (see the next section and the solution to the example in the problem set). Figure 11.7a show the axial foreshortening time history, and Figures 11.7b and 11.7c show the transverse tip deflections.

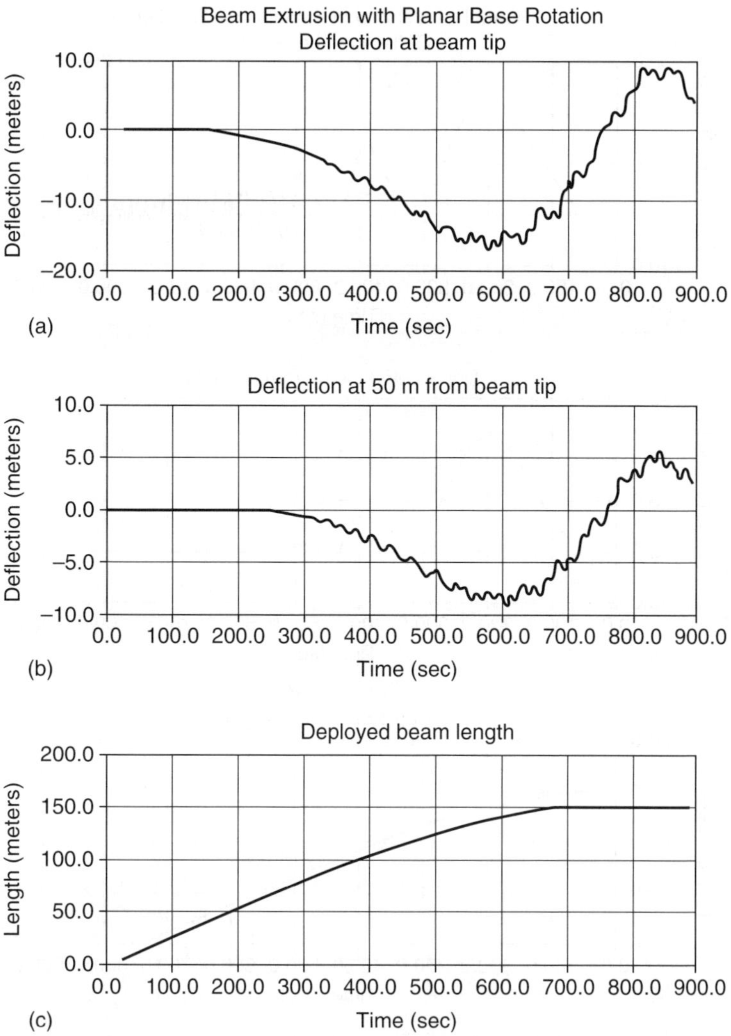

Figure 11.4 Deflection (m) at Beam Tip, 50 m from Tip, and Length (m) Deployed for Extrusion of a Beam from a Rotating Base. Banerjee [1]. Reproduced with permission of the American Institute of Aeronautics and Astronautics, Inc.

11.4 Deployment of a Cable from a Ship to a Maneuvering Underwater Search Vehicle

This section is based on Ref. [2] and deals with an application to searching for mines on the seafloor by means of a maneuvering vehicle connected to a ship by a cable. Figure 11.8 gives a sketch of the system, where an unmanned underwater vehicle (UUV) has its own control system, and the length of the cable connecting it to the ship above has to be extended or shortened to conform with a control command from the UUV; this is what constitutes the constraint. Efficient modeling of the dynamics of a cable being reeled out of a moving ship to

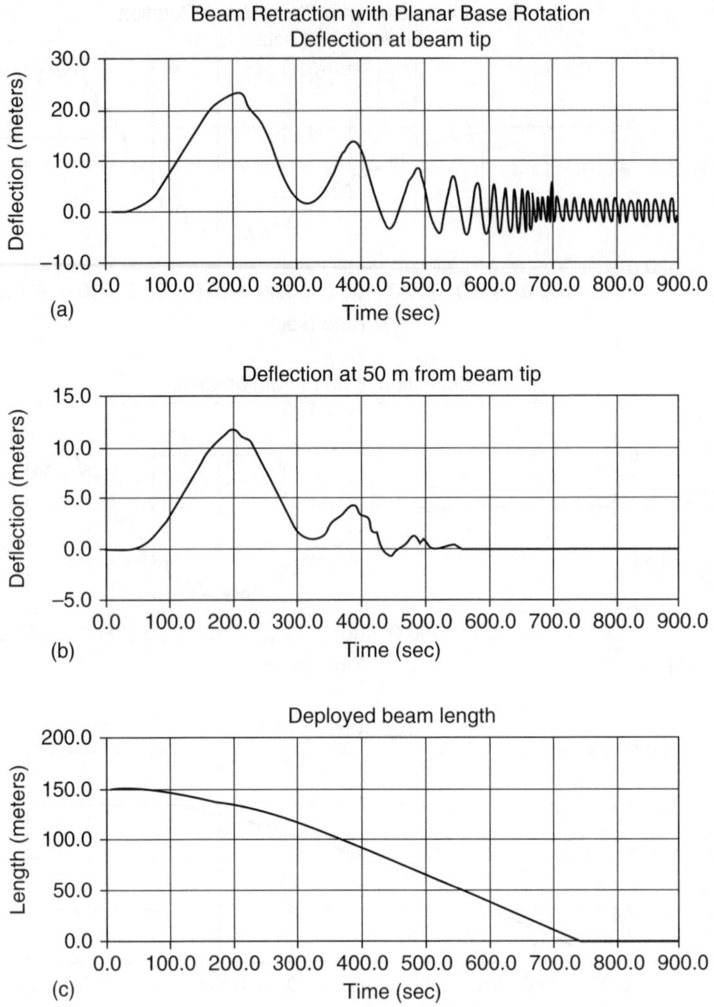

Figure 11.5 Deflection (m) at Beam Tip, and 50 m from Tip, and Length (m) Out for Retraction of a Beam into a Rotating Base {Frequency Increase Toward End Mimics Spaghetti-Eating !}. Banerjee [1]. Reproduced with permission of the American Institute of Aeronautics and Astronautics, Inc.

an underwater maneuvering vehicle requires a constrained order-n formulation that can handle both large three-dimensional curvature of the cable and end motion constraints. We consider the details in the following subsections.

*11.4.1 Cable Discretization and Variable-*n *Order-*n *Algorithm for Constrained Systems with Controlled End Body*

The algorithm used is essentially that given in Ref. [6], and the present concern being an application, only the relevant details of the constrained order-n algorithm are reviewed

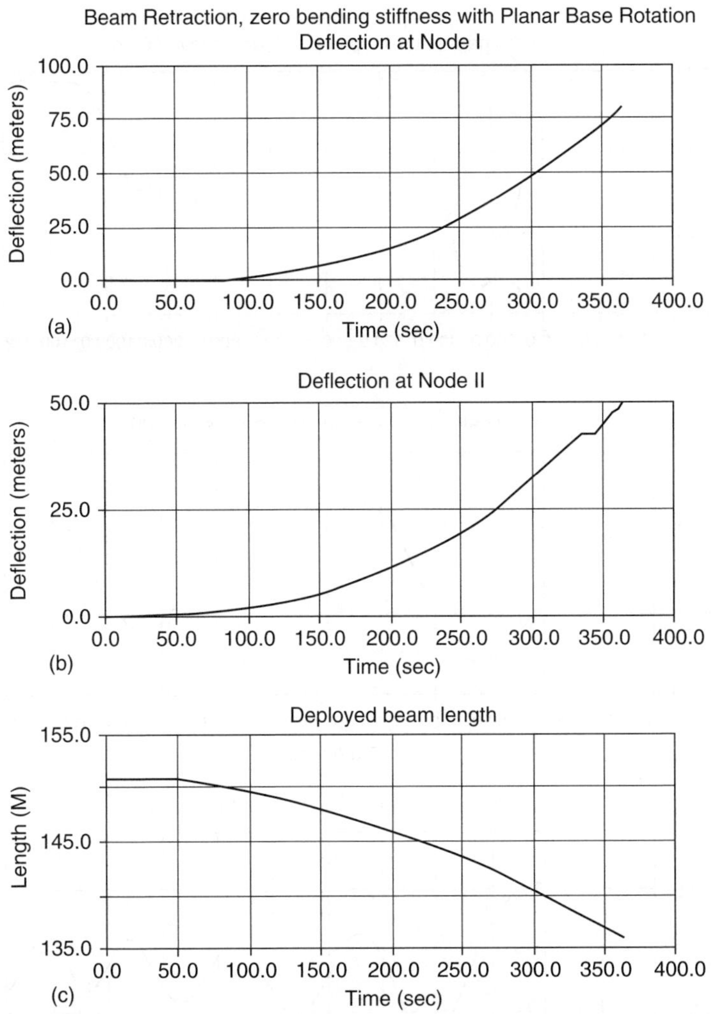

Figure 11.6 Deflection (m) of Zero Bending Stiffness Beam (Cable) at Tip, at 50 m from the Tip, and Length (m) Deployed of a Cable Being Retracted (at Half the Rate Used for Figure 11.5) with Planar Rotation. Banerjee [1]. Reproduced with permission of the American Institute of Aeronautics and Astronautics, Inc.

here. The cable is modeled as a chain of rods with 2 dof rotational joints having soft rotational stiffness (an assumption for a cable in water) to have the ability to produce three-dimensional shape of the cable. Figure 11.9 gives a sketch of the discretization of the cable.

Let λ be the unknown tension on the cable applied by the maneuvering underwater vehicle (UUV). Then considering the free-body diagram of the nth rod, connected to the UUV, and summing inertia and external forces on it, excluding the force and torque from the $(n-1)$th

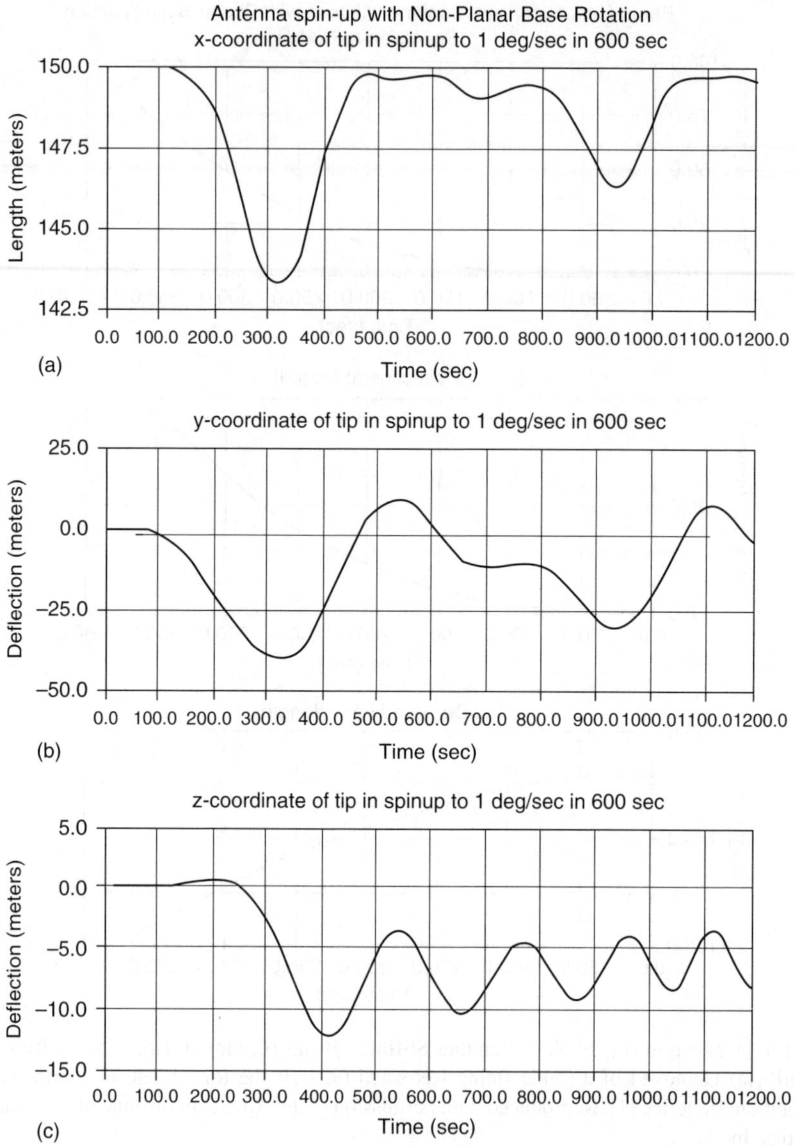

Figure 11.7 Three-Dimensional *X*-, *Y*-, *Z*- Deflection of Antenna due to Non-Planar Spin-up of Base to 1 Deg/sec; Figure 11.7 Shows Axial Fore-Shortening (Length Not Reaching 150 m). Banerjee [1]. Reproduced with permission of the American Institute of Aeronautics and Astronautics, Inc.

rod, and taking the moment of these forces about the hinge point Q_n, one has the following equilibrium equations:

$$\left\{ \begin{array}{c} f^{*n} - f_e^n - \lambda \\ t^{*n} - t_e^n - \tilde{L}\lambda \end{array} \right\} \equiv M^n \left\{ \begin{array}{c} a_0^{Q_n} \\ \alpha_0^n \end{array} \right\} + X^n - H^n \lambda \tag{11.23}$$

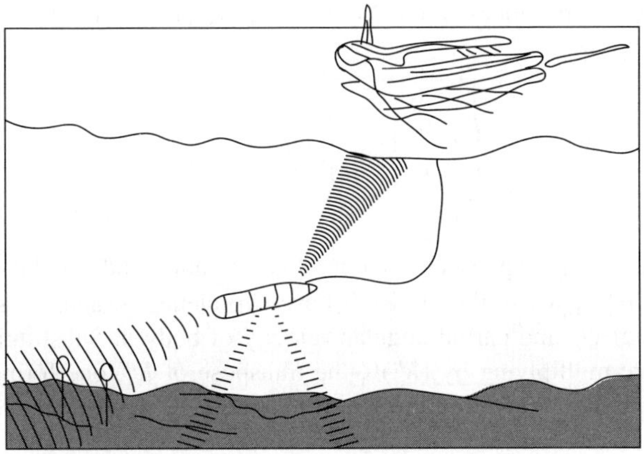

Figure 11.8 System Sketch of a Ship Towing an Underwater Search, Maneuvering Vehicle. Banerjee and Do [2].

Note that end-body force λ may not be aligned with the position vector from the end of link n to Q_n, and all vectors are resolved in body-n basis: In Eq. (11.23) f^{*n}, f_e^n are the inertia force and other external force as noted above, and t^{*n}, t_e^n are the inertia and external torques on the nth body, and we have defined:

$$M^n = \begin{bmatrix} m & -\tilde{S}^n \\ \tilde{S}^n & I^n \end{bmatrix}, \qquad H^n = \begin{Bmatrix} U \\ \tilde{L} \end{Bmatrix}$$
$$X^n = M^n \begin{Bmatrix} a_t^{Q_n} \\ \alpha_t^n \end{Bmatrix} + \begin{Bmatrix} \tilde{\omega}^n \, \tilde{\omega}^n \, S^n - f_e^n \\ \tilde{\omega}^n \, I^n \, \omega^n - t_e^n \end{Bmatrix} \tag{11.24}$$

As before, S^n and I^n are the first and second moments of inertia about Q_n, U is 3×3 unity matrix, and ω^n is the inertial angular velocity of body n in its own basis. Again, a tilde denotes a skew-symmetric matrix formed out (3×1) matrices, for matrix representation of vector cross products. Here, $\tilde{L}$ is formed out of a column matrix whose first element is L, the length of the rod, the other two element being zero. Symbols in Eq. (11.23) with zero subscript represent all terms that involve second derivatives of the general coordinates, and those in Eq. (11.24) with subscript t denote remainder acceleration terms in the expression for acceleration of the hinge point Q_n and the angular acceleration of body n, as has been used in earlier chapters:

$$\begin{Bmatrix} a^{Q_n} \\ \alpha^n \end{Bmatrix} = \begin{Bmatrix} a_0^{Q_n} \\ \alpha_0^n \end{Bmatrix} + \begin{Bmatrix} a_t^{Q_n} \\ \alpha_t^n \end{Bmatrix} \tag{11.25}$$

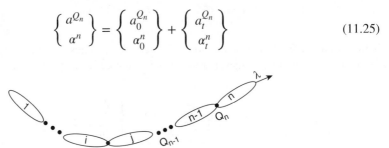

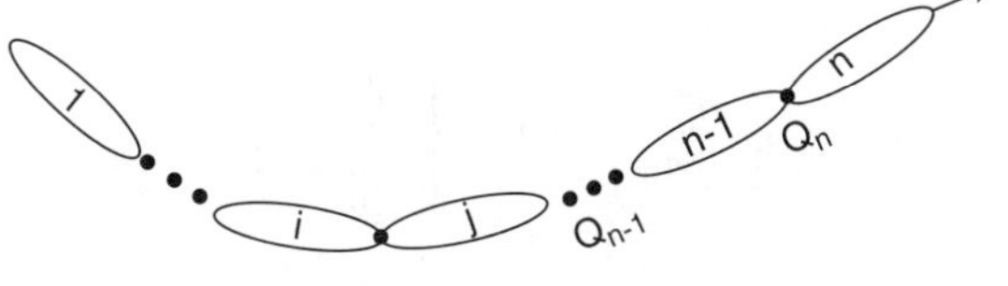

Figure 11.9 Discretization of the Cable. Banerjee and Do [2].

Again, exposing the contributions of the 2 dof relative rotation rates at the body n hinge, one can further write

$$\begin{Bmatrix} a_0^{Q_n} \\ \alpha_0^n \end{Bmatrix} = \begin{Bmatrix} \hat{a}_0^{Q_n} \\ \hat{\alpha}_0^n \end{Bmatrix} + R^n \ddot{\theta}^n \tag{11.26}$$

where the hat notation encompasses the group of acceleration and angular acceleration terms involving $\ddot{\theta}^1, \ldots\ldots, \ddot{\theta}^{n-1}$, where $\ddot{\theta}^i$ may itself have two elements, and R^n is the 6×2 matrix of partial velocity of Q_n and partial angular velocity of body n. Substituting Eq. (11.26) in Eq. (11.23) and pre-multiplying by $(R^n)^T$, the transpose of R^n, one forms Kane's dynamical equations with undetermined multipliers [8], associated with θ^n, in which non-working constraint forces are automatically eliminated (a property of Kane's equations),

$$\ddot{\theta}^n = -\nu_n^{-1} R^{n^T} \left\{ M^n \begin{pmatrix} \hat{a}_0^{Q_n} \\ \hat{\alpha}_0^n \end{pmatrix} + X^n - H^n \lambda \right\} + \nu_n^{-1} \tau^n \tag{11.27}$$

where

$$\nu_n = R^{n^T} M^n R^n \tag{11.28}$$

and τ^n is the two-component hinge torque applied by body $(n - 1)$ on body n. Using Eq. (11.27) in Eq. (11.23) gives D'Alembert equation of the zero-sum of forces and moments:

$$\begin{Bmatrix} f^{*n} - f_e^n - \lambda \\ t^{*n} - t_e^n - \tilde{L}\lambda \end{Bmatrix} \equiv M \begin{Bmatrix} \hat{a}_0^{Q_n} \\ \hat{\alpha}_0^n \end{Bmatrix} + X^n - H\,\lambda \tag{11.29}$$

Here the following notations have been used with U now representing a 6×6 identity matrix:

$$\begin{aligned} M &= PM^n, & P &= U - M^n R^n \nu_n^{-1} R^{n^T} \\ X &= PX^n, & H &= PH^n \end{aligned} \tag{11.30}$$

Now, as has been shown in previous chapters, kinematics relating the motion of inboard (n−1) to that of body n suggest a transfer of motion defined by a shift transform W^n:

$$\begin{aligned} \begin{Bmatrix} \hat{a}_0^{Q_n} \\ \hat{\alpha}_0^n \end{Bmatrix} &= W^n \begin{Bmatrix} a_0^{Q_{n-1}} \\ \alpha_0^{n-1} \end{Bmatrix} \\ W^n &= \begin{bmatrix} C_{n,n-1} & -C_{n,n-1}\tilde{L} \\ 0 & C_{n,n-1} \end{bmatrix} \end{aligned} \tag{11.31}$$

Here $C_{n,n-1}$ is the coordinate transformation relating the n-body basis vectors to $(n-1)$ body basis vectors, and $\tilde{L}$ has been defined before. The same shift transform serves to shift the forces and moments from the hinge Q_n to a statically equivalent system at the hinge $Q_{n\text{-}1}$:

$$\begin{Bmatrix} f^{*n} - f_e^n - \lambda \\ t^{*n} - t_e^n - \tilde{L}\lambda \end{Bmatrix} \equiv W^{n^T} \left\{ MW^n \begin{pmatrix} a_0^{Q_{n-1}} \\ \alpha_0^{n-1} \end{pmatrix} + X - H\lambda \right\} \tag{11.32}$$

Now considering the free-body diagram of the $(n-1)$th body and adding the inertia and all external forces, including hinge forces from body $n-2$ to body $n-1$ with the force system represented by Eq. (11.32), and recalling that non-working forces are eliminated, one gets

$$\begin{Bmatrix} f^{*n-1} - f_e^{n-1} \\ t^{*n-1} - t_e^{n-1} \end{Bmatrix} \equiv M^{n-1} \begin{pmatrix} a_0^{Q_{n-1}} \\ \alpha_0^{n-1} \end{pmatrix} + X^{n-1} - H^{n-1}\lambda \tag{11.33}$$

where one has made the following updates:

$$\begin{aligned} M^{n-1} &\leftarrow M^{n-1} + W^{n^T} M W^n \\ X^{n-1} &\leftarrow X^{n-1} + W^{n^T} X \\ H^{n-1} &= W^{n^T} H \end{aligned} \tag{11.34}$$

Equation (11.33) is just like Eq. (11.23). This completes a cycle that is repeated by going backward, considering bodies $n-2$, $n-3$, and so on, in succession, until one encounters body 1, for which the inboard motion – ship motion in this case – is assumed to be known. The equations for the hinge dof for the first link connected to the ship in this case can be written as follows:

$$\ddot{\theta}^1 = f_1^1 + c_1^1 \lambda \tag{11.35}$$

where

$$\begin{aligned} f_1^1 &= -\nu_1^{-1} R^{1^T} X^1 + \nu_1^{-1} \tau^1 \\ c_1^1 &= \nu_1^{-1} R^{1^T} H^1 \end{aligned} \tag{11.36}$$

In terms of Eqs. (11.35)–(11.36), Eq. (11.26), yields for body 1

$$\begin{Bmatrix} a_0^{Q_1} \\ \alpha_0^1 \end{Bmatrix} = g^1 + h^1 \lambda \tag{11.37}$$

where

$$\begin{aligned} g^1 &= R^1 f_1^1 \\ h^1 &= R^1 c_1^1 \end{aligned} \tag{11.38}$$

Equation (11.31) now yields, for a generic jth body,

$$\left\{ \begin{array}{c} \hat{a}_0^{Q_j} \\ \hat{\alpha}_0^j \end{array} \right\} = e^j + n^j \lambda \tag{11.39}$$

where the kinematical shift operation from body i, the inboard body of j, to body j produced:

$$\begin{aligned} e^j &= W^j g^i \\ n^j &= W^j h^i \end{aligned} \tag{11.40}$$

This defines the generic hinge rotation equations

$$\ddot{\theta}^j = f_1^j + c_1^j \lambda \tag{11.41}$$

where

$$\begin{aligned} f_1^j &= -v_j^{-1} R^{jT} \{M^j e^j + X^j\} \\ c_1^j &= -v_j^{-1} R^{jT} \{M^j n^j - H^j\} \end{aligned} \tag{11.42}$$

Use of Eqs. (11.41) and (11.39) in Eq. (11.27), along with new denotations,

$$\begin{aligned} g^j &= e^j + R^j f_1^j \\ h^j &= n^j + R^j c_1^j \end{aligned} \tag{11.43}$$

provides the general relation:

$$\left\{ \begin{array}{c} a_0^{Q_j} \\ \alpha_0^j \end{array} \right\} = g^j + h^j \lambda \tag{11.44}$$

The stage is now set for uncovering all the dynamical equations in a so-called second forward pass, going from body 2 to body n. This completes the review of the order-n algorithm for constrained systems. These equations have to be solved together with a consideration of the constraint, which is due to the underwater maneuvering vehicle having its own independent control system. Before doing that, however, we will take up the issue of specifying external forces on the underwater cable.

11.4.2 Hydrodynamic Forces on the Underwater Cable

For steady flow, the force exerted by a fluid on a submerged accelerating body is composed of two components, one depending on fluid friction and the other on inertia of the displaced fluid [8, 9]. The transverse oscillatory force on the cable due to vortex shedding is neglected here for simplicity. For cylindrical bodies, the force due to viscous drag is given by:

$$\mathbf{f}_d = -0.5\rho_w DL\left[\pi C_f\left|v_1\right|v_1\mathbf{b}_1 + C_{dn}(v_2\mathbf{b}_2 + v_3\mathbf{b}_3)\sqrt{v_2^2 + v_3^2}\right] \tag{11.45}$$

Here ρ_w is the density of seawater, D is the cable diameter, L is the cable length,C_f is the axial friction coefficient, C_{dn} is the nominal drag coefficient, and v_1, v_2, v_3 are the three components of the body relative to the water, taking into account water currents. The force required to move the displaced fluid is treated as an apparent added mass at the center of the link in the inertia computation [8],

$$m' = L(m + C_m\rho_w\pi D^2/4) \tag{11.46}$$

where m is the mass per unit length of the cable, L is the discretized segment length, and C_m is the fluid mass coefficient. Note that Eq. (11.46), while simplifying matters, does not rigorously treat the physics of fluid inertia in the sense that only acceleration components perpendicular to the cable will affect the apparent fluid mass. In addition to these effects, the force of buoyancy reduces the force of gravity of the rod element, giving rise to a force of "wet weight" approximation given by

$$f_w = -\pi\ DL(\rho_c - \rho_w)g\mathbf{n}_2 \tag{11.47}$$

where ρ_c, g, $\mathbf{n}_2$ are, respectively, the cable density, the acceleration due to gravity, and the inertial gravity direction unit vector. Finally, we consider the force at the end of the cable due to mass gain or loss associated with deployment or retraction, respectively. This force is in the nature of a thrust, and is represented as a force acting on the link closest to the ship, as follows:

$$\mathbf{f}_t^1 = -(\pi/4)D^2\rho_c\dot{L}\left[(v_1^1 - \dot{L})\,\mathbf{b}_1^1 + v_2^1\mathbf{b}_2^1 + v_3^1\mathbf{b}_3^1\right] \tag{11.48}$$

where $\dot{L}$, v_1^1, $\mathbf{b}_1^1$ are, respectively, the deployment rate, axial component of the velocity vector, and the unit vector along the axis for body *1* and so on.

11.4.3 Nonlinear Holonomic Constraint, Control-Constraint Coupling, Constraint Stabilization, and Cable Tension

With one end of the cable attached to the ship, the other end is constrained to move with the UUV, which has its own model and embedded propulsion control system as it goes in search of underwater mines. This implies the existence of a holonomic constraint, which means some of the generalized coordinates are dependent (indicated by the subscript d in the following) on the independent ones (denoted by subscript i). This in turn means that Eq. (11.41) can be written in column matrix form as:

$$\ddot{\theta}_i = F_i + C_i\lambda \tag{11.49}$$

$$\ddot{\theta}_d = F_d + C_d\lambda \tag{11.50}$$

The constraint force components represented in λ can be determined by solving Eqs. (11.49) and (11.50) simultaneously with the constraint equations. Now the constraint equation can be written in terms of an error in the position vectors from the inertial origin to the UUV measured in two ways: (1) once to the ship and then along the cable; and (2) to the UUV directly, as in the error vector equation:

$$\mathbf{e} = \left(\mathbf{p}^s + \sum_{i=1}^{n} L_i \mathbf{b}_1^i\right) - \mathbf{p}^{UUV} = 0 \tag{11.51}$$

Here $\mathbf{p}^s, \mathbf{p}^{UUV}$ stand for the position vectors from an inertial origin to the ship and to the UUV, respectively. Equation (11.51) is nonlinear in the sets of angles between the links, θ_i, θ_d. Although a differential-algebraic equation solver can be used to solve and satisfy Eqs. (11.49)–(11.51), this takes away from the efficiency of the order-n formulation that can be used in an online control system. One alternative is to differentiate Eq. (11.51) twice and solve the resulting equation together with Eqs. (11.49) and (11.50). This, of course, gives rise to the well-known problem of constraint violation with time, which necessitates constraint stabilization [11]. Here we use a proportional-integral-derivative correction of error, which is known to fight steady-state error in control. Using the scalar form of the equations in Eq. (11.51) in the inertial basis, and using their derivative and integral, one has:

$$\ddot{e} + k_d \dot{e} + k_p e + k_i \int_0^t e(\tau) d\tau = 0 \tag{11.52}$$

Equation (11.52) requires the computation of the velocity and acceleration of the point P of the cable that is attached to the UUV and comparing these, respectively, with the velocity and acceleration of the UUV. Writing the acceleration of P in matrix form as

$$a^P = J_1 \ddot{\theta}_i + J_d \ddot{\theta}_d + a_t^P \tag{11.53}$$

where J_i, J_d are Jacobian terms that are really partial velocities, with all vectors resolved in the inertial basis, and substituting Eqs. (11.49) and (11.50) in Eq. (11.53), one obtains from Eq. (11.52) the following equations to determine the cable tension λ:

$$[J_i C_i + J_d C_d]\ \lambda = a^{UUV} - \{J_i F_i + J_d F_d + a_t^P + k_p e + k_d (v^P - v^{UUV}) + k_i \int_0^t e(\tau) d\tau \tag{11.54}$$

where v^P and v^{UUV} are the velocities of the cable end points P and the UUV, respectively. The acceleration of the UUV, a^{UUV} in Eq. (11.54), is given as an output of the control system for depth and speed control of the UUV. This control system treats the force applied by the cable, λ, as a disturbance to the system, and thus the output acceleration becomes a function of λ. Because of the nonlinearities of the signal limiters used in an implementation of the control laws (see Figure 11.10), the functional dependence $a^{UUV}(\lambda)$ of the output acceleration on λ is

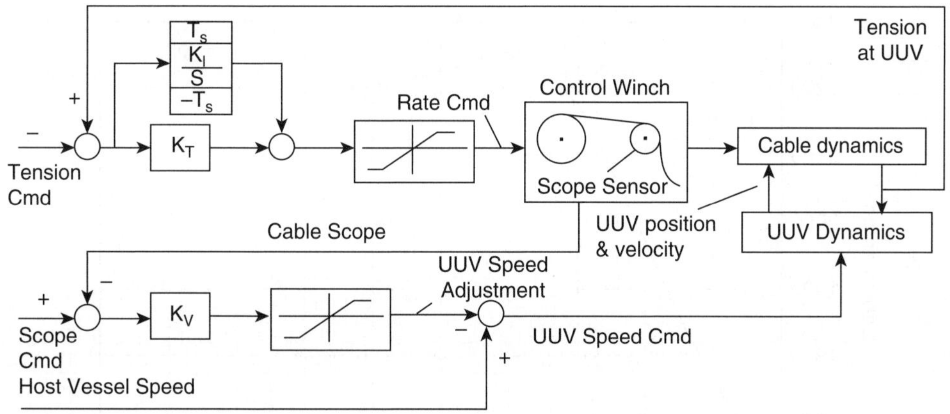

Figure 11.10 Cable Length Management Nonlinear Control Law Block Diagram. Banerjee and Do [2].

nonlinear. Thus, Eq. (11.54) are nonlinear equations in λ, and have to be solved iteratively, as follows:

$$[J_iC_i + J_dC_d]\lambda_{n+1} = a^{UUV}(\lambda_n) - [J_iF_i + J_dF_d + a_t^P + k_pe + k_d(v^P - v^{UUV}) + k_i \int_0^t e(\tau)d\tau] \tag{11.55}$$

Details of the control winch mechanization of nonlinear control with limiter itself are not described here, but are given in Ref. [2]. Figure 11.10 gives an overview of the control system.

For our purpose of describing the cable dynamics during deployment and retrieval, it suffices to say that it is a hardware realization of the cable rate control law given by:

$$\dot{L} \approx K_v(L_c - L) \tag{11.56}$$

Here L is the instantaneous length, L_c the commanded length, and K_v is a rate gain. This is a first order system that is stable at all values of K_v, and has a time constant of $1/K_v$. For the simulation, K_v is set to 0.01, so that for small errors in length (<150 ft), correction would be effective in 300 seconds. Effect of the rate limiter is felt beyond this 150 ft error. Simulation results of cable deployment and retrieval are described next.

11.5 Simulation Results

As has been said, the actual application of the cable dynamics algorithm given above is to an underwater vehicle doing seafloor mine detection, with the UUV having its own controller, while being attached to a ship by means of a cable. In these simulations Figures 11.11–11.13 show, respectively, the time-histories of cable length, cable tension, and constraint length violation error. Figure 11.14 shows the shape of the cable with deployment. Some typical results for retrieval of the cable, namely of cable length and cable tension, are given in Figures 11.15 and 11.16.

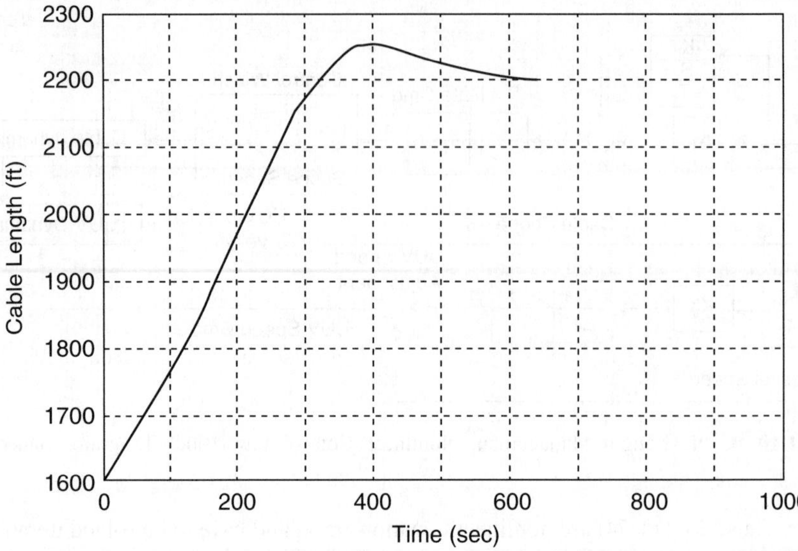

Figure 11.11 Cable Length vs. Time for Deployment of Underwater Vehicle. Banerjee and Do [2].

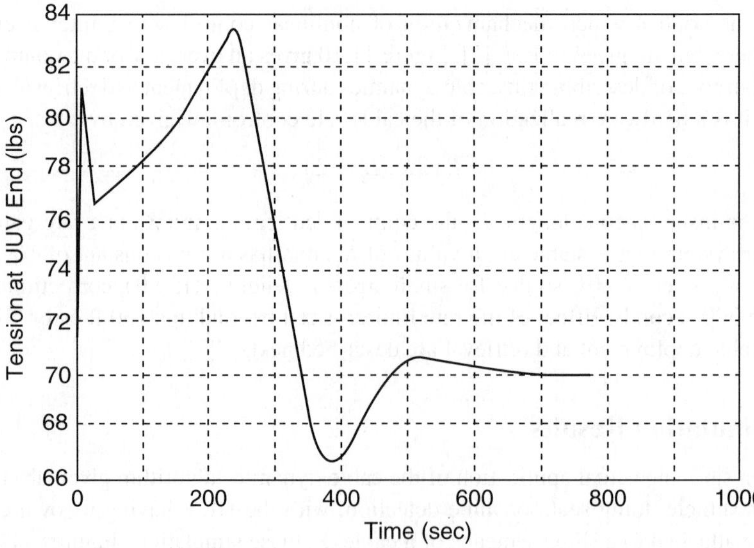

Figure 11.12 Cable Tension vs. Time for Deployment of Underwater Vehicle. Banerjee and Do [2].

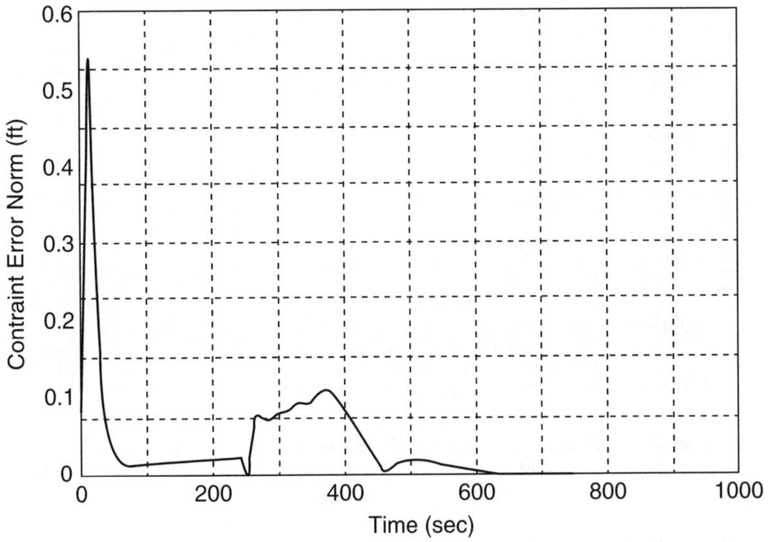

Figure 11.13 Constraint Length Error Norm vs. Time for Deployment of Underwater Vehicle. Banerjee and Do [2].

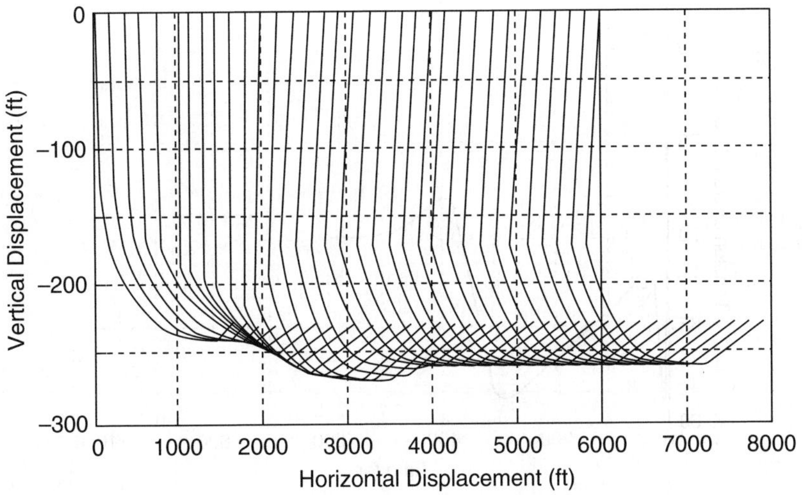

Figure 11.14 Cable Shape During Deployment of Underwater Vehicle. Banerjee and Do [2].

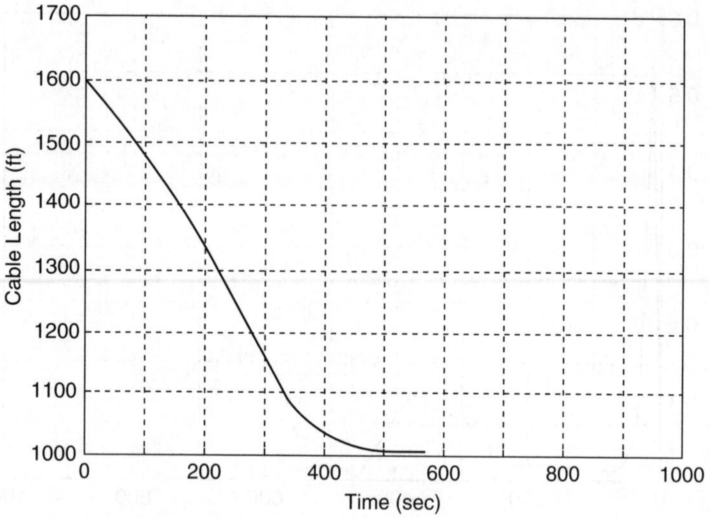

Figure 11.15 Cable Length During Retrieval of Underwater Vehicle. Banerjee and Do [2].

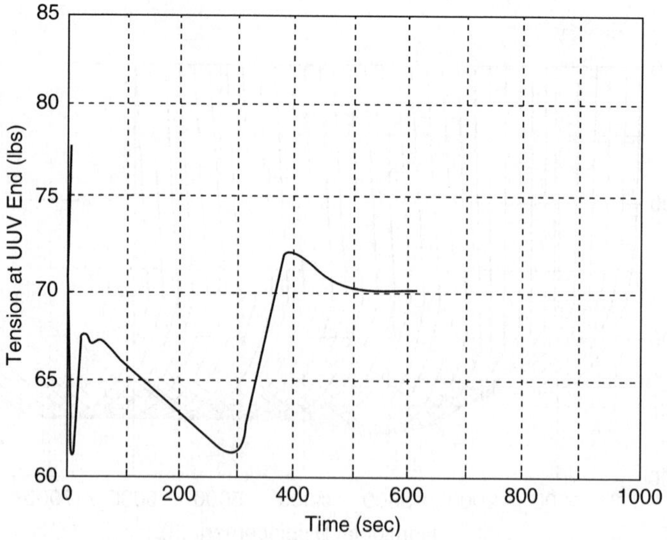

Figure 11.16 Cable Tension During Retrieval of Underwater Vehicle. Banerjee and Do [2].

Problem Set 11

The reader is encouraged to follow the steps in an alternative method of analysis for boom deployment or retrieval; the detailed solution is reported in Ref. [4]. The analysis is done on the following substitute system of a deploying/retracting boom coming out of a rotating base B. The links are massless, with lumped masses at end of links connected by spherical joints restrained by torsional springs, $KX_i, KY_i, (i = 1, \ldots, n)$. The idea is to work with n sets modal coordinates of vibration modes for the current number n of mass-spring sets that is out of the base, while the system is "convected" as a function of time $d(t)$, as shown in the Figure 11.16, to describe deployment / retrieval.

11.1 Represent the location of a particle P_k as: $x_k = \mathbf{r}^{OP_k}.\mathbf{b}_1; y_k = \mathbf{r}^{OP_k}.\mathbf{b}_2 \quad (k = 1, \ldots, n)$.

11.2 Use the following values for mass and springs, for each link of length and area moments of inertia *EI*, *EJ*:

$$m_1 = m_{n+2} = \frac{\rho L}{2(n+1)L}; \quad m_i = \frac{\rho L}{(n+1)L} \quad (i = 2, \ldots, n+1)$$

$$KX_i = EI/L; \quad KY_i = EJ/L \quad (i = 1, n-2)$$

$$KX_{n-1} = \frac{6(n-1)EI}{(3n-8)L}; \quad KY_{n-1} = \frac{6(n-1)EJ}{(3n-8)L}$$

$$KX_n = \frac{3nEI}{2(3n+1)L}; \quad KY_n = \frac{3nEJ}{2(3n+1)L}$$

11.3 Find vibration modes for n^* sets of modes for the maximum number n^* of particles and express $x_k = \sum_{j=1}^{\nu} A_{kj}q_j, k = 1, \ldots, n; y_k = \sum_{j=1}^{\nu} B_{kj}q_j, k = 1, \ldots, n$ where the modal matrix is:

$$\phi_n = \begin{bmatrix} A_{11} & A_{1\nu} \\ & \\ A_{n1} & A_{n\nu} \\ B_{11} & B_{1\nu} \\ & \\ B_{n1} & B_{n\nu} \end{bmatrix}; \quad q_n = \begin{Bmatrix} q_1 \\ \\ q_\nu \end{Bmatrix}; \quad \delta_n = \begin{Bmatrix} x_1 \\ \\ x_n \\ y_1 \\ \\ y_n \end{Bmatrix}; \quad v_n = \begin{Bmatrix} \dot{x}_1 \\ \\ \dot{x}_n \\ \dot{y}_1 \\ \\ \dot{y}_n \end{Bmatrix};$$

11.4 Show that the nonlinear expression for velocity of a generic particle P_k is:

$$
\begin{aligned}
{}^{N}\mathbf{v}^{P_k} = {}^{N}\boldsymbol{\omega}^{B} \times \left\{ \sum_{j=1}^{\nu} \left(A_{kj}\mathbf{b}_1 + B_{kj}\mathbf{b}_2 \right) q_j + \left[d + L \sum_{i=k}^{n} J_i^{0.5} \right] \mathbf{b}_3 \right\} \\
+ \sum_{j=1}^{\nu} \left(A_{kj}\mathbf{b}_1 + B_{kj}\mathbf{b}_2 \right) \dot{q}_j \\
+ \left(d - L \sum_{l=k}^{n} J_i^{-0.5} \sum_{i=1}^{\nu} \sum_{j=1}^{\nu} N_{lij}\dot{q}_i q_j \right) \mathbf{b}_3 \qquad (k = 1, \ldots, n)
\end{aligned}
$$

where

$$
J_i = 1 - \left(\frac{x_i - x_{i+1}}{L} \right)^2 - \left(\frac{y_i - y_{i+1}}{L} \right)^2 \qquad (i = 1, \ldots, n)
$$

$$
N_{lij} = \left(\frac{A_{li} - A_{l+1,i}}{L} \right) \left(\frac{A_{lj} - A_{l+1,j}}{L} \right) + \left(\frac{B_{li} - B_{l+1,i}}{L} \right) \left(\frac{B_{lj} - B_{l+1,j}}{L} \right)
$$

$(l, i, j = 1, \ldots, n)$

11.5 Extract the nonlinear partial velocity from above as:

$$
{}^{N}\mathbf{v}_i^{P_k} = A_{ki}\mathbf{b}_1 + B_{ki}\mathbf{b}_2 - L \sum_{l=k}^{n} \sum_{j=1}^{\nu} J_i^{-0.5} N_{lij} q_j \mathbf{b}_3 \qquad (k = 1, \ldots, n; \quad i = 1, \ldots, \nu)
$$

11.6 Linearize the partial velocity, then linearize the expression of velocity and differentiate that to get linear acceleration:

$$
\hat{J}_i = 1 \qquad (i = 1, \ldots, n)
$$

$$
{}^{N}\tilde{\mathbf{v}}_i^{P_k} = A_{ki}\mathbf{b}_1 + B_{ki}\mathbf{b}_2 - \sum_{j=1}^{\nu} C_{kij} q_j \mathbf{b}_3 \qquad (k = 1, \ldots, n; i = 1, \ldots, \nu)
$$

$$
C_{kij} = \sum_{l=k}^{n} L \left[\left(\frac{A_{li} - A_{l+1,i}}{L} \right) \left(\frac{A_{lj} - A_{l+1,j}}{L} \right) + \left(\frac{B_{li} - B_{l+1,i}}{L} \right) \left(\frac{B_{lj} - B_{l+1,j}}{L} \right) \right]
$$

$(k = 1, \ldots, n; i, j = 1, \ldots, \nu)$

$$
\begin{aligned}
{}^{N}\tilde{\mathbf{v}}^{P_k} = {}^{N}\boldsymbol{\omega}^{B} \times \left\{ \sum_{j=1}^{\nu} (A_{kj}\mathbf{b}_1 + B_{kj}\mathbf{b}_2) q_j + [d + L(n - k + 1)]\mathbf{b}_3 \right\} \\
+ \sum_{j=1}^{\nu} (A_{kj}\mathbf{b}_1 + B_{kj}\mathbf{b}_2)\dot{q}_j + \dot{d}\mathbf{b}_3 \qquad (k = 1, \ldots, n)
\end{aligned}
$$

11.7 Differentiate the linearized velocity to get linearized acceleration and form linearized generalized inertia force:

$$ {}^{N}\tilde{\mathbf{a}}^{P_k} = \frac{{}^{B}d^{N}\tilde{\mathbf{v}}^{P_k}}{dt} + {}^{N}\boldsymbol{\omega}^{B} \times {}^{N}\tilde{\mathbf{v}}^{P_k} \qquad (k = 1, \ldots, n) $$

$$ F_i^* = -\sum_{k=1}^{n} m_k {}^{N}\tilde{\mathbf{a}}^{P_k} . {}^{N}\tilde{\mathbf{v}}_i^{P_k} \qquad (i = 1, \ldots, \nu) $$

11.8 The linearized active force due to stiffness and damping is:

$$ \tilde{F}_i = -\Omega_i^2 q_i - 2\zeta_i \Omega_i $$

Then complete forming the linear equations of motion.

11.9 The dynamic bending moment at the root, R, of the beam is given by:

$$ M = \sqrt{M_1^2 + M_2^2} $$

$$ \begin{aligned} M_1 = &-\sum_{k=1}^{n} m_k [d - r + (n-k+1)L] \\ &\times \left\{ (\dot{\omega}_3 + \omega_1\omega_2) \sum_{j=1}^{\nu} A_{kj} q_j - \left(\omega_1^2 + \omega_3^2\right) \sum_{j=1}^{\nu} B_{kj} q_j + 2\omega_3 \sum_{j=1}^{\nu} A_{kj} \dot{q}_j \right\} \\ &-\sum_{k=1}^{n} m_k [d - r + (n-k+1)L] \\ &\times \left\{ \sum_{j=1}^{\nu} B_{kj} \ddot{q}_j - 2\omega_1 \dot{d} - (\dot{\omega}_1 - \omega_2\omega_3)[d + (n-k+1)L] \right\} \\ &-\sum_{k=1}^{n} m_k \left\{ (\omega_1^2 + \omega_2^2)[d + (n-k+1)L] - \ddot{d} \right\} \sum_{j=1}^{\nu} B_{kj} q_j \end{aligned} $$

$$ \begin{aligned} M_1 = &-\sum_{k=1}^{n} m_k [d - r + (n-k+1)L] \\ &\times \left\{ -(\dot{\omega}_3 + \omega_1\omega_2) \sum_{j=1}^{\nu} B_{kj} q_j - \left(\omega_2^2 + \omega_3^2\right) \sum_{j=1}^{\nu} A_{kj} q_j - 2\omega_3 \sum_{j=1}^{\nu} B_{kj} \dot{q}_j \right\} \\ &-\sum_{k=1}^{n} m_k [d - r + (n-k+1)L] \\ &\times \left\{ \sum_{j=1}^{\nu} A_{kj} \ddot{q}_j + 2\omega_1 \dot{d} + (\dot{\omega}_2 - \omega_3\omega_1)[d + (n-k+1)L] \right\} \\ &+\sum_{k=1}^{n} m_k \left\{ \left(\omega_1^2 + \omega_2^2\right) [d + (n-k+1)L] - \ddot{d} \right\} \sum_{j=1}^{\nu} A_{kj} q_j \end{aligned} $$

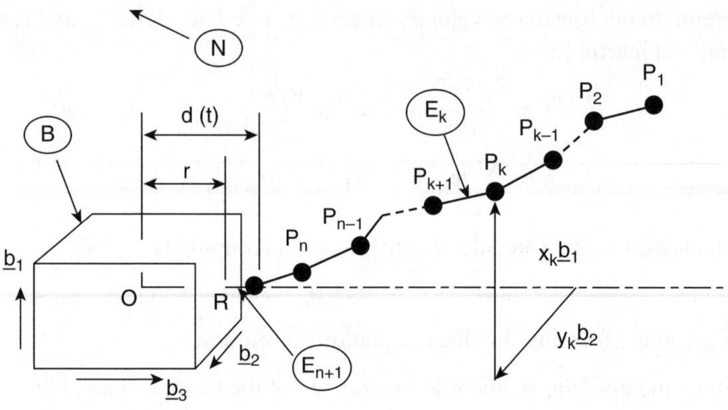

Figure 11.17 Discrete Particle System Convected to Model Deployment/Retrieval. Banerjee and Kane [4]. Reproduced with permission of the American Institute of Aeronautics and Astronautics, Inc.

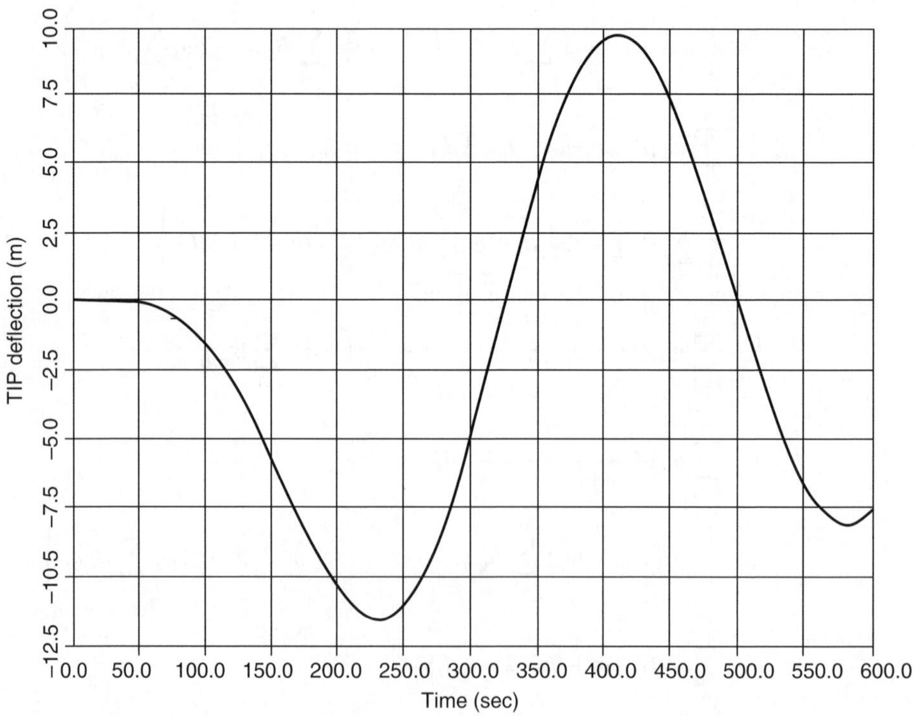

Figure 11.18 Spin-up Result Comparison between Present Discretized Beam Theory and a Continuum Beam Theory. Banerjee and Kane [4]. Reproduced with permission of the American Institute of Aeronautics and Astronautics, Inc.

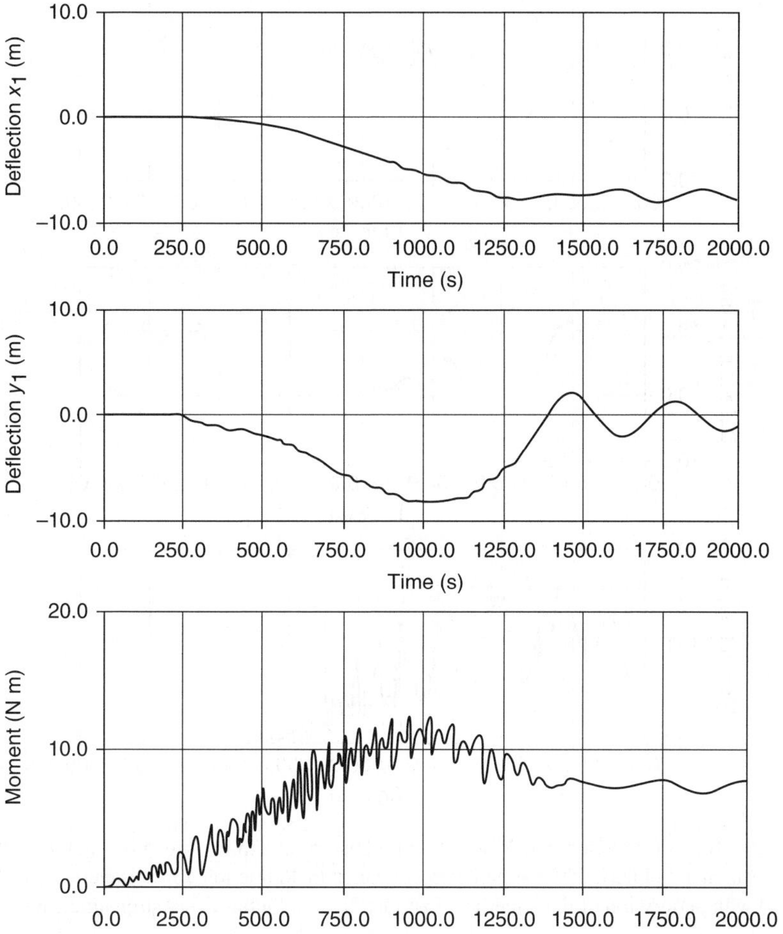

Figure 11.19 Beam Deflection at Tip in Two Axes and Bending Moment Time History with Prescribed Extrusion; Angular Velocity of Base Not Perpendicular to Extrusion Line. Banerjee and Kane [4]. Reproduced with permission of the American Institute of Aeronautics and Astronautics, Inc.

11.10 Code the theory and do a simulation with the following:

$$^N\boldsymbol{\omega}^B = \omega_1 \mathbf{b}_1 + \omega_2 \mathbf{b}_2 + \omega_3 \mathbf{b}_3,$$

$$\omega_1 = \frac{0.125}{57.3 \times 300}\left[t - \frac{150}{\pi}\sin\frac{\pi t}{150}\right] \text{ rad/sec} \quad t \le 300\,\text{sec}; \qquad \omega_2 = \omega_3 = 0.$$

Reproduce the expected result of Figure 11.18. It compares a continuum beam in spin-up (Ref. [12]) vs. the present theory, with $n^* = 29$, L = 5 m. As can be seen, the two curves agree very well.

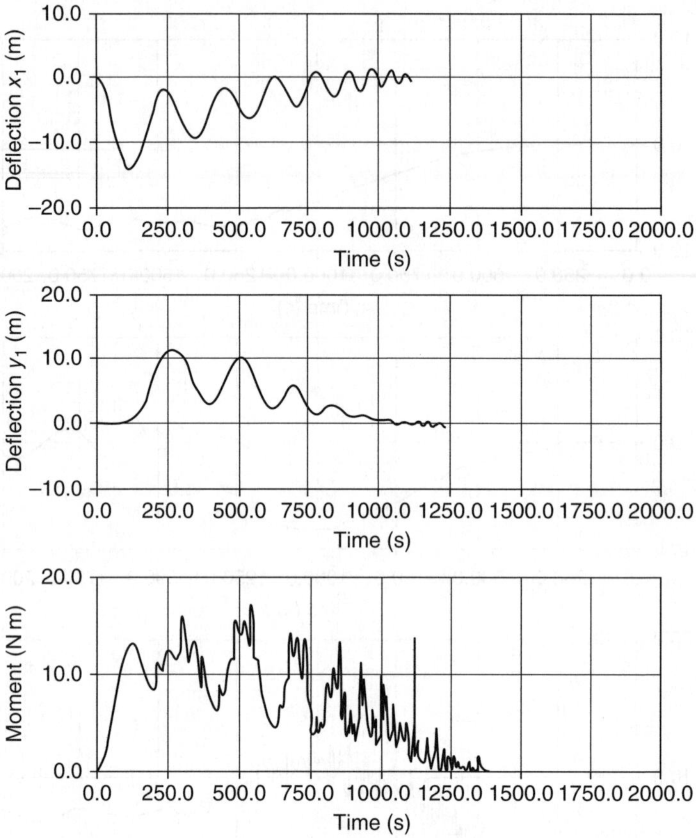

Figure 11.20 Beam Deflection and Moment Time History at Tip in Two Axes and with Prescribed Retraction; Angular Velocity of Base Not Perpendicular to Retraction Line. Banerjee and Kane [4]. Reproduced with permission of the American Institute of Aeronautics and Astronautics, Inc.

11.11 Reproduce the results of Figures 11.19 and 11.20 for extrusion and retraction time-history from 15 m to 150 m and vice versa, completed in 1387 sec, when the rate for extrusion/retraction is prescribed as:

$$\dot{d} = 0.133\left[1-\left(\frac{t}{1500}\right)^3\right] \text{ for extrusion;} \quad \dot{d} =$$

$$-0.133\left[1-\left(\frac{1500-t}{1500}\right)^3\right] \text{ for retraction}$$

Two vibration modes should be used in each plane, that is, $\nu = 4, \zeta_1 = \zeta_2 = 0.05$. Figure 11.18 applies to the case when the shuttle rotates at 1 degree/sec about an axis normal to the direction of extrusion.

References

[1] Banerjee, A.K. (1992) Order-n formulation of extrusion of a beam with large bending and rotation. *Journal of Guidance, Control, and Dynamics*, **15**(1), 121–127.

[2] Banerjee, A.K. and Do, V.N. (1994) Deployment control of a cable connecting a ship to an underwater vehicle. *Journal of Guidance, Control, and Dynamics*, **17**(6), 1327–1332.

[3] Banerjee, A.K. and Nagarajan, S. (1997) Efficient simulation of large overall motion of beams undergoing large deflection. *Multibody System Dynamics*, **1**, 113–126.

[4] Banerjee, A.K. and Kane, T.R. (1989) Extrusion of a beam from rotating base. *Journal of Guidance, Control, and Dynamics*, **12**(2), 140–146.

[5] Creamer, N.G. (1990) On the deployment of a Flexible Beam from an Oscillating Base. Proceedings of the AIAA Dynamic Specialist Conference, AIAA, Washington, DC, pp. 459–468.

[6] Banerjee, A.K. (1993) Block-diagonal equations of multibody elasto-dynamics with geometric stiffness and constraints. *Journal of Guidance, Control, and Dynamics*, **16**(6), 221–227.

[7] Banerjee, A.K. and Dickens, J.M. (1990) Dynamics of an arbitrary flexible body in large rotation and translation. *Journal of Guidance, Control, and Dynamics*, **13**(2), 140–146.

[8] Wang, J.T. and Huston, E.L. (1987) Kane's equations with undetermined multipliers – application to constrained multibody systems. *Journal of Applied Mechanics*, **54**, 424–429.

[9] Huston, R.L. and Kamman, J.W. (1981) A representation of fluid forces in finite segment cable models. *Computers and Structures*, **14**, 281–287.

[10] Myers, J.J., Holm, C.H., and McAllister, R.F. (eds) (1969) *Handbook of Ocean and Underwater Engineering*, McGraw-Hill.

[11] Baumgarte, J. (1972) Stabilization of constraint and integrals of motion in dynamical systems. *Computer Methods in Applied Mechanics and Engineering*, **1**, 1–16.

[12] Kane, T.R., Ryan, R.R. Jr., and Banerjee, A.K. (1987) Dynamics of a cantilever beam attached to a moving base. *Journal of Guidance, Control, and Dynamics*, **10**(2), 139–151.

12

Order-n Equations of Flexible Rocket Dynamics

12.1 Introduction

In this chapter we present a computationally efficient form of the equations of motion of a flexible rocket with a swiveling nozzle for directional control. Dynamics of systems with variable mass has been treated in the literature with a simplified particle approach, giving rise to the classical rocket equation [1], and the control volume approach [2] with flow of liquid propellant out of a solid base. It has been shown in Refs. [3, 4] that the equations derived with such flow considerations reduce to the equations given by the particle approach under the assumption of constant mass depletion rate. Ref. [5] considers the vibration problem of a rocket. In this chapter we derive equations of a flexible rocket by Kane's method in terms of motion variables that simplify the equations, with vibration characteristics of the rocket [6] approximated by time-updated system modes. The analysis is principally based on Refs. [7, 8]. Error in premature linearization, inherent in the use of modes [9], is compensated for by augmentation of the structural stiffness with geometric stiffness – softness in this case due to compressive nature of axial inertia and thrust; this feature enables the analyst to predict buckling of a slender rocket due to axial inertia force and thrust (see Chapter 5). Mode selection for rockets is an art, as has been noted for rockets and the space shuttle solid rocket booster in Ref. [10]. Here, for simplicity, we use free-free modes of a slender beam to characterize the vibration of a rocket. Mode selection strategies based on solving the modal eigenvalue problem for various "stages" of rocket mass, as discussed in Ref. [11], are strictly more appropriate. Equations developed here, however, should be useful as an adequate approximation of the dynamics of flexible multi-stage rockets, particularly for dynamics-based online nonlinear control design.

12.2 Kane's Equation for a Variable Mass Flexible Body

Consider a system of nodal particle P_k, $k = 1, \dots, \nu$, each losing mass, and let the motion of this system be described by ν generalized speeds. Nodal particle P_k of mass m^k at time t is

Flexible Multibody Dynamics: Efficient Formulations and Applications, First Edition. Arun K. Banerjee.

subjected to an external force $\mathbf{F}^k$, and at time $t + \Delta t$ it acquires a velocity $\mathbf{v}^k + \Delta\mathbf{v}^k$ by ejecting a particle of mass $(-\dot{m}^k\Delta t)$, where $\dot{m}^k$ is the mass loss rate, with an ejection velocity $\mathbf{v}_e^k$ relative to P_k. Acceleration of the ejected mass due to the changed velocity being $(\mathbf{v}_e^k/\Delta t)$, implies a force imparted to the ejected particle as:

$$\mathbf{F}_i^k = (-\dot{m}^k\Delta t)(\mathbf{v}_e^k/\Delta t) \tag{12.1}$$

By reaction, $-\mathbf{F}_i^k$ acts on nodal particle P_k, in addition to the external force $\mathbf{F}^k$. Now appealing to Newton's Law in the inertial frame N, one gets the vector form of the classical equation for a *particle losing mass*:

$$m^k\frac{{}^N d\mathbf{v}^k}{dt} = \mathbf{F}^k + \dot{m}^k\mathbf{v}_e^k \tag{12.2}$$

Ge and Cheng [5] dot-multiplied the vector equations, Eq. (12.2), by Kane's partial velocity [12] vector $\mathbf{v}_r^k$ of P_k with respect to the rth generalized speed of the system, to form what they called *extended Kane's equations* for the system of variable mass with n degrees of freedom:

$$F_r + F_r^* + F_r^{**} = 0 \qquad (r = 1, \dots, n) \tag{12.3}$$

Here, after writing $\frac{{}^N d\mathbf{v}^k}{dt}$ in the left hand side of Eq. (12.2) as $\mathbf{a}^k$, one has defined:

$$F_r = \sum_{k=1}^{\nu} {}^N\mathbf{v}_r^k \cdot \mathbf{F}^k \qquad (r = 1, \dots, n) \tag{12.4}$$

$$F_r^* = \sum_{k=1}^{\nu} {}^N\mathbf{v}_r^k \cdot (-m^k\,{}^N\mathbf{a}^k) \qquad (r = 1, \dots, n) \tag{12.5}$$

$$F_r^{**} = \sum_{k=1}^{\nu} {}^N\mathbf{v}_r^k \cdot (\dot{m}^k\,{}^N\mathbf{v}_e^k) \qquad (r = 1, \dots, n) \tag{12.6}$$

At this stage we consider a special case of a flexible rocket body with a rigid engine that swivels in a prescribed gimbal motion for directional control. The system with basis vectors fixed in the rocket undeformed configuration and the engine are shown in Figure 12.1.

Let O be the point where the engine joins the rocket, and the velocity of O in N, the angular velocity $\boldsymbol{\omega}^1$ of frame 1 in inertial frame N, and the velocity of a generic particle k in the flexible rocket body be given as follows in terms of $(6 + \mu)$ generalized speeds, where we have used μ number of vibration modes of the rocket body B:

$$\begin{aligned} {}^N\mathbf{v}^O &= u_1\mathbf{b}_1 + u_2\mathbf{b}_2 + u_3\mathbf{b}_3 \\ {}^N\boldsymbol{\omega}^1 &= u_4\mathbf{b}_1 + u_5\mathbf{b}_2 + u_6\mathbf{b}_3 \end{aligned} \tag{12.7}$$

$${}^N\mathbf{v}^k = {}^N\mathbf{v}^O + {}^N\boldsymbol{\omega}^1 \times (\mathbf{p}^k + \mathbf{d}^k) + \sum_{i=1}^{\mu} \boldsymbol{\Phi}_i^k u_{6+i}$$

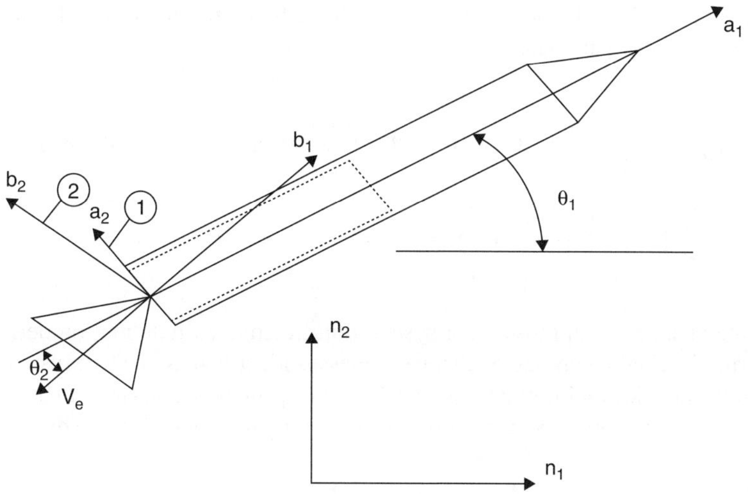

Figure 12.1 Planar View of a Rocket with a Gimbaled Engine. Banerjee [7].

Here we have expressed the deformation $\mathbf{d}^k$ at the location $\mathbf{p}^k$ from a fixed point O in frame-1 of the particle P_k, in terms of modes $\boldsymbol{\varphi}_i^k$, modal coordinates and associated generalized speeds as:

$$\mathbf{d}^k = \sum_{i=1}^{\mu} \boldsymbol{\phi}_i^k q_i; \quad u_{6+i} = \dot{q}_i \tag{12.8}$$

Equation (12.7) define the efficient generalized speeds that we have used previously in this book. Partial velocity of P_k for the rth generalized speed for $r = 1, \ldots, 6 + \mu$ follows from Eq. (12.7) as

$$^N\mathbf{v}_r^k = {}^N\mathbf{v}_r^O + {}^N\boldsymbol{\omega}_r^1 \times (\mathbf{p}^k + \mathbf{d}^k) + \delta_{ri}\boldsymbol{\phi}_i^k, \qquad (r = 1, \ldots, 6 + \mu) \tag{12.9}$$

where the Kronecker delta associates with the rth generalized speed, $\dot{q}_i$. Note that Eq. (12.7) uses linear vibration modes and this linearization is inherently premature, leading to error in results. As has been shown before, this error can be compensated *a posteriori* by including geometric stiffness due to associated inertia loads. Before we do that, we evaluate the acceleration of P:

$$^N\mathbf{a}^k = {}^N\mathbf{a}^O + {}^N\boldsymbol{\alpha}^1 \times (\mathbf{p^k} + \mathbf{d^k}) + \sum_{i=1}^{\mu} \boldsymbol{\phi}_i^k \ddot{q}_i + {}^N\boldsymbol{\omega}^1 \times \left[{}^N\boldsymbol{\omega}^1 \times (\mathbf{p^k} + \mathbf{d^k}) + 2 \sum_{i=1}^{\mu} \boldsymbol{\phi}_i^k \dot{q}_i \right] \tag{12.10}$$

Kane's equations for a flexible body losing mass can now be derived by substituting Eqs. (12.9) and (12.10) into Eqs. (12.3)–(12.6). Equation (12.3) can be separated into translation, rotation, and vibration equations depending on which r in the partial velocity equation, Eq. (12.9), is nonzero in the dot-multiplications of Eqs. (12.3)–(12.6). The translation equations are obtained for $r = 1,2,3$ by dot multiplying with the first term in the right hand side of Eq. (12.9) for $\mathbf{v}$ number of particles:

$$
{}^N\mathbf{v}_r^O.\sum_{k=1}^{\nu}\left\langle m^k\left\{{}^N\mathbf{a}^O + {}^N\boldsymbol{\alpha}^1\times(\mathbf{p}^k+\mathbf{d}^k) + \sum_{i=1}^{\mu}\boldsymbol{\phi}_i^k\ddot{q}_i + {}^N\boldsymbol{\omega}^1\times\left[{}^N\boldsymbol{\omega}^1\times(\mathbf{p}^k+\mathbf{d}^k)\right.\right.\right.
$$

$$
\left.\left.\left. + \; 2\sum_{i=1}^{\mu}\boldsymbol{\phi}_i^k\dot{q}_i\right]\right\} - \dot{m}^k\,{}^N\mathbf{v}_e^k\right\rangle = {}^N\mathbf{v}_r^O.\mathbf{F}^k \qquad r = 1,2,3 \tag{12.11}
$$

These equations agree with those for a system with relative particle motion derived in Ref. [1]. The term associated with rate of change of mass is identified as due to thrust. The rotation equations follow by dot-multiplying Eqs. (12.3)–(12.6) with the second term in the right side of Eq. (12.9), after using some vector identities representing moment of inertia dyadic about O:

$$
{}^N\boldsymbol{\omega}_r^1\cdot\left\langle\sum_{k=1}^{\nu}(\mathbf{p}^k+\mathbf{d}^k)\times\left\{m^k\left[{}^N\boldsymbol{a}^O + \sum_{i=1}^{\mu}\boldsymbol{\phi}_i^k\ddot{q}_i\right] - \dot{m}^k\,{}^N\mathbf{v}_e^k\right\} + \mathbf{I}.{}^N\boldsymbol{\alpha}^1 + {}^N\boldsymbol{\omega}^1\times\mathbf{I}.{}^N\omega^1\right\rangle
$$

$$
+{}^N\boldsymbol{\omega}_r^1\cdot\left[\sum_{k=1}^{\nu}2\,m^k(\mathbf{p}^k+\mathbf{d}^k)\times\left({}^N\boldsymbol{\omega}^1\times\sum_{i=1}^{\mu}\boldsymbol{\phi}_i^k\dot{q}_i\right)\right] = {}^N\boldsymbol{\omega}_r^1.\sum_{k=1}^{\nu}(\mathbf{p}^k+\mathbf{d}^k)\times\mathbf{F}^k \quad r = 4,5,6 \tag{12.12}
$$

This equation agrees with the basic form of the rotational equation given in Ref. [1], which after some vector-dyadic simplifications yields the more familiar result:

$$
{}^N\boldsymbol{\omega}_r^1\cdot\left\langle\sum_{k=1}^{\nu}(\mathbf{p}^k+\mathbf{d}^k)\times\left\{m^k\left[{}^{\mathrm{N}}\mathbf{a}^O + \sum_{i=1}^{\mu}\boldsymbol{\phi}_i^k\ddot{q}_i\right] - \dot{m}^k\,{}^N\mathbf{v}_e^k\right\} + \frac{\partial^B}{\partial t}(\mathbf{I}.{}^N\boldsymbol{\omega}^1)\right\rangle
$$

$$
+{}^N\boldsymbol{\omega}_r^1\cdot\left\langle{}^N\boldsymbol{\omega}^1\times\left[\sum_{k=1}^{\nu}m^k(\mathbf{p}^k+\mathbf{d}^k)\times\sum_{i=1}^{\mu}\boldsymbol{\phi}_i^k\dot{q}_i\right] - \sum_{k=1}^{\nu}(\mathbf{p}^k+\mathbf{d}^k)\times\dot{m}^k\left[{}^N\boldsymbol{\omega}^1\times(\mathbf{p}^k+\mathbf{d}^k)\right]\right\rangle
$$

$$
= {}^N\boldsymbol{\omega}_r^1\cdot\sum_{k=1}^{\nu}(\mathbf{p}^k+\mathbf{d}^k)\times\mathbf{F}^k \qquad r = 4,5,6 \tag{12.13}
$$

Here use has been made of the following identity for differentiation in the body B- or 1-frame (in this case), of the deformation-dependent vector of angular momentum about O, extending the result given by Thomson on p. 237 of Ref. [1]:

$$
\frac{\partial^B}{\partial t}(\mathbf{I}\cdot{}^N\boldsymbol{\omega}^1) = \sum_{k=1}^{\nu}\left\{\sum_{i=1}^{\mu}\begin{array}{l} m^k\boldsymbol{\phi}_i^k\dot{q}_i\times\left[{}^N\boldsymbol{\omega}^1\times(\mathbf{p}^k+\mathbf{d}^k)\right] + (\mathbf{p}^k+\mathbf{d}^k)\times\left[{}^N\boldsymbol{\omega}^1\times\sum_{i=1}^{\mu}m^k\boldsymbol{\phi}_i^k\dot{q}_i\right] \\ +(\mathbf{p}^k+\mathbf{d}^k)\times\dot{m}^k\left[{}^N\boldsymbol{\omega}^1\times(\mathbf{p}^k+\mathbf{d}^k)\right] \end{array}\right\}
$$

$$
+\mathbf{I}\cdot{}^N\boldsymbol{\alpha}^1 + {}^N\boldsymbol{\omega}^1\times\mathbf{I}\cdot{}^N\boldsymbol{\omega}^1 \tag{12.14}
$$

There are two terms involving the rate of change of mass in Eq. (12.13). Meirovitch [2] labels the first term in Eq. (12.13) as the moment due to jet damping and the second term as the moment due to thrust misalignment. The vibration equation for the flexible rocket given below account for thrust effects and is believed to be new in the literature, and it follows from Kane's equations by dot-multiplying the third-term in Eq. (12.9) with Eqs. (12.3)–(12.6):

$$\boldsymbol{\varphi}_r^k \cdot \sum_{k=1}^{\nu} \left\langle m^k \left\{ {}^N\mathbf{a}^O + {}^N\boldsymbol{\alpha}^1 \times (\mathbf{p}^k + \mathbf{d}^k) + \sum_{i=1}^{\mu} \boldsymbol{\phi}_i^k \ddot{q}_i + {}^N\boldsymbol{\omega}^1 \times \left[{}^N\boldsymbol{\omega}^1 \times (\mathbf{p}^k + \mathbf{d}^k) + 2\sum_{i=1}^{\mu} \boldsymbol{\phi}_i^k \dot{q}_i \right] \right\} - \dot{m}^k\, {}^N\mathbf{v}_e^k \right\rangle = \boldsymbol{\phi}_i^k . \mathbf{F}^k \quad r = 7, \ldots, 6+\mu \tag{12.15}$$

Here, to compensate for premature linearization endemic with the use of vibration modes, $\boldsymbol{\phi}_i^k$, one must consider geometric stiffness – softness in this case because of a compressive system of loads in equilibrium of axial inertia and thrust – as well as aerodynamic drag. Ignoring the drag for now, we compute the generalized force associated with load-dependent geometric stiffness as the matrix,

$$S^g = \Phi^t K_g \Phi f \tag{12.16}$$

where K_g is the geometric stiffness matrix due to unit load, and f is the thrust force; along the axis given by

$$f = \sum_{k=1}^{\nu} \dot{m}^k\, {}^N\mathbf{v}_e^k \cdot \mathbf{b}_1 \tag{12.17}$$

the right-hand side of Eq. (12.15) changes, including geometric stiffness, structural stiffness and damping, as:

$$\sum_{k=1}^{\nu} \boldsymbol{\phi}_i^k \cdot \mathbf{F}^k \equiv \left[\sum_{j=1}^{\mu} S_{rj}^g q_j - m\Omega_r^2 q_r \right] + 2\xi_r m \Omega_r \dot{q}_r + \sum_{k=1}^{\nu} \boldsymbol{\phi}_i^k . \mathbf{f}_{\text{ext}}^k \tag{12.18}$$

Mass loss and thrust cause changes in the assumed modes and frequencies. While the frequencies increase with mass loss, the effect of axial thrust is to lower the frequencies [5]. Here we make an approximation that the structural stiffness stays the same while the mass varies according to some interpolation function as assumed later. In practice, one can work with interpolated values of mass and modal integrals that are stored for modes for various fill levels of the fuel, as is reported in Ref. [11]. Before closing this section, we note some of the limitations of the particle-based formulation given above. The most obvious omission is the consideration of the flow of the combustion gases inside the rocket, which gives rise to reaction forces and moments due to Coriolis inertia effect and unsteady flow. However, for solid-propellant launch vehicles, these effects are negligible, and even for liquid-engine rockets the major flow effect can be approximated as that due to the burnt gases expelled through the nozzle.

12.3 Matrix Form of the Equations for Variable Mass Flexible Body Dynamics

Equations written above are in vector notation. For implementation of these in an algorithm we need the scalar version; for this we express all vectors in the frame-*1* basis. This process results in the representation of the translation and rotation equations, Eqs. (12.11) and (12.12) as:

$$M_1 \begin{Bmatrix} {}^N a^O \\ {}^N \alpha^1 \end{Bmatrix} + A_1^T \ddot{q} + X_1 = \begin{Bmatrix} 0 \\ 0 \end{Bmatrix} \tag{12.19}$$

$$M_1 = \begin{bmatrix} mU & -\tilde{s} \\ \tilde{s} & I \end{bmatrix} \tag{12.20}$$

$$A_1^T = \begin{Bmatrix} b \\ g \end{Bmatrix} \tag{12.21}$$

$$X_1 = \begin{Bmatrix} {}^N\tilde{\omega}_1 ({}^N\tilde{\omega}_1 s + 2b\dot{q}) - \dot{m}\, v_e - f_{\text{ext}}^1 \\ {}^N\tilde{\omega}_1\, I\, {}^N\omega_1 + 2 \sum_{i=1}^{\mu} N_i^T \dot{q}_i {}^N\tilde{\omega}_1 - \dot{m}\tilde{r}_c v_e - t_{\text{ext}}^1 \end{Bmatrix} \tag{12.22}$$

Here U stands for a 3×3 unity matrix, m, s, I are the instantaneous mass, mass moment and moment of inertia about O, and ω_1 is a 3×1 matrix of angular velocity; a tilde sign over a symbol denotes a skew-symmetric matrix representation of the corresponding vector in a cross product. Furthermore, we have used the following summations, where ϕ^k, p^k, d^k are 3×1 matrices of mode shape, position vector, and deformation vector of Eq. (12.8), respectively, and N_i, I_i, q_i are the ith modal dyadic, inertia dyadic, and modal coordinate, respectively:

$$b = \sum_{k=1}^{\nu} m^k \phi^k \tag{12.23}$$

$$g = \sum_{k=1}^{\nu} m^k \tilde{p}^k \phi^k \tag{12.24}$$

$$s = \sum_{k=1}^{\nu} m^k p^k + bq \tag{12.25}$$

$$N_i = \sum_{k=1}^{\nu} m^k \left[\left(p^{k^T} \phi_i^k \right) U - p^k \left(\phi_i^k \right)^T \right] \qquad i = 1, \ldots, \mu \tag{12.26}$$

$$I_1 = I_1^0 + \sum_{i=1}^{\mu} \left(N_i + N_i^T \right) q_i \tag{12.27}$$

$$\tilde{r}_c = \frac{1}{\nu} \sum_{k=1}^{\nu} (\tilde{p}_k + \tilde{d}_k) \tag{12.28}$$

It is assumed here that all nodal bodies P_k have the same mass reduction rate and ejection velocity as in a process of sublimation. The vibration equations, Eq. (12.14), have the matrix representation, with associated matrices,

$$A_1 \left\{ \begin{matrix} {}^N a^O \\ {}^N \alpha^1 \end{matrix} \right\} + E_1 \ddot{q} + Z_1 = 0 \tag{12.29}$$

$$E_1 = \sum_{k=1}^{\nu} \phi^{k^T} m^k \phi^k \tag{12.30}$$

$$Z_1 = Y_1 + (m\Omega^2 + K^g f)q + 2m\xi\Omega\dot{q} - \Phi^T f_{\text{ext}} \tag{12.31}$$

$$Y_1 = \left\{ \begin{matrix} -[{}^N\omega^1]^T \left(D_1^N \omega^1 - 2 \sum_{j=1}^{\mu} d_{1j}\dot{q}_j \right) \\ \vdots \\ \vdots \\ \vdots \\ -[{}^N\omega^1]^T \left(D_\mu{}^N\omega^1 - 2 \sum_{j=1}^{\mu} d_{\mu j}\dot{q}_j \right) \end{matrix} \right\} - \dot{m} \begin{bmatrix} \sum_{k=1}^{\nu} \phi_1^{k^T} \\ \vdots \\ \vdots \\ \vdots \\ \sum_{k=1}^{\nu} \phi_\mu^{k^T} \end{bmatrix} \{v_e\} \tag{12.32}$$

where we have allowed for a three-component exit velocity $\{v_e\}$ in the B-basis and used the notations:

$$D_r = N_r + \sum_{k=1}^{\nu} \sum_{i=1}^{\mu} q_i \left[\phi_i^{k^T} \phi_r^k U - \phi_i^k \phi_r^{k^T} \right] m^k \qquad (r = 1, \ldots, \mu) \tag{12.33}$$

$$d_{rj} = \sum_{k=1}^{\nu} m^k \tilde{\phi}_r^k \phi_j^k \qquad (r, j = 1, \ldots, \mu) \tag{12.34}$$

12.4 Order-n Algorithm for a Flexible Rocket with Commanded Gimbaled Nozzle Motion

A flexible rocket with a gimbaled engine represents two articulated bodies, with the rocket being a flexible body and the gimbaled engine usually modeled as a rigid body. The overall model being of $(8 + \mu)$ degrees of freedom (dof), with the rocket having $(6 + \mu)$ dof and the engine assigned 2 dof of rotation, a dense matrix formulation may be expensive for online control design using a number (μ) of modes. A block-diagonal mass matrix formulation can be revisited for this purpose, and is summarized in the Appendix. When the nozzle or engine motion is prescribed (as is realizable by a high-bandwidth controller), this algorithm has to be modified. In preparation for that we express each kinematical vector in its body basis. Thus, if the attachment point O of the rocket and nozzle in Figure 12.1 has a velocity vector ${}^N\mathbf{v}^O$ with generalized speeds u_1, u_2, u_3, the acceleration of O can be written in matrix notation as:

$${}^N a^O = {}^N \dot{v}^O + {}^N \tilde{\omega}^1 \, {}^N v^O \tag{12.35}$$

With u_4, u_5, u_6 being the generalized speeds for the angular velocity ${}^N\boldsymbol{\omega}^1$ of the rocket body frame-1, angular acceleration of frame-1 has matrix components:

$$^N\alpha^1 = \begin{bmatrix} \dot{u}_4 & \dot{u}_5 & \dot{u}_6 \end{bmatrix}^T \tag{12.36}$$

The nozzle has its reference frame, frame-2, with a fixed point that coincides with O of the flexible body. Frame-2 is oriented with respect to frame-1 by a body 2-3 rotation through given time histories of θ_1, θ_2. Angular velocity of frame-2 is given in 2-basis by the matrix:

$$^N\omega^2 = C_{12}^T\, {}^N\omega^1 + R\begin{Bmatrix} \dot{\theta}_1 \\ \dot{\theta}_2 \end{Bmatrix} \tag{12.37}$$

$$R = \begin{bmatrix} \sin\theta_2 & 0 \\ \cos\theta_2 & 0 \\ 0 & 1 \end{bmatrix} \tag{12.38}$$

For prescribed relative rotations θ_1, θ_2, angular acceleration of frame-2, the engine in Figure 12.1 is split into two terms, as done before, one involving derivatives of the generalized speeds of body-1 basis rotation, and the other a remainder term involving prescribed rotation rate derivatives:

$$^N\alpha^2 = {}^N\alpha_0^2 + {}^N\alpha_t^2 \tag{12.39}$$

$$\alpha_0^2 = C_{12}^T\alpha^1 \tag{12.40}$$

$$\alpha_t^2 = R\begin{Bmatrix} \ddot{\theta}_1 \\ \ddot{\theta}_2 \end{Bmatrix} + \left[\dot{R} + C_{12}^T\tilde{\omega}^1 C_{12}R\right]\begin{Bmatrix} \dot{\theta}_1 \\ \dot{\theta}_2 \end{Bmatrix} \tag{12.41}$$

The order-n algorithm for prescribed nozzle motion is started by first applying Kane's equations to the dynamics of the nozzle. To this end we first write the equilibrium system of inertia and external forces and torques about O on body 2 in its own basis, including unknown forces and torques of interaction from body 1, as follows, with s^2, I^2 meaning first and moment of inertia of the nozzle about the hinge:

$$\begin{Bmatrix} f^{*2} - f^{\text{ext}2} \\ t^{*2} - t^{\text{ext}2} \end{Bmatrix} \equiv M_2\begin{Bmatrix} {}^Na^O \\ {}^N\alpha^2 \end{Bmatrix} + \begin{Bmatrix} {}^N\tilde{\omega}^2\, {}^N\tilde{\omega}^2 s^2 - f^{\text{ext}2} \\ {}^N\tilde{\omega}^2 I^2\, {}^N\omega^2 - t^{\text{ext}2} \end{Bmatrix} \tag{12.42}$$

M_2 is given by Eq. (12.20) for a rigid body. One can write, in view of Eqs. (12.39) and (12.40),

$$\begin{Bmatrix} {}^Na^O \\ {}^N\alpha^2 \end{Bmatrix} = W\begin{Bmatrix} {}^Na^O \\ {}^N\alpha^1 \end{Bmatrix} + \begin{Bmatrix} 0 \\ {}^N\alpha_t^2 \end{Bmatrix} \tag{12.43}$$

$$W = \begin{bmatrix} C_{12}^T & 0 \\ 0 & C_{12}^T \end{bmatrix} \tag{12.44}$$

so that Eq. (12.42) is rewritten, for body-2 in basis 2 as:

$$\begin{Bmatrix} f^{*2} - f^{\text{ext2}} \\ t^{*2} - t^{\text{ext2}} \end{Bmatrix} \equiv M_2 W \begin{Bmatrix} {}^N a^O \\ {}^N \alpha^1 \end{Bmatrix} + X_2 \tag{12.45}$$

$$X_2 \equiv M_2 \begin{Bmatrix} 0 \\ {}^N \alpha_t^2 \end{Bmatrix} + \begin{Bmatrix} {}^N \tilde{\omega}^2 \tilde{\omega}^2 s^2 - f^{\text{ext2}} \\ {}^N \tilde{\omega}^2 I^2 \omega^2 - t^{\text{ext2}} \end{Bmatrix} \tag{12.46}$$

Equation (12.45) is now expressed in body-1 basis by using the W-transformation in Eq. (12.44) as:

$$\begin{Bmatrix} f^{*2} - f^{\text{ext2}} \\ t^{*2} - t^{\text{ext2}} \end{Bmatrix} = W^T M_2 W \begin{Bmatrix} a^O \\ \alpha^1 \end{Bmatrix} + W^T X_2 \tag{12.47}$$

In preparation for considering inertia forces and torques on the flexible body, body 1, we first write the solution of Eq. (12.29) formally as

$$\ddot{q} = -E_1^{-1} \left\{ A_1 \begin{pmatrix} {}^N a^O \\ {}^N \alpha^1 \end{pmatrix} + Z_1 \right\} \tag{12.48}$$

where the interaction forces and moments do not appear on the right-hand side due to the assumption that the elastic deformation at O is zero. It may be noted that E_1 in Eq. (12.30) is a diagonal matrix of variable mass. Now the D'Alembert equilibrium system of inertia and external forces on body 1, including interaction forces and torques from body 1, can be expressed in body 1 basis as per Eq. (12.19):

$$\begin{Bmatrix} f^{*1} - f^{\text{ext1}} \\ t^{*1} - t^{\text{ext1}} \end{Bmatrix} \equiv M_1 \begin{Bmatrix} {}^N a^O \\ {}^N \alpha^1 \end{Bmatrix} + A_1^T \ddot{q} + X_1 \tag{12.49}$$

Use of Eq. (12.48) in (12.49) and the introduction of the notations,

$$\hat{M}_1 = M_1 - A_1^T E_1^{-1} A_1 \tag{12.50}$$

$$\hat{X}_1 = X_1 - A_1^T E_1^{-1} Z_1 \tag{12.51}$$

lead to the resultant of all inertia and external forces and torques on body 1 in basis 1 to be:

$$\begin{Bmatrix} f^{*1} - f^{\text{ext1}} \\ t^{*1} - t^{\text{ext1}} \end{Bmatrix} \equiv \hat{M}_1 \begin{Bmatrix} {}^N a^O \\ {}^N \alpha^1 \end{Bmatrix} + \hat{X}_1 \tag{12.52}$$

Finally, the result of the inertia and external forces for the two-body system, with interaction forces/torques cancelled, is obtained by adding, for two bodies, Eqs. (12.47) and (12.52):

$$\left[W^T M_2 W + \hat{M}_1 \right] \begin{Bmatrix} {}^N a^O \\ {}^N \alpha^1 \end{Bmatrix} + \left\{ W^T X_2 + \hat{X}_1 \right\} = 0 \tag{12.53}$$

Solution of Eq. (12.53) for a^O, α^1 leads to the explicit statement of the kinematical differential equations, Eqs. (12.35) and (12.36) of the frame, and the vibration equation, Eq. (12.48) for the flexible rocket. This completes the order-n algorithm for a flexible rocket with prescribed nozzle rotation.

12.5 Numerical Simulation of Planar Motion of a Flexible Rocket

The order-n algorithm is numerically implemented here for a planar motion of a flexible rocket. For simplicity, the rocket body is represented by the continuum model, as against a discrete set of particles, of a free-free beam including the aft motor dome where the nozzle connects to the main body. This makes the modal sums as integrals of the mode shape functions for each λ_i, or ith root of the characteristic equation of a cantilever beam of length L. Only two modes are kept in the simulation:

$$\phi_i(x) = \cosh(\lambda_i x/L) + \cos(\lambda_i x/L) - \sigma_i[\sinh(\lambda_i x/L) + \sin(\lambda_i x/L)] \qquad i = 1,2 \quad (12.54)$$

$$\sigma_i = \frac{\cosh \lambda_i - \cos \lambda_i}{\sinh \lambda_i - \sin \lambda_i} \qquad i = 1,2$$

One can evaluate the following time-varying modal integrals:

$$m_i = \int_0^{m(t)} \phi_i^2 dm \qquad i = 1,2 \quad (12.55)$$

$$b_{2i} = \int_0^{m(t)} \phi_i dm = 2m_i\sigma_i/\lambda_i \qquad i = 1,2 \quad (12.56)$$

$$g_i = \int_0^{m(t)} x\phi_i dm = 2m_i L/\lambda_i^2 \qquad i = 1,2 \quad (12.57)$$

Here $m(t) = m_0 - \dot{m}t$ can be taken for simplicity, or by Hermite interpolation from final mass m_f at time t_f where the instantaneous mass is interpolated as: $m(t) = m_f f(t)$, as in Ref. [8].

$$f(t) = \left\{ \frac{m_0}{m_f} - \frac{\dot{m}_0 t}{m_f} + \left(\frac{t}{t_f}\right)^2 \left[3 + 2\frac{\dot{m}_0 t_f}{m_f} - 3\frac{m_0}{m_f}\right] - \left(\frac{t}{t_f}\right)^3 \left[2 + \frac{\dot{m}_0 t_f}{m_f} - 2\frac{m_0}{m_f}\right] \right\} \quad (12.58)$$

Before passing, it should be noted that one can also work with a mean value of the mass, and use modes for the rocket structure with this mean mass value [11]. The matrices A_1, E_1, M_1, X_1 given by Eqs. (12.21), (12.30), (12.20), and (12.22) are as follows:

$$A_1 = \begin{bmatrix} 0 & b_{21} & g_1 \\ 0 & b_{22} & g_2 \end{bmatrix} \quad (12.59)$$

$$E_1 = \begin{bmatrix} m_1 & 0 \\ 0 & m_1 \end{bmatrix} \quad (12.60)$$

$$M_1 = \begin{bmatrix} m_1 & 0 & -s_2 \\ 0 & m_1 & s_1 \\ -s_2 & s_1 & J_1 \end{bmatrix} \tag{12.61}$$

$$X_1 = \left\{ \begin{array}{c} -\dot{\theta}_1^2 s_1 - 2\dot{\theta}_1 \sum_{i=1}^{2} b_i \dot{q}_i - \dot{m} v_e \cos\theta_2 - f^{\text{ext1}} \\ -\dot{\theta}_1^2 s_2 - \dot{m} v_e \sin\theta_2 - f^{\text{ext2}} \\ \dot{m} v_e \left[\sum_{i=1}^{2} \phi_i \left(\frac{L}{2}\right) q_i \cos\theta_2 - \frac{L}{2} \sin\theta_2 \right] - t^{\text{ext1}} \end{array} \right\} \tag{12.62}$$

where θ_1, θ_2 are the prescribed nozzle angles, and Eq. (12.25) defines:

$$\left\{ \begin{array}{c} s_1 \\ s_2 \end{array} \right\} = \left\{ \begin{array}{c} m_1 L/2 \\ b_{21} q_1 + b_{22} q_2 \end{array} \right\} \tag{12.63}$$

The two components of the external force and torque due to aerodynamics and gravity are expressed in the rocket body frame as:

$$\left\{ \begin{array}{c} f^{\text{ext2}} \\ f^{\text{ext2}} \\ t^{\text{ext}} \end{array} \right\} = 0.5\rho\left(\dot{x}^2 + \dot{y}^2\right) A_s \left\{ \begin{array}{c} C_x \\ C_y \\ L_m C_m + 0.0212\, L\alpha \end{array} \right\} - m_1 g \left\{ \begin{array}{c} \sin\theta_1 \\ \cos\theta_1 \\ 0.5L\cos\theta_1 \end{array} \right\} \tag{12.64}$$

Here A_s, L_m, C_x, C_y, C_m are the aerodynamic surface area, characteristic length for moment, and three aerodynamic coefficients, respectively, given as functions of angle of attack α. Matrix Z_1 is given by Eq. (12.31), incorporating Eq. (12.18):

$$Z_1 = \left\{ \begin{array}{c} m_1\left(\omega_1^2 - \dot{\theta}_1^2\right) q_1 \\ m_2\left(\omega_2^2 - \dot{\theta}_2^2\right) q_2 \end{array} \right\} + \frac{6\dot{m}V_e}{5L} \left\{ \begin{array}{c} [\phi_1(L)]^2 q_1 + \phi_1(L)\phi_2(L) q_2 \\ [\phi_2(L)]^2 q_2 + \phi_1(L)\phi_2(L) q_1 \end{array} \right\} - f_{\text{ext}}^{(2)} \left\{ \begin{array}{c} \phi_1(L/2) \\ \phi_2(L/2) \end{array} \right\} \tag{12.65}$$

Here the second-column matrix, involving the multiplier $\dot{m}$, represents the geometric stiffness effect due to axial thrust on a cantilever beam [7]. For the nozzle oriented as in Figure 12.1, matrices M_2, X_2 follow from Eqs. (12.20) and (12.22) for rigid bodies:

$$M_2 = \begin{bmatrix} m_2 & 0 & 0 \\ 0 & m_2 & -m_2 r_2 \\ 0 & -m_2 r_2 & J_2 \end{bmatrix} \tag{12.66}$$

$$X_2 = \left\{ \begin{array}{c} 0 \\ -m_2 r_2 \\ J_2 \end{array} \right\} \ddot{\theta}_2 + \left\{ \begin{array}{c} m_2(\dot{\theta}_1 + \dot{\theta}_2)^2 r_2 + m_2 g \sin(\theta_1 + \theta_2) \\ m_2 g \cos(\theta_1 + \theta_2) \\ -m_2 g r_2 \cos(\theta_1 + \theta_2) \end{array} \right\} \tag{12.67}$$

The direction cosine matrix of Eq. (12.44), referring to Figure 12.1, becomes:

$$W = \begin{bmatrix} \cos\theta_2 & \sin\theta_2 & 0 \\ -\sin\theta_2 & \cos\theta_2 & 0 \\ 0 & 0 & 1 \end{bmatrix} \tag{12.68}$$

Open-loop control in the form of prescribed gimbal motion for the nozzle is taken in this example as:

$$\begin{aligned} &\theta_2 = 2.618\times 10^{-3}t && t < 10 \\ &\theta_2 = 2.618\times 10^{-3}t\left[1 - \frac{t-10}{60}\right] + 5.236\times 10^{-3}\sin\frac{\pi(t-10)}{30} && 10 < t < 70 \end{aligned} \tag{12.69}$$

Now matrices in Eqs. (12.50) and (12.51) can be formed. Equation (12.53) is set up with two acceleration components, where:

$$\left\{ \begin{matrix} a^O \\ \alpha^1 \end{matrix} \right\} = \left\{ \begin{matrix} \ddot{x}_1 - \dot{y}_1\dot{\theta}_1 \\ \ddot{y}_1 + \dot{x}_1\dot{\theta}_1 \\ \ddot{\theta}_1 \end{matrix} \right\} \tag{12.70}$$

Finally, gimbal torque required to realize the prescribed motion is obtained by reducing Eq. (12.78) given in the Appendix:

$$\tau_c = -J_2\ddot{\theta}_2 - \begin{bmatrix} 0 & -m_2r_2 & J_2 \end{bmatrix} W \left\{ \begin{matrix} a^O \\ \alpha^1 \end{matrix} \right\} + m_2gr_2\cos(\theta_1 + \theta_2) \tag{12.71}$$

Numerical simulation was done with the following data for an actual system:

$$\begin{aligned} &m_1 = 4400 \text{ slug},\ m_2 = 90 \text{ slug},\ L = 60 \text{ ft},\ \dot{m} = -39 \text{ slug/ sec},\ v_e = -I_{sp}g;\ I_{sp} = 267; \\ &J_2 = 200 \text{ slug} - \text{ft}^2,\ r_2 = 0.5 \text{ ft},\ Q = 2425 \text{ lb/ft}^2,\ A_s = 46 \text{ ft}^2,\ L_m\ 21.4 \text{ ft}; \\ &\text{beam } EI = 1.14\times 10^9 \text{ lb} - \text{ft}^2,\ \Omega_1 = 1.05 \text{ Hz},\ \Omega_2 = 6.6 \text{ Hz}; \\ &C_x = 0.0115\alpha - 0.420,\ C_y = -0.0425\alpha,\ C_m = 0.149\alpha, \\ &\rho = .002377\exp\left(-4.15x10^{-5}\ y_i\right) \end{aligned} \tag{12.72}$$

Three separate simulations were done: for a rigid rocket, for a rocket with low flexibility (36 times the value of the EI factor quoted above), and for actual flexibility. Results are shown in Figures 12.3–12.9, following the gimbal angle prescribed motion of Figure 12.2. Figures 12.3 and 12.4 show the differences in heights reached and horizontal distance, with implications on possible landing differences. The pitch angle history of Figure 12.5 shows that a gyro placed on a flexible rocket would read different value of the pitch from that of a rigid rocket, and thus would control differently. How bending deformations vary along the length of a highly flexible rocket is shown in Figure 12.6; note that deflection is negative corresponding to a positive nozzle angle. Figures 12.7 and 12.8 show the effect on natural frequency of the thrust, highlighting the softening effect of thrust. Naturally frequencies increase with mass reduction, with our assumption that structural stiffness remains unchanged. Other results are also relevant and interesting. Figure 12.9 shows the gimbal torque required to produce the prescribed gimbal rotation in Eq. (12.69).

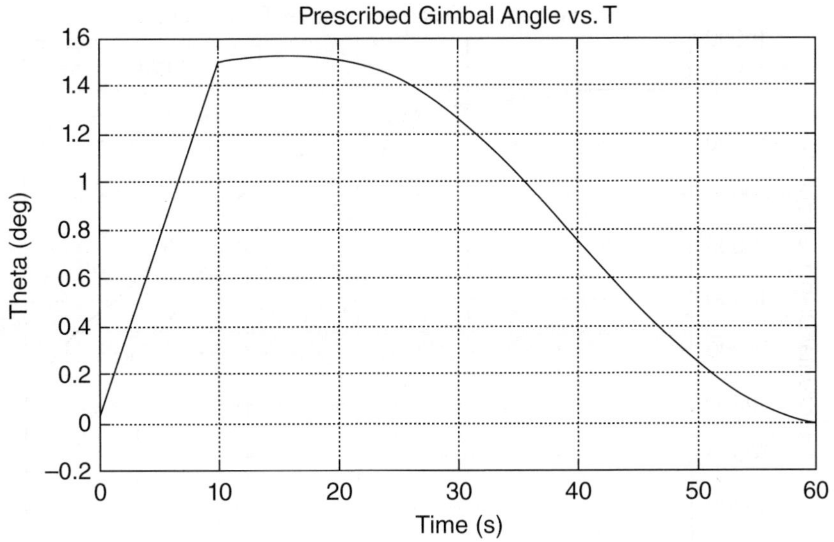

Figure 12.2 Prescribed Motion of Rocket Nozzle Angle. Banerjee [7].

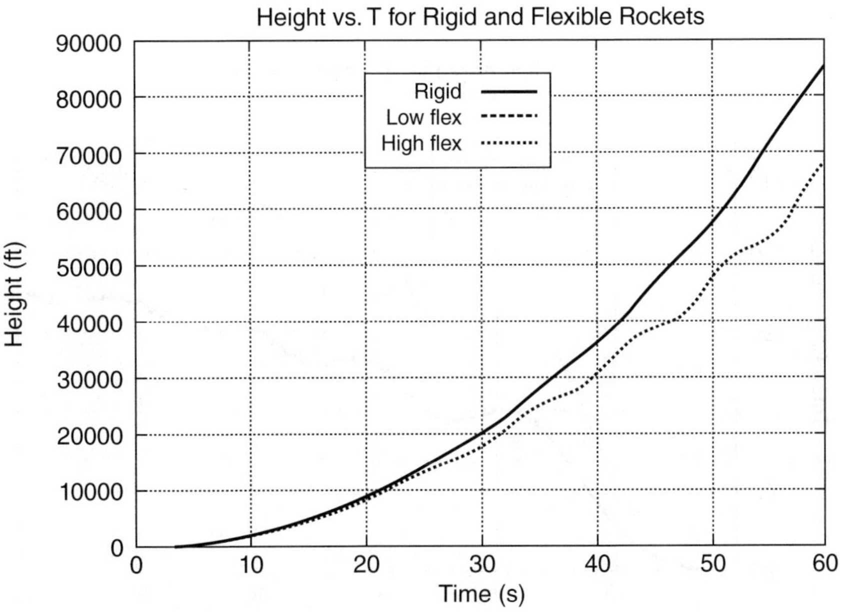

Figure 12.3 Rocket Altitude vs. Time for Three Models. Banerjee [7].

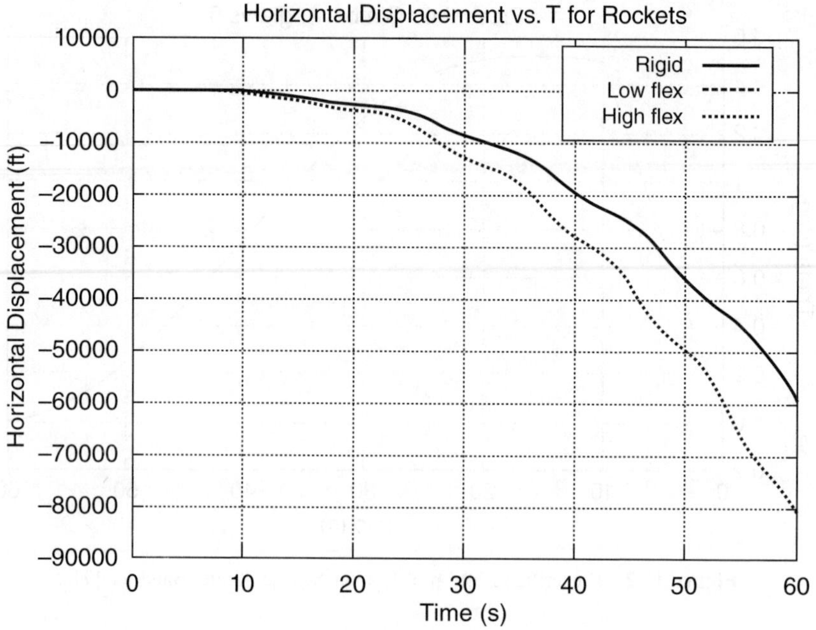

Figure 12.4 Rocket Horizontal Displacement vs. Time for Three Models. Banerjee [7].

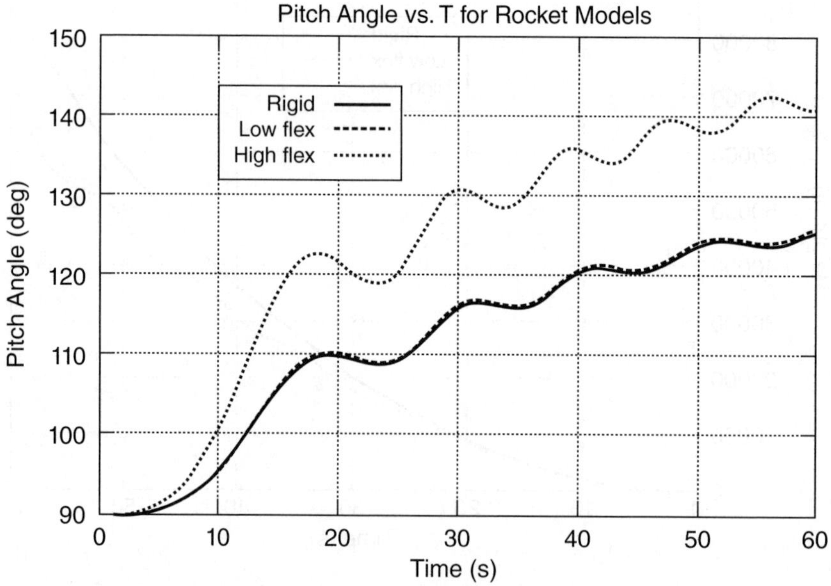

Figure 12.5 Rocket Pitch Angle vs. Time for Three Models. Banerjee [7].

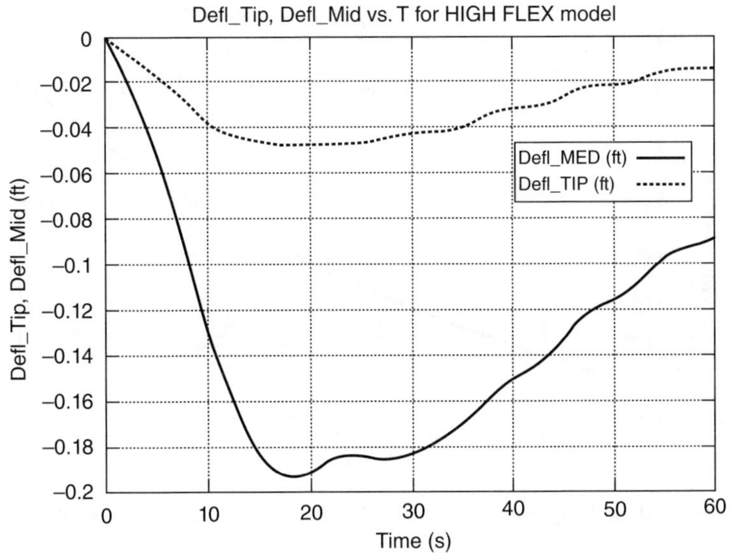

Figure 12.6 Tip and Mid-point Deflection of the High-Flexibility Rocket Model. Banerjee [7].

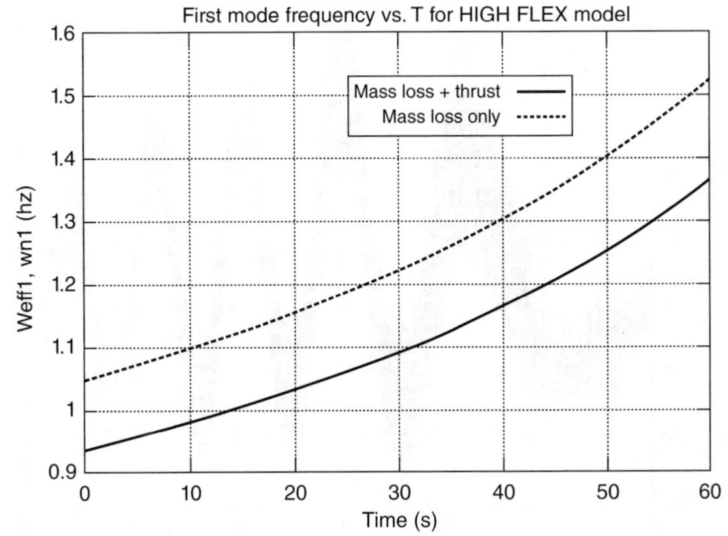

Figure 12.7 First Mode Frequency vs. Time for High-Flexibility Rocket with/without Thrust Effect. Banerjee [7].

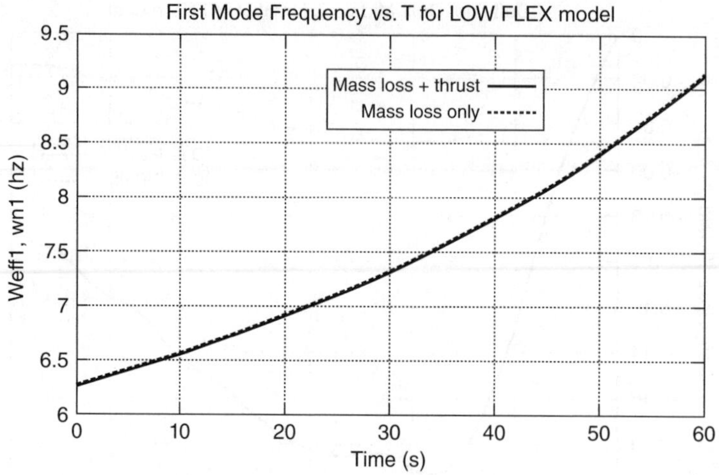

Figure 12.8 First Mode Frequency vs. Time for Low-Flexibility Rocket with/without Thrust Effect. Banerjee [7].

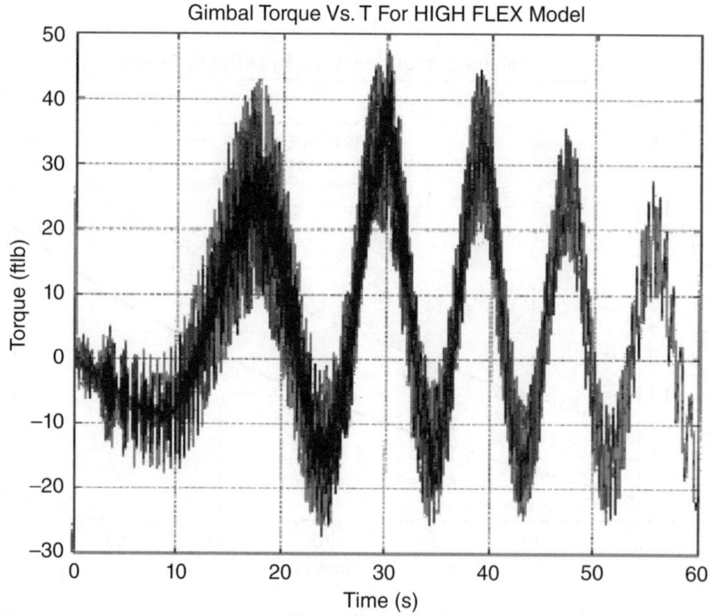

Figure 12.9 Gimbal Torque for High-Flexibility Rocket with Prescribed Nozzle Motion. Banerjee [7].

12.6 Conclusion

Results show substantial effect of flexibility on flight behavior for control of a rocket, with the nozzle angle time history prescribed as realizable by a high-gain control system. In particular, Figures 12.3–12.5 show that the horizontal and vertical motion and the pitch angle are quite different for a flexible body model from those for a rigid body model. Fidelity thus demands the need in rocket control simulations for having a model of flexible body with mass loss. The order-*n* algorithm lends itself to a computationally efficient online nonlinear dynamics model for control design of the rocket.

Acknowledgment

The author is grateful to Dr. C. J. Chang of Lockheed Martin Missiles and Space Systems Company, Sunnyvale, California, for providing the data for an actual rocket test flight. Results reported here compared quite well with the case study of a low-flexibility test rocket flight measurements, and to Dr. Tushar Ghosh of L3 Communications current practice in updating modes.

Appendix 12.A Summary Algorithm for Finding Two Gimbal Angle Torques for the Nozzle

Summary of an order-*n* algorithm for a rocket driven by gimbal torques τ_1, τ_2 driving two gimbal degrees of freedom in pitch-yaw nozzle angles θ_1, θ_2 is given below.

1. Compute $M_1, X_1, A_1, M_2, X_2, Z_1$ as defined.
2. Introduce in terms of Eq. (12.38) the 6×2 partial velocity matrix

$$\Pi = \begin{bmatrix} 0 \\ R \end{bmatrix} \tag{12.73}$$

   and define for the inertia matrix I_2 of the gimbal,

$$\mu = R^t I_2 R \tag{12.74}$$

3. Compute for a 6×6 unity matrix U:

$$M = \left[U - M_2 \Pi \mu^{-1} \Pi^T\right] M_2 \tag{12.75}$$

$$X = \left[U - M_2 \Pi \mu^{-1} \Pi^T\right] X_2 - M_2 \Pi \mu^{-1} \begin{Bmatrix} \tau_1 \\ \tau_2 \end{Bmatrix} \tag{12.76}$$

4. Solve the linear algebraic equation, using Eqs. (12.50), (12.51):

$$\left[W^T M W + \hat{M}_1\right] \begin{Bmatrix} a^O \\ \alpha^1 \end{Bmatrix} + \left\{W^t X + \hat{X}_1\right\} = 0 \tag{12.77}$$

5. The gimbal dynamics equations are as follows. Given the gimbal motion of the engine, the torque components requied can be computed from this:

$$\begin{Bmatrix} \ddot{\theta}_1 \\ \ddot{\theta}_2 \end{Bmatrix} = -\mu^{-1} \left\{ \Pi^T \left[M_2 W \begin{pmatrix} a^O \\ \alpha^1 \end{pmatrix} + X_2 \right] - \begin{Bmatrix} \tau_1 \\ \tau_2 \end{Bmatrix} \right\} \tag{12.78}$$

Problem Set 12

12.1 When a non-spinning rocket rotates about a transverse axis, the ejected gas gives it a reaction torque that is opposite to the direction of rotation. If the moment of inertia of the rocket about its mass center is mk^2, and if the external torque is zero, show by equating torque to rate of change of angular momentum that the angular velocity of the rocket is given by $\omega = \omega_0[m/m_0]^{(L^2/k^2-1)}$ where ω_0, m_0, L, k are, respectively, the initial angular velocity, initial mass, length from mass center to nozzle exit, and radius of gyration. This is a simple analysis of jet damping for a rigid rocket.

12.2 Simulate the motion of a variable mass *single* flexible body for relevant data in Eq. (12.72).

12.3 Assume that the rocket is a variable-mass single rigid body, and simulate motion for the relevant data in Eq. (7.2), and overlay the results against those in Problem 12.2.

12.4 Recall that Newton's law is strictly $\mathbf{F} = m\mathbf{a}$ for constant mass. In Ref. [4], Kane shows using Newton's law that the rocket equation considering the mass inside *and* ejected, for a body filled with a liquid propellant, modeled as a continuum, is $\mathbf{F} = (M + m_0 - \mu t)\,{}^R\mathbf{a}^{P*} + \mu\,\mathbf{v}_e$, where $\mathbf{F}$ is the external force on the rocket body, M is the mass of the rocket body, m_0 is the mass of the propellant at $t = 0$, μ is the mass flow rate, and ${}^R\mathbf{a}^{P*}$, $\mathbf{v}_e$ are the acceleration of the rocket body and the exit velocity of the ejected mass. Derive this equation of motion.

References

[1] Thomson, W.T. (1963) *Introduction to Space Dynamics*, John Wiley & Sons, Inc., pp. 230–236.

[2] Meirovitch, L. (1970) General motion of a variable mass flexible rocket with internal flow. *Journal of Spacecraft and Rockets*, **7**(2), 186–195.

[3] Eke, F.O. and Wang, S.-M. (1994) Equations of motion of two-phase variable mass systems with solid base. *Journal of Applied Mechanics*, **61**, 855–860.

[4] Kane, T.R. (1961) Variable mass dynamics? *Bulletin of Mechanical Engineering Education*, **2**(20), 62–65.

[5] Ge, Z.M. and Cheng, Y.H. (1982) Extended Kane's equations for nonholonomic variable mass system. *Journal of Applied Mechanics*, **49**, 429–431.

[6] Joshi, A. (1995) Free vibration characteristics of variable mass rockets having large axial thrust/acceleration. *Journal of Sound and Vibration*, **187**(4), 727–736.

[7] Banerjee, A.K. (2000) Dynamics of a variable-mass, flexible-body system. *Journal of Guidance, Control, and Dynamics*, **23**(3), 501–508.

[8] Banerjee, A.K. and Lemak, M.E. (2008) Dynamics of a Flexible Body with Rapid Mass Loss. Proceedings of the International Congress on Theoretical and Applied Mechanics, Australia.

[9] Banerjee, A.K. and Dickens, J.M. (1990) Dynamics of an arbitrary flexible body in large rotation and translation. *Journal of Guidance, Control, and Dynamics*, **13**(6), 221–227.

[10] Craig, R.R. Jr. (1981) *Structural Dynamics: An Introduction to Computer Methods*, John Wiley & Sons, Inc.

[11] Quiocho, L.J., Ghosh, T.K., Frenkel, D., and Huynh, A. (2010) Mode Selection Techniques in Variable Mass Flexible Body Modeling. Paper No. AIAA 2010-7607. AIAA Modeling and Simulation Technologies Conference, 2–5 August, Toronto, Canada.

[12] Kane, T.R. and Levinson, D.A. (1985) *Dynamics, Theory and Applications*, McGraw-Hill.

A

Efficient Generalized Speeds for a Single Free-Flying Flexible Body

When there is just one flexible body free-flying in space, like a spacecraft or an airplane, it is possible to simplify the equations even further with an efficient choice of generalized speeds. This is achieved by defining a new set of elastic generalized speeds replacing Eq. (5.4), while retaining the same rigid body generalized speeds, Eqs. (5.2) and (5.3). Thus, in view of Eq. (5.5), define generalized speeds, $\boldsymbol{u}_7, \ldots, \boldsymbol{u}_{6+\nu}$ as follows [17]:

Refer again to the picture and define as the rigid body generalized speeds,

$$u_i = {}^N\mathbf{v}^O \cdot \mathbf{b}_i \quad (i = 1, 2, 3) \tag{A.1}$$

$$u_{3+i} = {}^N\boldsymbol{\omega}^B \cdot \mathbf{b}_i \quad (i = 1, 2, 3) \tag{A.2}$$

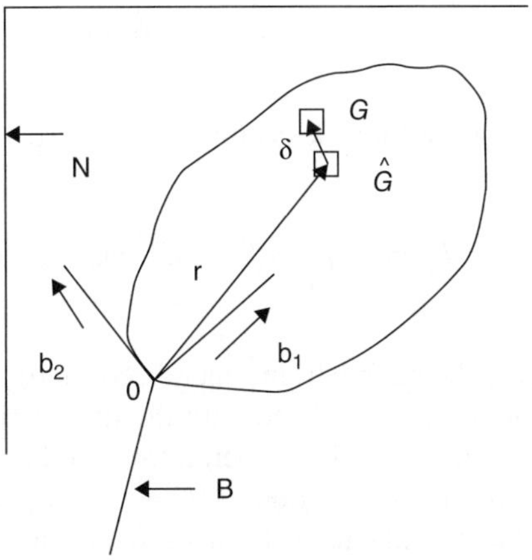

Figure A.1 Single Flexible Body with "Small" Elastic Displacements in a Flying Reference Frame *B* Undergoing "Large" Rotation and Translation in a Newtonian Frame *N*. Banerjee and Dickens [1].

Now define a new set of elastic generalized speeds

$$\sum_{k=1}^{\nu} \boldsymbol{\phi}_k u_{6+k} = \boldsymbol{\omega}^j \times \sum_{k=1}^{\nu} \boldsymbol{\phi}_k q_k + \sum_{k=1}^{\nu} \boldsymbol{\phi}_k \dot{q}_k \tag{A.3}$$

What it does is that velocity of the particle at $\boldsymbol{G}^*$ in N can be written as

$${}^N\mathbf{v}^{G^*} = \sum_{i=1}^{3} u_i \mathbf{b}_i + \sum_{i=1}^{3} u_{3+i} \mathbf{b}_i \times \mathbf{r} + \sum_{k=1}^{\nu} \boldsymbol{\phi}_k u_{6+k} \tag{A.4}$$

The acceleration of G* may be written from Eq. (A.4) as

$$\begin{aligned} {}^N a^{G^*} &= \sum_{i=1}^{3} \dot{u}_i b_i + \sum_{i=1}^{3} \dot{u}_{3+i} b_i \times r + \sum_{i=1}^{\nu} \boldsymbol{\phi}_k \dot{u}_{6+i} + \sum_{i=1}^{3} u_{3+i} b_i \\ &\times \left[\sum_{i=1}^{3} u_i b_i + \sum_{i=1}^{3} u_{3+i} b_i \times r + \sum_{i=1}^{\nu} \boldsymbol{\phi}_k u_{6+i} \right] \end{aligned} \tag{A.5}$$

Partial velocities of $\boldsymbol{G}^*$ follow as

$$\begin{aligned} {}^N\mathbf{v}_i^{G^*} &= \mathbf{b}_i, \quad i = 1, 2, 3 \\ &= \mathbf{b}_i \times \mathbf{r}, \quad i = 4, 5, 6 \\ &= \boldsymbol{\phi}_i \quad i = 7, \ldots, 6 + \nu \end{aligned} \tag{A.6}$$

Now form the generalized inertia force neglecting rotary inertia, that is,

$$F_i^* = -\int_D {}^N\mathbf{v}_i^{G^*} \cdot {}^N\mathbf{a}^{G^*} dm \tag{A.7}$$

Note that the partial velocities being free from generalized coordinates, the terms of the mass matrix, which arise only from terms associated with the time-derivatives of the generalized speeds, will not contain any term involving generalized coordinates. That makes the mass matrix constant in the matrix dynamical equations, $M\dot{U} = C$, resulting in significant increase in speed of simulations. Regarding the kinematical equations, note that dot-multiplying Eq. (A.3) by φ_i and integrating over the body, we get

$$\sum_{j=1}^{\nu} (\dot{q}_j - u_{6+j}) \int_0^m \boldsymbol{\phi}_i \cdot \boldsymbol{\phi}_j dm = \omega \cdot \sum_{j=1}^{\nu} q_j \int_0^m \boldsymbol{\phi}_i \times \boldsymbol{\phi}_j dm \qquad i = 1, \ldots, \nu \tag{A.8}$$

where for generality we have written $\boldsymbol{\omega} = \sum_{i=1}^{3} u_{3+i}\mathbf{b}_i$. Defining $\sigma = [u_7, \dots, u_{6+\nu}]^t$, and Eq. (A.8) is expressed in matrix form with the $\nu \times \nu$ matrices of generalized modal mass $\boldsymbol{\mu}$, and unity U.

$$\dot{q} = [-\mu^{-1}G_q \quad U]\begin{Bmatrix} \omega \\ \sigma \end{Bmatrix} \tag{A.9}$$

Here $G_q\omega$ represents gyroscopic stiffness rate, just as $\boldsymbol{\sigma}$ is a modal rate, defined in terms of the $n \times 3$ matrix shown below, where $\tilde{\varphi}_k$ is a 3×3 skew-symmetric matrix corresponding to vector cross products with the 3×1 vector φ_k, $k = 1, \dots, \nu$.

$$G_q = \begin{bmatrix} \sum_j q_j \int_0^m \{\tilde{\varphi}_j \varphi_1\}^t \, dm \\ \vdots \\ \sum_j q_j \int_0^m \{\tilde{\varphi}_j \varphi_n\}^t \, dm \end{bmatrix} \tag{A.10}$$

Reference

[1] Banerjee, A.K. and Dickens, J.M. (1990) Dynamics of an arbitrary flexible body in large rotation and translation. *Journal of Guidance, Control, and Dynamics*, **13**(2), 221–227.

B

A FORTRAN Code of the Order-*n* Algorithm: Application to an Example

```
C     ORDER-N FORMULATION FOR  EXTRUSION OF A BEAM WITH LARGE
C    FRAME ROTATION, BENDING AND TORSION, CHAPTER 11
C    CODE LINES SHOULD START AT 6TH COLUMNS. THIS IS ONLY A GUIDE
C    IT WILL NEED ADJUSTMENT FOR USER'S COMPUTER
      EXTERNAL EQNS
      LOGICAL STPSZ
      DIMENSION X(144),Q(72),DFLN(72)
      COMMON/DFQLST/T,STEP,RELERR,ABSERR,NCUTS,NEQNS,STPSZ
      COMMON/PRESCR/OMG,OMGDT,DIST,DVEL,DACC
      COMMON/PARAM/NPMAX,NP,EM(72),EL(72),AI(3,3,72),AK(72)
      DATA T,TMAX,STEP/0.0,120.0,0.1/
      DATA NCUTS,ABSERR,RELERR,STPSZ/20,1.0E-6,1.0E-6,.FALSE./
      DATA NPMAX,EM1,EL1,AK1,AI3/72, 3.124352E-3, 5.42, 0.02,
4.0357/
C
      ATIME=SECOND( )
      NP=NPMAX
      DO 1 K=1,NPMAX
      EM(K)=EM1
      EL(K)=EL1
      AK(K)=AK1
      DO 2 I=1,3
      DO 2 J=1,3
2     AI(I,J,K)=0.0
```

Flexible Multibody Dynamics: Efficient Formulations and Applications, First Edition. Arun K. Banerjee.

```
1      AI(3,3,K)=AI3
       NEQNS=2*NPMAX
       DO 5 I=1,NEQNS
5       X (I)=0.0
       X(31)=1.0
       NSTEPS=INT(TMAX/STEP+0.1)+1
       DO 50 JLOOP =1,NSTEPS
       IF(MOD(JLOOP,2).EQ.1)THEN
       WRITE(6,20) T,X(NEQNS)
       END IF
       IF (JLOOP.EQ.NSTEPS)GO TO 50
       CALL DEQS(EQNS,X,*99)
50 CONTINUE
       BTIME=SECOND( )
       CPU=BTIME-ATIME
       WRITE(6,101) CPU
101 FORMAT (1X,'CPU TIME FOR ORDER-N FORMULATION',E13.5)
20    FORMAT(1X,6E12.4)
99    WRITE(6,100)T
100   FORMAT(1X,'STEPSIZE HALVED 20 TIMES AT T=',E13.5)
101   STOP
       END
CCCCCCCCCCCCCCCCCCCCCCCCCCCCCCCCCCCCCCCCCCCCCCCCCCC
       SUBROUTINE EQNS(T,X,XDT)
       SAVE
       COMMON/PARAM/NPMAX,NP,EM(72),EL(72),AI(3,3,72),AK(72)
       COMMON/PRESCR/OMG,OMGDT,DIST,DVEL,DACC
       DIMENSION X(144),XDT(144)
       DIMENSION W0(3),ALF0(3),A0(3)
       DIMENSION
WKK(3,72),WKI(3,72),CKI(3,3,73),WK(3,72),ALFKT(3,72), 2
BIGAKT(3,3,72),AKT(3,72),VKK(3,72),FSTKK(3,72), 3
TSTKK(3,72),FXTK(3,72),TXTK(3,72),TAUK(72),FSTKT(3,72). 4
TSTKT(3,72),BIGMK(6,6,72),VEC(6),A(6),B(6), 5
BIGXK(6,72),BIGM(6,6),BIGX(6),YK(6,72),ZK(6,72),EMKK(72), 6
                       FK(72),PK(6),WC(6,6,72),DUM(6,6)
DATA I1ST/0/
EL2=EL(1)/2.0
DO 5 I=1,NPMAX
NI=NPMAX+I
XDT(NI)=X(I)
5      XDT(I)=0.0
       DO 100 K=1,NP
       WKK(1,K)=0.0
       WKK(2,K)=0.0
       WKK(3,K)=0.0
```

```
      WKI(1,K)=0.0
      WKI(2,K)=0.0
      WKI(3,K)=X(K)
C     CKI
      NPK=NPMAX+K
      SK=SIN(X(NPK))
      CK=COS(X(NPK))
      CKI(1,1,K)=CK
      CKI(1,2,K)=SK
      CKI(1,3,K)=0.0
      CKI(2,1,K)=-SK
      CKI(2,2,K)=CK
      CKI(2,3,K)=0.0
      CKI(3,1,K)=0.0
      CKI(3,2,K)=0.0
      CKI(3,3,K)=1.0
      IF(K.EQ.1)THEN
      WK(1,K)=WKI(1,K)
      WK(2,K)=WKI(2,K)
      WK(3,K)=WKI(3,K)
C
      ALFKT(1,K)=0.0
      ALFKT(2,K)=0.0
      ALFKT(3,K)=0.0
      AKT(1,K)= -WK(3,K)**2*EL2
      AKT(2,K)=0.0
      AKT(3,K)=0.0
      BIGAKT(1,1,K)= -(WK(3,K)**2+WK(2,K)**2)
      BIGAKT(2,2,K)= -(WK(1,K)**2+WK(3,K)**2)
      BIGAKT(3,3,K)= -(WK(1,K)**2+WK(2,K)**2)
      BIGAKT(1,2,K)= -ALFKT(3,K)+WK(1,K)*WK(2,K)
      BIGAKT(1,3,K)=  ALFKT(2,K)+WK(1,K)*WK(3,K)
      BIGAKT(2,1,K)=  ALFKT(3,K)+WK(2,K)*WK(1,K)
      BIGAKT(2,3,K)= -ALFKT(1,K)+WK(2,K)*WK(3,K)
      BIGAKT(3,1,K)= -ALFKT(2,K)+WK(3,K)*WK(1,K)
      BIGAKT(3,2,K)=  ALFKT(1,K)+WK(3,K)*WK(2,K)
      END IF
C
      IF(K.GT.1)THEN
      I=K-1
      WK(1,K)=WKI(1,K)+CK*WK(1,I)+SK*WK(2,I)
      WK(2,K)=WKI(2,K)-SK*WK(1,I)+CK*WK(2,I)
      WK(3,K)=WKI(3,K)+WK(3,I)
      ALFKT(1,K)=0.0
      ALFKT(2,K)=0.0
      ALFKT(3,K)=0.0
```

```
      BIGAKT(1,1,K)= -(WK(3,K)**2+WK(2,K)**2)
      BIGAKT(2,2,K)= -(WK(1,K)**2+WK(3,K)**2)
      BIGAKT(3,3,K)= -(WK(1,K)**2+WK(2,K)**2)
      BIGAKT(1,2,K)= -ALFKT(3,K)+WK(1,K)*WK(2,K)
      BIGAKT(1,3,K)=  ALFKT(2,K)+WK(1,K)*WK(3,K)
      BIGAKT(2,1,K)=  ALFKT(3,K)+WK(1,K)*WK(2,K)
      BIGAKT(2,3,K)= -ALFKT(1,K)+WK(3,K)*WK(2,K)
      BIGAKT(3,1,K)= -ALFKT(2,K)+WK(1,K)*WK(3,K)
      BIGAKT(3,2,K)=  ALFKT(1,K)+WK(3,K)*WK(2,K)
      VEC(1)=AKT(1,I)+BIGAKT(1,1,I)*EL2
      VEC(2)=AKT(2,I)+BIGAKT(2,1,I)*EL2
      VEC(3)=AKT(3,I)+BIGAKT(3,1,I)*EL2
      AKT(1,K)=CKI(1,1,K)*VEC(1)+ CKI(1,2,K)*VEC(2)+CKI(1,3,K)*
     2                   VEC(3)+BIGAKT(1,1,K)*EL2
      AKT(2,K)=CKI(2,1,K)*VEC(1)+ CKI(2,2,K)*VEC(2)+CKI(2,3,K)*
     2                   VEC(3)+BIGAKT(2,1,K)*EL2
      AKT(3,K)=CKI(3,1,K)*VEC(1)+ CKI(3,2,K)*VEC(2)+CKI(3,3,K)*
     2                   VEC(3)+BIGAKT(3,1,K)*EL2
      END IF
      VKK(1,K)=0.0
      VKK(2,K)=WKK(3,K)*EL2
      VKK(3,K)=0.0
      CALL EXTFOR (FXTK,TXTK,K)
      FXTK(1,K)=EM(K)*AKT(1,K)-FXTK(1,K)
      FXTK(2,K)=EM(K)*AKT(2,K)-FXTK(2,K)
      FXTK(3,K)=EM(K)*AKT(3,K)-FXTK(3,K)
C
      Z1=AI(1,1,K)*WK(1,K)+AI(1,2,K)*WK(2,K)+AI(1,3,K)*WK(3,K)
      Z2=AI(2,1,K)*WK(1,K)+AI(2,2,K)*WK(2,K)+AI(2,3,K)*WK(3,K)
      Z3=AI(3,1,K)*WK(1,K)+AI(3,2,K)*WK(2,K)+AI(3,3,K)*WK(3,K)
      TSTKT(1,K)=AI(1,1,K)*ALFKT(1,K)+ AI(1,2,K)*ALFKT(2,K)+AI(1,3,K)*
     2           ALFKT(3,K)-WK(3,K)*Z2+ WK(2,K)*Z3-TXTK(1,K)
      TSTKT(2,K)=AI(2,1,K)*ALFKT(1,K)+ AI(2,2,K)*ALFKT(2,K)+AI(2,3,K)*
     2           ALFKT(3,K)+WK(3,K)*Z1- WK(1,K)*Z3-TXTK(2,K)
      TSTKT(3,K)=AI(3,1,K)*ALFKT(1,K)+ AI(3,2,K)*ALFKT(2,K)+AI(3,3,K)*
     2           ALFKT(3,K)-WK(2,K)*Z1+ WK(1,K)*Z2-TXTK(3,K)
      DO 10 I=1,6
      DO 10 J=1,6
10    BIGMK(I,J,K)=0.0
      BIGMK(1,1,K)=EM(K)
      BIGMK(2,2,K)=EM(K)
      BIGMK(3,3,K)=EM(K)
      DO 15 I=4,6
      I3=I-3
      BIGXK(I3,K)=FSTKT(I3,K)
      BIGXK(I,K)=TSTKT(I3,K)
```

```
      DO 15 J=4,6
15    BIGMK(I,J,K)=AI(I3,J-3,K)
      DO 20 I=1,3
      YK(I,K)=VKK(I,K)
20    YK(I+3,K)=WKK(I,K)
100   CONTINUE
C   BEGIN BACKWARD PASS
      K=NP
      DO 110 I=1,6
      BIGX(I)=BIGXK(I,K)
      DO 110 J=1,6
110 BIGM(I,J)=BIGMK(I,J,K)
200   CONTINUE
      DO 215 I=1,6
      ZK(I,K)=0.0
      DO 215 J=1,6
215   ZK(I,K)=ZK(I,K)+BIGM(I,J)*YK(J,K)
      EMKK(K)=YK(1,K)*ZK(1,K)+ YK(2,K)*ZK(2,K)+ YK(3,K)*ZK(3,K)
     2        +YK(4,K)*ZK(4,K)+ YK(5,K)*ZK(5,K)+ YK(6,K)*ZK(6,K)
      CALL HINGE(TAUK,X,K)

FK(K)=YK(1,K)*BIGX(1)+YK(2,K)*BIGX(2)+YK(3,K)*BIGX(3)+YK(4,K)
     2       *BIGX(4)+YK(5,K)*BIGX(5)+YK(6,K)*BIGX(6)-TAUK(K)
      IF(K.EQ.1)GO TO 300
      DO 220 I=1,6
220   PK(I)=ZK(I,K)/EMKK(K)
      DO 230 I=1,6
      BIGX(I)=BIGX(I)-PK(I)*FK(K)
      DO 230 J=1,6
230 BIGM(I,J)=BIGM(I,J)-ZK(I,K)*PK(J)
C  SHIFT TRANSFORM
      DO 240 I=1,3
      I3=I+3
      DO 240 J=1,3
      WC(I,J,K)=CKI(I,J,K)
      WC(I3,J,K)=0.0
      J3=J+3
      WC(I,J3,K)=0.0
240 WC(I3,J3,K)=WC(I,J,K)
      VEC(1)=EL2+CKI(1,1,K)*EL2
      VEC(2)=CKI(2,1,K)*EL2
      VEC(3)=0.0
      WC(1,4,K)=0.0
      WC(2,4,K)=CKI(3,1,K)*VEC(1)
      WC(3,4,K)= -CKI(2,1,K)*VEC(1)+CKI(1,1,K)*VEC(2)
      WC(1,5,K)=0.0
```

```
      WC(2,5,K)=CKI(3,2,K)*VEC(1)
      WC(3,5,K)= -CKI(2,2,K)*VEC(1)+CKI(1,2,K)*VEC(2)
      WC(1,6,K)= -CKI(3,3,K)*VEC(2)
      WC(2,6,K)= CKI(3,3,K)*VEC(1)
      WC(3,6,K)= -CKI(2,3,K)*VEC(1)
      DO 250 I=1,6
      DO 250 J=1,6
      DUM(I,J)=0.0
      DO 250 L=1,6
250 DUM(I,J)=DUM(I,J)+BIGM(I,L)*WC(L,J,K)
      KM1=K-1
      DO 260 I=1,6
      DO 260 J=1,6
      BIGM(I,J)=BIGMK(I,J,KM1)
      DO 260 L=1,6
260 BIGM(I,J)=BIGM(I,J)+WC(L,I,K)*DUM(L,J)
      DO 270 I=1,6
      VEC(I)=0.0
      DO 270 J=1,6
270 VEC(I)=VEC(I)+WC(J,I,K)*BIGX(J)
      DO 280 I=1,6
280  BIGX(I)=BIGXK(I,KM1)+VEC(I)
      K=KM1
      GO TO 200
300  CONTINUE
C    BEGIN FORWARD PASS
      XDT(1)=-FK(1)/EMKK(1)
      IF(NP.EQ.1) RETURN
      DO 310 I=1,6
310 A(I)=YK(I,1)*XDT(1)
      K=2
320 CONTINUE
     DO 330=0 I=1,6
     B(I)=0.0
     DO 330 J=1,6
330 B(I)=B(I)+WC(I,J,K)*A(J)
     XDT(K)= -(ZK(1,K)*B(1) + ZK(2,K)*B(2)+ ZK(3,K)*B(3)
    2            +ZK(4,K)*B(4) +ZK(5,K)*B(5) +ZK(6,K)*B(6)
    3            +FK(K))/EMKK(K)
     IF(K.EQ.NP) RETURN
     DO 340 I=1,6
340 A(I)=B(I)+YK(I,K)*XDT(K)
     K=K+1
     GP TO 320
     END
```

```
CCCCCCCCCCCCCCCCCCCCCCCCCCCCCCCCCCCCCCCCCCC
    SUBROUTINE EXTFOR (FXTK,TXTK,K)
    SAVE
    DIMENSION FXTK(3.72),TXTK(3,72)
C  SPECIFY EXTERNAL FORCES & TORQUES; BELOW IS A PLACE HOLDER
    DO 10 I=1,3
    FXTK(I,K)=0.0
10 TXTK(I,K)=0.0
     RETURN
     END
CCCCCCCCCCCCCCCCCCCCCCCCCCCCCCCCCCCCCCCCCCCC
    SUBROUTINE HINGE (TAUK,X,K)
    SAVE
    COMMON/PARAM/NPMAX,NP,EM(72),EL(72),AI(3,3,72),AK(72)
    DIMENSION TAUK(1),X(1)
    TAUK(K)= -AK(K)*X(NPMAX+K)
       RETURN
       END
CCCCCCCCCCCCCCCCCCCCCCCCCCCCCCCCCCCCCCCCCCCCCCCCCCCCCCC
     SUBROUTINE DEQS(F,Y,*)
C   VARIABLE STEP, FOURTH ORDER RUNGE-KUTTA-MERSON INTEGRATOR
     IMPLICIT REAL (A-Z)
     INTEGER I,NCUTS,NEQ
     LOGICAL DBL,STPSZ
     SAVE
     EXTERNAL F
     COMMON/DFQLST/T,STEP,REL,ABSERR,NCUTS,NEQ,STPSZ
     DIMENSION F0(200),F1(200),F2(200),Y1(200),Y2(200),Y(NEQ)
     DATA HC /0.0/
C   CHECK FOR INITIAL ENTRY AND ADJUST HC, IF NECESSARY
     IF (NEQ.NE.0) GO TO 10
     HC=STEP
     RETURN
10   IF(STEP.EQ.0.0) RETURN 1
C   CHANGE DIRECTION, IF REQUIRED
     IF (HC*STEP) 20,30,40
20  HC=-HC
     GO TO 40
30  HC=STEP
C  SET LOCAL VARIABLES
40   EPSL=REL
     FINAL=T+STEP
     H=HC
     TT=T+H
     T=FINAL
     H2=H/2.0
```

```
      H3=H/3.0
      H6=H/6.0
      H8=H/8.0
C   MAIN KUTTA-MERSON LOOP
50   IF ((H.GT.0.0 .AND. TT.GT.FINAL) .OR.
      2 (H.LT.0.0 .AND. TT.LT.FINAL)) GO TO 190
60    CALL F(TT-H,Y,F0)
       DO 70 I=1,NEQ
70     Y1(I)=F0(I)*H3+Y(I)
       CALL F(TT-2.0*H3,Y1,F1)
       DO 80 I=1,NEQ
80     Y1(I)=(F0(I)+F1(I))*H6+Y(I)
       CALL F(TT-2.0*H3,Y1,F1)
       DO 90 I=1,NEQ
90     Y1(I)=(F1(I)*3.0+F0(I))*H8+Y(I)
       DO 100 I=1,NEQ
100    Y1(I)=(F2(I)*4.0-F1(I)*3.0+F0(I))*H2+Y(I)
       CALL F(TT,Y1,F1)
       DO 110 I=1,NEQ
110    Y2(I)=(F2(I)*4.0+F1(I)+F0(I))*H6+Y(I)
C ** DOES THE STEPSIZE H NEED TO BE CHANGED
       IF(EPSL.LE.0.0) GO TO 170
       DBL=.TRUE.
       DO 160 I=1,NEQ
       ERR=ABS(Y1(I)-Y2(I))*0.2
       TEST=ABS(Y1(I))*EPSL
       IF (ERR.LT.TEST .OR. ERR.LT.ABSERR) GO TO 150
C ** HALVE THE STEPSIZE
       H=H2
       TT=TT-H2
       IF (.NOT.STPSZ) GO TO 120
       TEMP=TT-H2
       WRITE(6,200)H,TEMP
C      HAS THE STEPSIZE BEEN HALVED TOO MANY TIMES ?
120    NCUTS=NCUTS-1
       IF (NCUTS.GE.0) GO TO 130
       T=TT-H2
       WRITE(6,210)T
       RETURN 1
C ** IF STEPSIZE IS TOO SMALL RELATIVE TO TT TAKE RETURN 1
130    IF(TT+H .NE. TT) GO TO 140
       T=TT
       RETURN 1
140    H2=H/2.0
       H3=H/3.0
       H6=H/6.0
```

```
      H8=H/8.0
      GO TO 60
150   IF (DBL .AND. 64.0*ERR .GT. TEST
     2    .AND. 64.0*ERR .GT. ABSERR) DBL=.FALSE.
160   CONTINUE
C     DOUBLE THE STEPSIZE, MAYBE.
      IF(.NOT. DBL .OR. ABS(2.0*H) .GT. ABS(STEP) .OR.
     2 ABS(TT+2.0*H) .GT. ABS(FINAL) .AND.
     3 ABS(TT-FINAL) .GT. ABS(FINAL)*1.0E-7) GO TO 170
       H2=H
       H=H+H
       IF (STPSZ) WRITE(6,200)H,TT
       H3=H/3.0
       H6=H/6.0
       H8=H/8.0
       NCUTS=NCUTS+1
170    DO 180 I=1,NEQ
180    Y(I)=Y2(I)
       TT=TT+H
       GO TO 50
190    IF(EPSL.LT.0.0) RETURN
C  *  NOW BE SURE TO HAVE T=FINAL
       HC=H
       H=FINAL-(TT-H)
       IF (ABS(H).LE.ABS(FINAL)*1.0E-7)RETURN
       TT=FINAL
       EPSL= -1.0
       H2=H/2.0
       H3=H/3.0
       H6=H/6.0
       H8=H/8.0
       GO TO 60
200    FORMAT(1X,'THE STEPSIZE IS NOW',1E12.4,'AT T=',1E12.4)
210    FORMAT(1X,'THE STEPSIZE HAS BEEN HALVED TOO MANY TIMES;',
      2                    'T=',1E12.4)
       END
```

Index

Flexible Multibody Dynamics: Efficient Formulations and Applications, First Edition. Arun K. Banerjee.